D1155882

The Persisting Ecological Constraints of Tropical Agriculture

The Persisting Ecological Constraints of Tropical Agriculture

Wolfgang Weischet
and
Cesar N. Caviedes

Longman
Scientific &
Technical
Copublished in the United States with
John Wiley & Sons, Inc., New York

AUDISIANA COLLEGE Library

Longman Scientific & Technical,
Longman Group UK Ltd,
Longman House, Burnt Mill, Harlow,
Essex CM20 2JE, England
and Associated Companies throughout the world.

Copublished in the United States with
John Wiley & Sons, Inc., 605 Third Avenue, New York, NY 10158

© Longman Group UK Limited 1993

All rights reserved; no part of this publication may be reproduced, stored in a retrieval system, or transmitted in any form or by any means, electronic, mechanical, photocopying, recording, or otherwise without either the prior written permission of the Publishers or a licence permitting restricted copying in the United Kingdom issued by the Copyright Licensing Agency Ltd, 90 Tottenham Court Road, London W1P 9HE.

First published 1993

ISBN 0–582–05692–6

British Library Cataloguing in Publication Data
A catalogue record for this book is available from the British Library

Library of Congress Cataloging-in-Publication Data
A catalogue record for this book is available from the Library of Congress

Set in Bembo
Printed in Singapore

S
481
.W45
1993
27171963

Augustana College Library
Rock Island, Illinois 61201

Contents

Preface

Since the 1950s there has been an increasing world-wide awareness of the uneven economic and socio-political development and degree of progress between the industrialized nations of the North and the nations of the tropical belt.

Initially, the disparity was believed to have been caused by an inherent under-development of the tropical regions. For many people this underdevelopment was simply the result of the exploitation of the tropical environment and the repression of aboriginal societies by the colonial powers of the eighteenth and nineteenth centuries. Major efforts in foreign assistance were designed to stimulate development in many ways and to compensate for past neglect. The notion of 'underdeveloped' nations was upgraded to 'developing' nations.

Meanwhile, the original public awareness has grown into concern: in most parts of the developing world the increase in gross national product and especially food production cannot keep up with actual population growth, and the situation is getting worse. Only a few tropical countries have managed to break the vicious concurrence of population growth and production decline, at least for the time being, by means of Green Revolution technology. Most nations of the tropical world have been unable to bridge the gap, in spite of receiving similar foreign aid and technical assistance. This is especially true in the regions of the humid inner tropics where large areas of the lush tropical rain forest had remained virtually untouched until the first decades of this century. But since food production could not be increased on traditional farmland people began to explore the forest, first by turning valuable timber reserves into instant profit, and then by attempting to transform the degraded, burnt areas into agriculturally productive lands. However, sooner in some cases and later in others, the idea of converting rain-forest lands into a 'food basket' turned out to be an illusion. Notwithstanding, the destruction of the forest reserves goes on at an accelerated pace.

In view of the fact that the squandering of this unique and largely unexplored reserve (gene pool) and the threatening transformation of the 'green hell' into a 'red desert' (Goodland and Irwin, 1975) will affect not only the people of the tropics, but all mankind, the leaders of the industrialized nations – one after another – have felt obliged to voice publicly their concern about these ruinous processes and to demand counteractive measures.

However, as long as the decision-makers of those countries which are shared by population growth and food decline believe in the possibility of reaping lasting economic benefits from their reserves or of utilizing them as transmigration areas for their own overflowing rural populations, and as long as they cling to the illusion that these measures will help solve their problems, they will continue using up their natural reserves. Dennis Mahar's recent (1988) statement that in Brazil the key role in the deforestation process was played by government also holds true for the other nations in the tropical forest belt.

To stem this tide of destruction a convincing campaign is needed concerning the questionability of the whole enterprise. Indeed, a decisive change in the general attitude towards the so-called developing regions of our world is called for.

Many experts in the geosciences are convinced that certain basic characteristics of the North–South disparity, such as the discrepancy in agricultural productivity, have to do with incisive geoecological differences between the tropics and the extratropical regions. If they want to see a change in the attitude towards the tropical world, they must start by explaining these differences to the concerned public.

The objective of this book is to pass on to the interested non-experts, in the form of substantial but easy-to-follow deductions, the findings that experts of the various geosciences have arrived at in the course of the last three decades. The difficulty with this approach is that the geoecological viewpoint includes contributions from a wide range of disciplines, such as geology, petrography, mineralogy, climatology, hydrology, limnology, chemical and physical soil sciences, and geobotany. However, by presenting data and by showing how to put them into a geoecological context, we hope to equip the reader to reach his or her own conclusions about not yet fully understood characteristics of the world.

The philosophy behind our approach to the problem of the North–South disparity is first to find out whether or not there are coincidences in space or time between a certain socio-cultural phenomenon and certain geoecological conditions. If such coincidences can be detected, the second step will be to substantiate them through facts and figures and carefully to point out the interrelations so as to enable the reader to check them critically.

We do not merely want to convey a message but, more importantly, we are interested in passing on material for the reader's own evaluation.

Such a procedure should be a matter of course, but experience shows that there is a great gap between what the experts know and what transpires to the general public. Normally, the language and technical vocabulary used among the various groups of experts make nearly everyone else an outsider who will find it impractical to replicate the experiments or verify the arguments. When it becomes necessary to inform the public, who usually foot the bill for the experts' investigations, this happens mostly in the form of condensed statements at conferences or via the media. In both instances, a great deal of trust in the exactness and objectivity of the reporting agency is needed.

Acknowledgements

This book owes much to the valuable assistance of several institutions and persons. Most of the manuscript was written in the Hume Library of the University of Florida, Gainesville. The Deutsche Forschungsgemeinschaft sponsored in part the visits to Gainesville and the senior author's trip to Yurimaguas Experimental Station, eastern Peru, where Dr José Benites and his crew provided field-work support.

Mrs Christiana Caviedes read and patiently revised lengthy passages of the original manuscript. Librarian Dolores Jenkins, Professors Robert B. Marcus and David L. Niddrie, all of the University of Florida, Professor Franz H. Link, University of Freiburg, and Consul Adolf Charles McCarthy, Freiburg, kindly advised on the style of the text. Mrs Charlotte Grabowski and Anna-Maria Feser as well as cartographers Klaus-Dieter Lickert and Walter Hoppe, all of the Institut für Physische Geographie, University of Freiburg, devoted much time to the completion of the typescript and figures.

B.G. Teubner Verlag, Stuttgart, gave permission to reprint figures from previously published books.

To all of them the authors express their sincere appreciation. We also thank the staff of Longman Higher Education and Reference and an anonymous English colleague for their valuable co-operation.

Last but not least, the authors feel deeply indebted to all those scholars (see references) from different disciplines who provided the fundamental data that made this geoecological synopsis possible.

We are grateful to the following for permission to reproduce copyright material: *Amazonia* for Table 4.3 (Anon/Klinge, 1972a); American Association for the Advancement of Science for Table 7.2 (Pimentel *et al.*, 1973) Copyright 1973 by the AAAS; American Institute of Biological Sciences for Fig. 8.5 (Nicholaides *et al.*, 1985a); the author D. E. Bandy for Table 8.6 (Bandy & Benites, 1983); Blackwell Scientific Publications for Tables 3.7 (Swift *et al.*, 1979) & 3.8 (Anderson & Swift, 1983); Butterworth Heinemann Ltd. and the author, Dr. P. A. Sanchez for Table 2.2 (Seubert *et al.*, 1977) © Butterworth Heinemann Ltd.; Cambridge University Press for Table 7.1 (Grigg, 1984); Columbia University Press for Tables 9.1 & 9.2 (Fearnside, 1986); the editorial office, *Die Erde* for Fig. 1.8 (Weischet, 1969); Elsevier Science Publishers

BV and the author, H. Linnemann for Fig. 1.14 & Table 1.6 (Linnemann *et al.*, 1979); Ferdinand Enke Verlag for Table 3.5 (Scheffer & Schachtschabel, 1976); Food and Agriculture Organization of the United Nations for Tables 1.7 (Higgins *et al.*, 1982), 2.3 (Sommer, 1976), 2.4 (Lanly, 1982) & Appendix 1 (FAO, 1986); Gebrüder Borntraeger for Figs 6.3 (Büdel, 1982) & 6.4 (Louis, 1964); Walter de Gruyter Publishers for Figs 1.1 & 1.3 (Blüthgen & Weischet, 1980): Institut für Wissenschaftliche Zusammenarbeit for Fig. 6.1 (Mensching & Ibrahim, 1977); International Wheat Council for Table 7.5 (International Wheat Council, 1980–81); Methuen & Co. for Fig. 1.11 (Morgan & Pugh, 1969); Methuen Educational for Fig. 1.2 (Ledger, 1969); Nathan for Fig. 5.1 (Benchetrit *et al.*, 1971) © Benchetrit and Alü, 1971; North Carolina State University (Soil Science Dept.) for Tables 8.3–8.5 (Bandy *et al.*, 1980), Table 8.6 (Bandy & Benites, 1983); Oxford University Press for Fig. 3.10 (Juo, 1981) Oxford University Press Inc. for Figs 3.5, 3.11 and Tables 3.6 (Birkeland, 1984) copyright © 1984 by Oxford University Press Inc. Reprinted by Permissions; Routledge for Figs 2.1 (Morgan, 1969a), 2.4 (De Schlippe, 1956), 2.9 (Morgan, 1969b) & 2.10 (De Schlippe, 1956); the author, Dr. P. A. Sanchez for Fig. 2.2 (Sanchez & Cochrane, 1980), Fig. 8.6 (Sanchez *et al.*, 1985a); Verlag Ferdinand Schöningh for Fig. 7.1 (Weischet, 1978); Springer-Verlag and the author, Prof. Dr. K. H. Wedepohl for Tables 3.2, 3.3 & 5.1 (Wedepohl, 1969); Franz Steiner Verlag for Figs 9.1 & 9.2 (Kohlhepp, 1976) & Table 1.5 (Carol, 1973); chief editor, *Tropical Ecology* for Table 4.4 (Klinge, 1973a); Unesco for Fig. 6.7 (Unesco, 1971) & Tables 2.1 (Rodin & Basilivic, 1968) & 4.1 (Van Baren, 1961) © 1961, 1968 & 1971 Unesco; The Regents of the University of California and the University of California Press for Fig. 2.3 (Spencer, 1966) copyright © 1978 The Regents of the University of California; University of Hawaii Press for Fig. 5.5 & Table 5.8 (Kawaguchi & Kyuma, 1977); UNRISD for Tables 5.12 & 7.4 (Palmer, 1976); Weltforum Verlag for Fig. 5.6 & Table 5.13 (Lagemann, 1977); Westermann Schulbuch Verlag GmbH for Fig. 1.12 (Schmidt & Mattingly, 1966); John Wiley & Sons Inc. for Fig. 2.5 & Tables 3.9 & 5.7 (Sanchez, 1976) copyright © 1976 John Wiley & Sons Inc.; Yale University Press for Fig. 5.2 (Pelzer, 1964) copyright © 1964 Yale University Press.

Whilst every effort has been made to trace the owners of copyright material, in a few cases this has proved impossible and we take this opportunity to offer our apologies to any copyright holders whose rights we may have unwittingly infringed.

Part I

The ecological fundamentals of the constraints

1
Introducing the problem

1.1 The theses to be proven

One of the pressing problems of world-wide significance is the marked difference in the socio-economic state of development between industrialized nations and Third World countries. The commonly accepted formulation of the North–South disparity reflects the geographic reality that almost all the so-called Third World countries lie in the tropics.

At present, among the numerous symptoms attributed to underdevelopment, insufficient food production, coupled with fast population growth, is one of the most aggravating. The countries involved include not only those of the semiarid outer tropics (for the definition of this term see section 1.2), periodically beset by catastrophic droughts, but also the countries of the humid inner tropics where climatic conditions permit lush natural vegetation.

The insufficient food production in those countries is not principally caused by a lack of arable land since, with a few exceptions, the ratio of population to arable land is very low and every year millions of hectares of virgin forest are cleared for agricultural purposes. The exceptional areas are usually very limited isolated pockets with extremely high population densities surrounded by sparsely inhabited areas of great size. Moreover, the high density of population correlates with high cultural standards in comparison with the thinly populated areas. Remarkable examples are the islands of Java and Bali as opposed to Borneo and large areas of Sumatra, or the *agreste* of northern Brazil compared to the Amazon basin. The reasons for spatial limitation of high population densities and their cultural advance will be discussed, along with other similar cases, in Chapter 5. First, it is necessary to deepen our understanding of the spatially dominating (nearly 80 per cent of the tropics) phenomenon of scant food production in areas of low man-per-hectare ratio (Chs 2–4).

The FAO's technical report (see also p. 35) *Potential Supporting Capacities of Lands in the Developing World* (Higgins *et al.*, 1982) indicates for the agroclimatic zone of the warm humid tropics, with a growing period longer than 330 days, such as the Amazon basin, an actual population density of 0.02–0.05 persons per hectare. It suggests that the potential population density can range from 1.02 to 1.39 persons per hectare (102–139 people per square kilometre), even under the still prevailing low-level input circumstances, i.e. 'assuming only hand labor, no fertilizer and no pesticide applications, no soil conservation measures, . . . and cultivation of the presently grown mixture of crops and potentially cultivable rainfed lands'. Under conditions of intermediate or high-level inputs the report ventures to state that the estimated densities would be 219–408, or 831–985 persons per square kilometre, respectively. High-level input means 'complete mechanization, full use of optimum genetic material, necessary farm chemicals and soil conservation measures, and cultivation of only the most calorie-protein productive crops on potentially cultivable rainfed lands' (Higgins *et al.*, 1982, p. 32).

This specialized publication by FAO experts expresses the still widely held view that the agricultural production modes and the land-use systems found today in those tropical countries where food production is scarce can first be improved and ultimately be replaced by modern technology. As a consequence of this attitude one would have to assume that the still dominating traditional agricultural practices are underdeveloped and that the difficulties that the world is confronted with today would not have occurred if better agricultural systems had been adopted in the tropics sooner.

The scientific shortcoming of this FAO report and of other likewise optimistic assessments is that the authors do not seek the reasons for the persistence of the traditional land-use systems and do not bother to prove whether modern agrotechnical methods, with the implied use of fertilizers, are applicable to the regions, where despite all technical assistance slash-and-burn practices and shifting cultivation dominate even now. We will see that precisely the difficulties with the application of inorganic fertilizers are the main obstacle in all efforts to replace the so-called underdeveloped traditional land-use systems by modern and more efficient agrotechnologies (Chs 3, 7 and 8).

The dominating traditional farming practices in the tropics outside of the densely populated pockets comprise those rotational land-use systems that alternate a cropping period with annuals, biennials or short-lived perennials with a long-lasting fallow period during which the abandoned crop area is recolonized by a succession of native vegetation. According to FAO calculations (Hauck, 1974), about 36 million km², or roughly 30 per cent of the total croplands of the world, are thus cultivated.

As these systems are evidently land-demanding, forest-squandering and increasingly failing 'to satisfy the requirements for higher production per unit area' (Hauck, 1974), the first question to arise is why shifting cultivation or land rotation practices, which were abandoned more than 1000 years ago in the subtropical and temperate middle latitudes, persisted in the tropics? And second, why did 30 years of international assistance with the know-how of modern agricultural technology accomplish improvements on only a local level without changing the detrimental effects of shifting cultivation in general?

The problems connected with the first question will be dealt with in the first part of the book (Chs 2–6). In this part we wish to provide the proof, or at least the elements of judgement, of why the largest part of the tropics (80–85 per cent) suffers from a handicap that is rooted in the natural conditions that render agricultural production more difficult than in subtropical or mid-latitudinal regions. This ecological disadvantage is of such magnitude that it suffices to explain the historically documented North–South disparity more convincingly than all socio-economic reasoning.

The second part of the book considers the efforts that have been undertaken in international collaboration in order to overcome decisive ecological constraints and to specify alternative farming practices. In so doing it will be established that agro-ecological disadvantages restrict agricultural production even under modern technological conditions (Chs 7–9).

In order to define the key ideas of this book particular theses are proposed in the following paragraphs. Before spelling out the theses, a preliminary differentiation must be made first between the highlands and lowlands, and then within the tropical lowlands between the permanently humid inner tropics, where the growing period lasts at least 330 days, and the semihumid and semiarid outer tropics, where rainfed agriculture is restricted to a period of half a year.

For the permanently humid tropical warm lowlands the thesis to be proven is: it seems that shifting cultivation or land rotation with bush-fallowing has evolved as a specialized technique in response to certain restrictive soil properties. In 80–85 per cent of that area, even with the application of modern agrotechnics, it is not possible at the scale of normal farm units to replace the extremely biomass-rich forest by continuous and fertilizer-intensive cropping systems for the production of cereals or tubers as man's main staples.

The key arguments for proving this thesis are that the inorganic mineral substances of the warm humid tropics' dominant soils (oxisols, ultisols, xanthic and orthic ferralsols) are the result of a lengthy and intensive allitic weathering process (see Ch. 3). The clay minerals of the leached and desilified soils consist almost exclusively of 1 : 1 layered kaolinites. This is why the mineral substance is extremely poor, not only in nutrients but also in cation exchange capacity (CEC). Under an undisturbed natural forest, soil nutrients and CEC are bound mainly to organic matter, i.e. humic acids. With the mineralization rate of organic matter being at least five times higher than in temperate climates, the main carriers of nutrients and CEC disappear from the soil within 2 or 3 years after removal of the natural forest and its replacement by a biomass-exporting system for crop production. Modern agrotechnology, with its intensive use of common fertilizers, does not work on these soils owing to their low CEC and also because of the high amounts of rain and the frequent occurrence of intensive downpours.

With regard to the *semihumid* and *semiarid outer tropics* the following thesis applies:

in general, the soils of that part of the tropics are well endowed with nutrients and have a good CEC, conditions which are principally conducive to continuous cultivation. Their peculiar disadvantage is conditioned by the region-related problem of developing suitable water management systems and irrigation in order to bridge the seasonal dryness and to lessen the consequences of droughts which are notorious for the regions near the poleward boundary of rainfed agriculture.

In the semiarid regions most of the soils are sialitic, which means they possess a relatively high nutrient content and are rich in 2 : 1 layered clay minerals of high CEC. Irrigation could guarantee the production of enough food for a large population. The actual situation, however, is that in the semiarid regions of Africa and South America, flood control and irrigation systems are restricted to a few small areas. In the Deccan Plateau, 29 out of the 40 dams which existed in 1961 were built after 1951. Why then, in such a drought- and famine-prone region, did it take so long to build these installations, while in the surrounding lowlands large-scale water management and irrigation systems have been in use for generations?

The main reason for this inconsistency lies in the particular configuration of land surfaces in the semiarid outer tropics (see Ch. 6), which are part of a geomorphological zone characterized by excessive planation surfaces (peneplains). The rivers do not cut deep valleys into the basement; they flow in extreme shallow troughs, regardless of their altitude above sea-level. A consequence of this particular geomorphological condition is that the average width of the Deccan Plateau dams exceeds that of the longest of all Mediterranean dams. The construction of the extended dams is further complicated by the seasonal floods of the monsoon streams that peak at 30 000 or even 40 000 m^3 5^{-1}. It requires a transportation technology that permits closing the inner parts of the dam between two floods, a technological capability that did not exist prior to its development during the Second World War.

Summing up what has been stated so far: in the wet inner tropics with their optimum climate for plant growth, the transition from land rotation with bush fallowing to continuous cropping was prevented in most areas by an ecological barrier caused by pedological conditions. This resulted in relatively low agricultural population densities, low labour inputs, and low yields per surface unit. We assume that cultural progress stagnated, due to the lack of minimum population density required for social and cultural diversification.

In the wet-and-dry outer tropics the advance to more intensive permanent agriculture was possible, provided sufficient irrigation water was available. In this regard, there are two completely different environments: the intensively used cultural landscapes under irrigation since early times on one side, and the hazardous realm of rainfed agriculture on the other (see Chs 6 and 7).

The theses presented have differing weights regarding tropical Africa, South America or southern Asia. Especially in Africa, but also in South America, the ecologically disadvantaged realms predominate absolutely, while in southern Asia they are more restricted.

In order to substantiate the postulates and to put across the basic reasoning, one needs a great deal of special knowledge drawn from different geosciences, such as climatology, pedology, hydrology, mineralogy, geobotanics and geoecology. We are aware of the difficulties of such an endeavour, but hope that, in the end, a convincing deduction is presented that can be grasped by the reader.

1.2 Climatic delimitations and characterization of the tropics

1.2.1 *Radiation conditions*

A simple but effective delimitation criterion of the tropics is the radiative–climatic one which defines them as the area between the Tropic of Cancer (23° 27′ N) and the Tropic of Capricorn (23° 27′ S). Accordingly, the tropical belt is the part of the world where, at local noontime, the sun is, at least once a year, directly overhead and even during the period of low sun, never below 66° at the Equator and 43° at the outer margins of the tropics. Within this belt there are 12 hours of daylight all year round near the Equator, and between 10.5 and 13.5 hours in the outer margins.

For many reasons it is convenient to differentiate between the inner tropical belt (inner tropics) as the latitudes on either sides of the Equator and the adjacent outer tropical zones (outer tropics) which lie beyond this belt, but within the tropical circles.

The highest annual energy input on the earth's surface occurs near the tropical margins due to their proximity to the subtropical dry belts, where the atmosphere is practically cloud-free and low in water vapour. Conditions are different in the humid and cloudy inner tropics where a large part of the solar radiation is reflected and absorbed by the atmosphere. Even there, the sum of direct and indirect short-wave radiation (global radiation) has throughout the year the same magnitude as in the mid-latitudes during a few midsummer days. Evidence of the ensuing energy surplus is the fact that, in the tropical rain forest, despite several storeys of trees and a lush display of epiphytes, climbers and lianas, a dense undergrowth of broad-leaved herbs thrives in the dim light near the ground.

1.2.2 *Thermal conditions*

Closely related to the radiant energy intake are the thermal conditions in the tropical lowlands. In climatological classifications, the tropics are generally limited by the 18° isotherm of the coldest month. In this way, a climatic belt is defined in which the temperatures during the coldest month are comparable to those experienced in the maritime-influenced parts of the mid-latitudes at the height of the summer, or in more continental parts in early summer. This means that throughout the year growing conditions in the thermally defined tropics are similar to those governing the agriculturally highly productive countries of the mid-latitudes during the summer months only.

Within the tropical lowlands regional as well as temporal differences in thermal conditions are relatively slight. To illustrate the combined effect of both, the seasonal and diurnal variations of air temperature (thermo-isopleths) of an equatorial station (Singapore) and an outer tropical station (Nagpur) are shown (see Fig. 1.1).

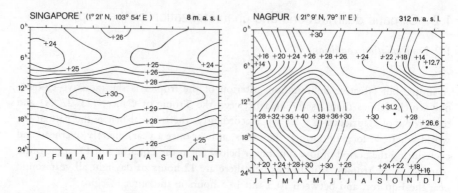

Fig. 1.1   Seasonal and diurnal variations of mean air temperature (thermo-isopleth diagrams) of an equatorial (Singapore) and an outer tropical lowland station (Nagpur, India). (Source: Blüthgen and Weischet, 1980)

In the equatorial region temperature conditions are characterized by seasonal uniformity and diurnal variations. At marine locations the latter are particularly small, but also in the interior of the Amazon or Congo basins the daily range does not exceed 8–10 °C. These conditions resemble those of a greenhouse. Sufficient humidity provided, the inner tropics offer optimal climatic conditions for plant growth. Unfortunately, however, the high energy input, continuous high temperature and humidity result in intensive chemical weathering of mineral substances and rapid decomposition of organic matter, with all its unfavourable ecological consequences.

The Nagpur diagram is typical for a continental station near the outer margins of the tropics. The daily and annual range of mean temperatures and diurnal variations is generally larger than in the inner tropics. Maximum temperatures occur shortly before the onset of the rainy season, which brings about a reduction of radiation input by clouds, haze and rain. Temperature minima are reached after the rains during the low-sun season. The diurnal range is considerably larger than in the inner tropics.

1.2.3   Thermal effects of elevation

The relative horizontal uniformity of thermal conditions that prevails in tropical lowlands is superimposed by a strong vertical differentiation. In the free atmosphere the rate of decrease (lapse rate) is usually around 5 °C per 1000 m elevation. In mountainous regions it varies in response to radiation and topography. Under the cloudy skies of the humid inner tropics or during the rainy season of the outer tropics the decrease approaches that of the free atmosphere. Under clear skies, however, solar radiation raises the air temperature over extended highlands during the day and the resulting temperature difference between lowlands and highlands is smaller than in the free atmosphere. At night the heat loss through terrestrial irradiation is greater than over the lowlands, but the cooled air can flow down the mountain slopes so that the lapse rate may increase during night hours, but not enough to equalize the positive effect of the daytime upland heating by solar radiation.

The result is that the mean air temperature over dry tropical highlands is generally some degrees higher than in the free atmosphere at the same altitude and that the daily range is greater than in the adjacent lowlands. Calculating with a lapse rate of 5 °C per 1000 m, the resulting theoretical temperature is always lower than the recorded temperatures. Variations are due to special topographic conditions, to seasonal cloud cover and rainfall, to prevailing winds, and to local circulations. The overriding rule for the decline of temperature with elevation is that the mean temperature of the coldest month reaches the thermal limit of the tropics (18 °C) at the 1000–1200 m level. At 2500 m the lowest monthly mean temperatures equal those of cool summer days in temperate latitudes. Above this level, night frosts are already to be reckoned with during the season of low sun.

The change of temperature has its obvious impact on the environmental conditions. Of interest here are the consequences with respect to agricultural possibilities: in all tropical high mountains a vertical succession of mountain belts is recognized. In Latin America they are called *tierra caliente*, *tierra templada*, *tierra fria* and *tierra helada*.

The *tierra caliente*, which reaches up to 1000 m above sea-level at most, is characterized by an annual shift of monthly mean temperatures between 20 and 30 °C. The characteristic tropical lowland crops are cultivated here, cacao, oil-palms and bananas being the most significant.

In the *tierra templada* between 1000 and 2000 m (2500 m at the most) monthly mean temperatures vary between 15 and 21 °C. Typical *tierra templada* crops are coffee and tea as well as a number of subtropical crops, such as maize, sugar-cane and citrus.

The *tierra fria* is situated between 2500 m and the altitudinal boundary of agriculture. This upper limit lies near 3500 m in the humid equatorial tropics and extends up to a maximum height of 4000 m above sea-level near Lake Titicaca in the Bolivian *altiplano*. Thermal conditions are similar to those in the oceanic parts of mid-latitudes with monthly means ranging from 8 to 18 °C. Characteristic crops are those of the mid-latitudes, such as wheat, barley and potatoes. In the ever-humid high mountains of Colombia, potatoes can be cultivated all year round, but they need 11 months to mature.

The 200 or 300 m above the upper limit of agriculture are often called the *páramo* or *puna* belt, named according to plant formations of tall herbs and low tussock grass which prevail at these heights in the equatorial Andes or the altiplano, respectively. Only extensive grazing is possible.

The *tierra helada* reaches up to the limit of permanent snow and ice. The greater part of the year is dominated by strong winds and crisp weather with fog and snowfall, and a great number of days with freeze–thaw action. This fringe is unsuitable for any kind of land use.

1.2.4 *Soil temperatures*

Records of soil temperatures are very scarce. Apart from the atmospheric conditions, they depend in a very complicated way upon the vegetation cover and physical characteristics of the soil. Measurements under grass at different depths near Jakarta (Indonesia) by C. Braak (1929) may serve as an example (Table 1.1). The values indicate that, under humid tropical conditions, the mean yearly temperature of the soil exceeds that of the atmosphere by about 3 °C and that the extreme values range between 38 and 23 °C. The discrepancy between air and soil temperature decreases with increasing vegetation density. In the middle of tropical forests the mean soil temperatures are almost the same as those of the air.

Table 1.2 gives values of the mean and of the absolute extreme temperatures from the dry savanna region of West-Africa.

In the wet-and-dry outer tropics, experiments conducted by Ramdas and Dravid (1936) at the Central Agricultural Meteorological Observatory of Poona (India) showed that during the dry season, with its relatively low sun, the temperatures at a depth of 10–20 cm in different soils oscillate between 22 and 24 °C during the morning hours and between 24 and 26 °C in the afternoon. On the surface itself 45–50 °C were registered. During high sun, the means rise to 58–65 °C at the soil

Table 1.1   Soil temperature (°C) at Jakarta (Indonesia) (after Braak, 1929. Taken from Lockwood, 1976)

Depth (cm)	Mean daily maximum temperature	Yearly mean of temperature 24 hours	Absolute maxima of temperature	Absolute minima of temperature
Air	29.97	26.11	35.8	18.3
5	32.11	29.30	37.9	23.1
10	31.09	29.48	35.6	23.6
30	29.56	29.44	31.5	23.4
90	29.51	29.49	30.1	26.9
110	29.49	29.47	29.9	26.3

Table 1.2   Soil surface temperatures (°C) at Niamey (Rep. of Niger) (Values taken from Janke, 1972)

	Jan.	Feb.	Mar.	Apr.	May	June	July	Aug.	Sept.	Oct.	Nov.	Dec.
Abs. maximum	51	52	58	65	61	60	60	56	58	62	56	54
Mean	27	31	34	37	36	34	32	31	33	34	30	27
Abs. minimum	6	8	9	9	13	12	13	16	12	8	7	6

surface and to 30–32 °C in the morning and to 30–37 °C in the afternoon at a depth of 10–20 cm. From all these observed results, it is found that at the ecologically decisive depth of 10–30 cm the soil temperatures lie within the 25–30 °C range everywhere in the tropics.

1.2.5 *Hygro-climatic conditions*

Overlaying the optimal radiative and thermal conditions, hygro-climatic factors play the decisive role with respect to plant growth and soil formation. With evapotranspiration being so high, the quantity and temporal distribution of the rainfall are crucial for the soil, flora and fauna as well as for the subsistence of man.

It is common knowledge that, with increasing distance from the Equator, there is a decrease in quantity of annual rainfall and in length of the rainy season (Fig. 1.2 for West Africa). In a general overview the following zones have been distinguished (Figs 1.3 and 1.4):

1. The wet equatorial climates, where rainfall is copious in at least 11 months of the year and mean annual totals range between 2000 and 3000 mm normally and up to several metres in extreme cases. No month receives less than 50 mm, and precipitation exceeds evapotranspiration in at least 11 months out of 12.
2. The tropical wet-and-dry climates (approx. 5° to 20°N and S), which represent

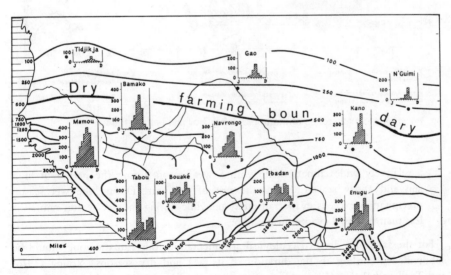

Fig. 1.2 Mean annual amounts and course of monthly rainfall in West Africa (after Ledger, 1969)

11

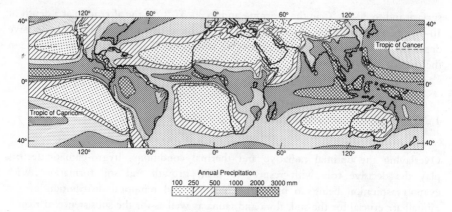

Fig. 1.3   Mean annual precipitation in the tropics (after Blüthgen and Weischet, 1980)

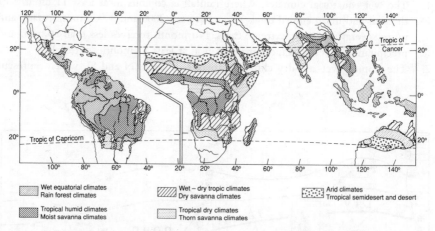

Fig. 1.4   Hygro-climatic zones of the tropics (after Troll and Paffen, 1964)

a relatively wide climatic zone with a wet season during the period of high sun and a dry season during the rest of the year. During the wet season precipitation exceeds evapotranspiration, while during the dry season rainfall is less than potential evapotranspiration.

For the sake of a detailed ecological argumentation, this climatic zone should be subdivided according to the duration of its wet and dry seasons. Young (1976) and Troll and Paffen (1964) differentiate the following subzones within the tropical wet-and-dry climates (since both subdivisions are based on the regional zonation of

12

the natural vegetation belts, they differ only in their nomenclature, hardly in their climatological criteria):

2a The rainforest–savanna transition climates (Young) or tropical rainy climates (Troll and Paffen) in successive bands at both sides of the wet equatorial climate, with 11 to 9.5 humid months and annual precipitation amounts between 1500 and 800 mm.

2b The moist savanna climates (Young) or tropical humid climates (Troll and Paffen) with 9.5 to 7 humid months and precipitation amounts between 1000 and 700 mm.

2c The dry savanna climates (Young) or genuine wet-and-dry tropical climates (Troll and Paffen) with 7 to 4.5 humid months and precipitation amounts between 900 and 600 mm. In the central part of this belt, the wet and dry spells are of equal length.

2d The semiarid climates (Young) or tropical dry climates (Troll and Paffen) with 4.5 to 2 humid months and annual precipitation amounts between 600 and 250 mm.

3. The tropical dry climates with less than 2 humid months and annual precipitation means below 250 mm. This type includes the 'tropical semi-desert and desert climates' of Troll and Paffen and the 'arid climates' of Young.

Apart from quantity and seasonal distribution, the regional and temporal variability and the intensity of tropical rainfall are also of profound ecological importance. Both characteristics are consequences of the fact that tropical precipitation occurs in the form of showers that result from convectional clouds in tropical warm and moist air masses.

It is a general climatological rule that the interannual variability of the amounts of rainfall as well as its temporal sequence increase with decreasing annual totals and with shortening of the rainy reason. Figure 1.5 illustrates these facts for the African tropics. The consequences of the variability are grave, especially in the critical zone near the boundary of rainfed agriculture in the semiarid climate. With long-term means ranging between 600 and 450 mm and a mean annual deviation of 30 per cent, there may be 800 to 600 mm of precipitation in one year and only 400 to 300 mm in the next. Unfortunately, negative deviations tend to coincide with a delayed onset of the rainy season, a very irregular shower activity, and concentration in a few downpours.

Typical of this situation are the measurements near Niamey (Republic of Niger) reported by Janke (1972) (Figs 1.6a and b). Figure 1.6(a) illustrates the sequence of yearly totals and the repetitious tendency of 'meagre and good years' over periods of 10–12 years. Figure 1.6(b) depicts the actual rainfall sequences during a dry and during a wet year. In 1949, delayed onset, scarce frequency and high intensity of rain showers caused a crop failure in the whole area around Niamey and produced a retreat of rainfed agriculture far to the south of Niamey. In contrast, during the humid year of 1952, the dry limit of agriculture moved some 200–250 km north into the semi-desert (Janke, 1972). The far-reaching implications of this problem

13

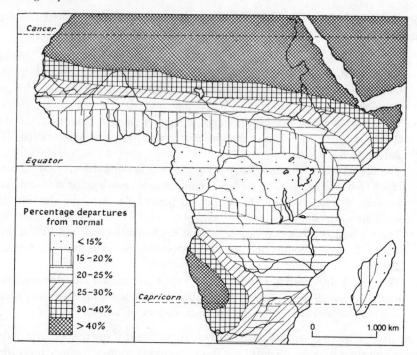

Fig. 1.5   Variability of annual rainfall amounts (after Biel, 1929, taken from Gregory, 1969)

will become more evident in a later paragraph where the relatively high population density, land demand, ineffective water management, and shortage of irrigation water are addressed.

The 1952 data of Fig. 1.6(b) show another property of tropical convective precipitation which has severe agricultural implications, namely the occurrence of particularly heavy downpours. The 129 mm registered on 28 August at the end of the growing season probably did no serious damage to the established plants. But what if a downpour of comparable intensity had occurred shortly after planting or seeding? The possible effects can be illustrated with data from the experimental station in Yurimaguas, in the Peruvian part of the Amazon basin (Fig. 1.7).

During the relatively wet year of 1984, daily precipitation values surpassed 50 mm on 13 out of 173 rainy days. On three occasions (February, April and September) amounts close to 100 mm in 24 hours were recorded. The drier year of 1985 had only 135 days with rainfall, 8 of which had more than 50 mm. On 3 days the 100 mm level was surpassed, and on 21 March a maximum of 226 mm was registered. Noteworthy is an almost even distribution of the downpour events over the year with a slight increase only between October and May.

Looking synoptically at the five downpours of more than 50 mm that occurred in October 1983 and the corn planting and fertilization dates during the same month, it

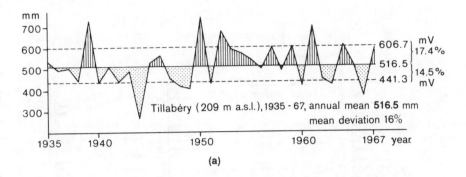

Tillabéry (209 m a.s.l.), 1935 - 67, annual mean **516.5** mm
mean deviation 16%

(a)

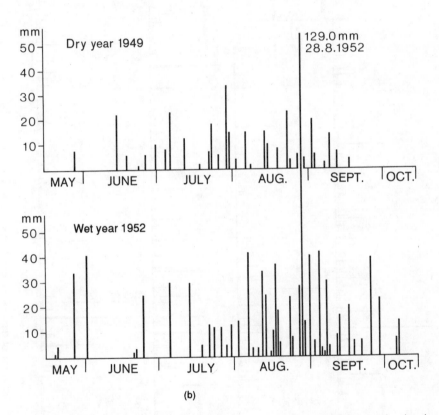

(b)

Fig. 1.6 Irregularity of rainfall in the outer tropics: (a) the sequence of yearly totals, showing variability of annual precipitation in west Niger; (b) the actual rainfall sequence during a dry and during a wet year in Niamey (Rep. of Niger) (after Janke, 1972)

15

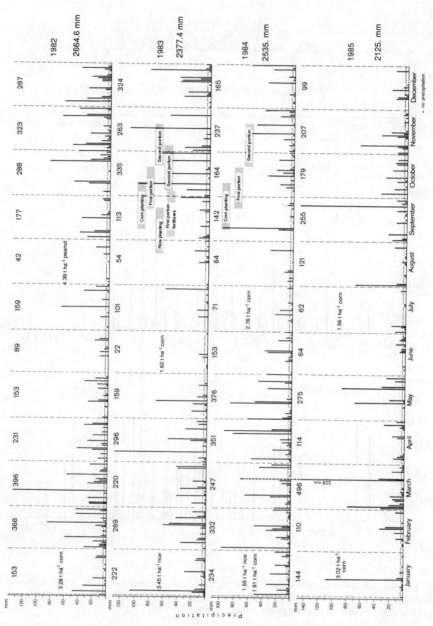

Fig. 1.7 Sequence of daily rainfall amounts for rainy and less rainy years in the wet inner tropics of the Amazon basin (Yurimaguas, Peru).

is easy to imagine what happened to the crops that were drenched five times with 50 or even 100 l m^{-2} of water. In this context it must be taken into account that the precipitation values are given for a time unit of 24 hours for the purpose of observation practices only. In reality a rainstorm lasts 'rarely more than about two hours' (Nieuwolt, 1977). This means that in the case of agricultural land use most of the ash or the fertilizer can be washed away, and the young seedlings possibly as well. Since rains of the mentioned intensities are inherent of tropical precipitation regimes it becomes evident that in an ecological context argumentation with monthly or yearly values only is insufficient. Utilization of at least daily values becomes a necessity. For more details see sections 8.1 and 8.3.

Except for the very limited areas of cultivation and of road surfaces, the natural soil surfaces in the wet equatorial tropics are well protected by natural vegetation cover. Consequently, erosional hazards are minor in that part of the tropics. They become a major problem, though, in the wet-and-dry climates of the savannas. Especially where land use alternates between pastures and crop cultivation, soil erosion leads rapidly to the formation of badlands (see section 8.3).

1.2.6 *Hygric effects of elevation*

In extratropical mountains it is a rule that, over a longer period of time, the amount of precipitation increases to an altitude of approximately 3500 m. This is not so in tropical areas.

As early as 1950, while interpreting the very dense network of climatological stations in Java, de Boer noticed an increase in precipitation up to 1100–1200 m only. Higher up, there was a significant decrease in annual precipitation amount (Table 1.3). In later years values for maximum precipitation amounts at heights between 900 and 1300 m were given by Leopold (1951) for Hawaii, Schmidt (1952) and Trojer (1959) for Colombia, and Hastenrath (1967) for Central America. The general rule for this phenomenon, the detailed topographic dependency, and the reasons for the difference with regard to the conditions outside the tropics, were elaborated by Weischet (1966, 1969) on the basis of several stations in the coffee-growing areas of the Colombian Andes. For a cross-section through the Andes near 5° N. Fig. 1.8 gives the location and altitude of the stations in correlation with the mean annual amounts of precipitation. The following facts emerge: tropical high mountains are relatively dry in comparison with the surrounding lowlands. In the Andes, precipitation increases from the valley bottoms over the slopes of the various ranges (occidental, central and oriental) up to an altitude of some 1000–1100 m. Above this level there is a distinct decrease in rainfall. In vast high plains or basins surrounded by mountain ranges, such as the 'Sabana de Bogotá' in the Cordillera Oriental, orographic air lift causes a secondary increase in rainfall (stations Arrayan and El Verjón).

In the meantime, the general validity of the rule was proven in other publications (Flohn, 1968; Domroes, 1968; Hastenrath, 1968). Differences of secondary importance

17

Table 1.3 Vertical distribution of precipitation in Java (de Boer, 1950)

	Intervals of height (m)																
	Sea-level	50–150	150–250	250–350	350–450	450–550	550–650	650–750	750–900	900–1100	1100–1300	1300–1500	1500–1750	1750–2000	2000–2500	2500–3000	3000–3500
West Java																	
Number of stations	36	87	55	63	46	48	45	44	39	67	47	25	20	11	7	—	2
Mean height (m)	1	100	200	300	400	496	600	700	809	989	1181	1411	1607	1827	2221	—	3050
Mean yearly rainfall (mm)	1990	3038	3375	3237	3573	3454	3546	3151	3695	3389	4460	3529	2908	3492	3738	—	3070
Central Java																	
Number of stations	39	201	137	67	49	46	36	26	17	32	14	11	7	—	11	3	3
Mean height (m)	2	100	200	300	400	500	600	700	806	999	1175	1382	1605	—	2133	2582	3191
Mean yearly rainfall (mm)	2313	2554	2483	3044	3007	3139	3487	3895	4062	3815	4094	3496	3294	—	3267	2481	2473
East Java																	
Number of stations	36	208	104	85	85	68	42	31	19	24	10	6	3	7	5	6	3
Mean height (m)	2	100	200	300	400	500	600	700	821	990	1186	1377	1663	1877	2239	2720	3392
Mean yearly rainfall (mm)	1526	1930	2231	2443	2442	2708	2942	3186	3391	3073	3401	2840	2195	2176	1949	1818	1933

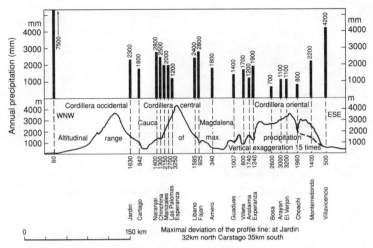

Fig. 1.8 Vertical distribution of yearly rainfall amounts for a hygric cross-section of the Colombian Andes at 5°N. (Source: Weischet, 1969)

relate to findings concerning the altitude of the maximum precipitation-level only: it varies between 900 and 1500 m.

With precipitation diminishing drastically in tropical mountains above 1000 m, the environmental conditions that prevail in the agroclimatic belts of *tierra templada* and *tierra fria* correspond not only in thermal, but also in hygric respects to those which are characteristic for the lowlands of oceanic extratropical climates.

The significant difference between tropical and extratropical vertical rainfall distribution types is due to the origin of the precipitation. Within the tropics, most of the rainfall originates from convectional clouds, while in the extratropics the greater part results from frontogenetic clouds. In the case of predominance of convective cloud formation by vertical air movement, the large decrease in water vapour content from the warm humid lower layers to the cooler and drier upper layers of the troposphere causes the clouds above the mountains to have a smaller water supply than those originating over the lowlands. Therefore, in the higher parts of the tropical mountains, rains are generally less in quantity and intensity than in the lowlands and foothills of high mountains (Weischet, 1966).

1.3 Natural vegetation regions

Since the classification and regionalization of tropical climates are generally made in accordance with the natural vegetation distribution, it is not surprising that the boundaries found on vegetation maps coincide with those on climatic maps. In a transect from the permanently humid regions near the Equator to the semi-deserts at the outer fringes of the tropics, the following natural vegetation belts are distinguished, with minor deviations among authors (Fig. 1.9):

19

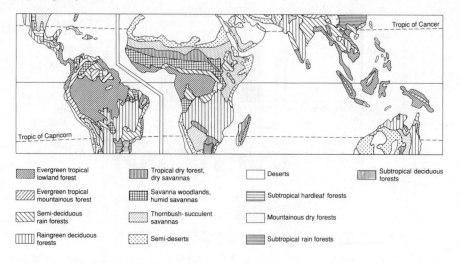

Tropic of Capricorn

Evergreen tropical lowland forest	Tropical dry forest, dry savannas	Deserts	Subtropical deciduous forests
Evergreen tropical mountainous forest	Savanna woodlands, humid savannas	Subtropical hardleaf forests	
Semi-deciduous rain forests	Thornbush-succulent savannas	Mountainous dry forests	
Raingreen deciduous forests	Semi-deserts	Subtropical rain forests	

Fig. 1.9   Map of the natural vegetation belts in the tropics (after Schmithüsen, 1976)

1. *The evergreen tropical lowland forests* (essentially synonymous with 'equatorial rain forests' or 'tropical broadleaf evergreen forests'). The term 'evergreen' refers to the permanently green appearance due to the low proportion of deciduous vegetation and the fact that different trees shed their leaves at different times during the year, replacing them at once by new ones.

Concerning the annual quantity of dead leaves and litter, Anderson and Swift (1983) have produced the following table (t ha^{-1} yr^{-1}):

Malaya	6.3 leaves	10.6 litter
Ivory Coast	7.8–8.1 leaves	—
Zaïre	—	12.3 litter
Nigeria	7.2 leaves	—
Ghana	7.4 leaves	9.7 litter
Colombia	—	8.5 litter
Sarawak	6.6 leaves	9.4 litter

For the Manaus area of Brazil Mabberly (1983) estimated the leaf-fall at some 7 t ha^{-1} yr^{-1}.

These values are extraordinarily high as compared with those of the temperate deciduous forests, where values of 3 t of leaves and 5–5.5 t of litter are recorded.

The *equatorial rain forest* consists of tall, close-set trees with relatively small crowns which tend to be arranged in three storeys. The tallest trees reach heights of 40 m or more, their crowns towering above the rest. The second storey is continuous and occupies the level between 15 and 30 m. The third storey is made up of smaller trees between 5 and 15 m with narrow crowns. The undergrowth consists of tree saplings, shrubs and herbs. All layers are thickly interlaced with wooden lianas; trunks and branches of the trees host a variety of epiphytes, especially orchids, ferns, mosses and lichens. The floor of the rain forest is covered by dead leaves and a tangle of branches and trunks at different stages of decomposition. Litter and dead wood are covered with all kinds of fungi.

Regarding the standing phytomass, experts have come up with different values. Klinge (1976b) has compiled the following data (t ha^{-1} of dry matter):

New Guinea	Mountain rain forest	571
Manaus	Rain forest	473
Panama	Rain forest	360
Ghana	Humid forest	287
Colombia	Rain forest	181

Another important aspect of the rain forest is the extreme floral diversity: more than 1000 different species can be counted on 1 km^2 and it is hard to find two identical large trees within 1 ha.

2. The *semi-deciduous rain forests*, located in a transition zone between the equatorial rain forests and the raingreen tropical deciduous forests. They are composed of the more xerophytic of the rain forest species and of raingreen deciduous trees and shrubs.

3. *The raingreen deciduous forests* (essentially synonymous with 'monsoon forests' in Asia, '*miombo* forest' or 'tree steppe' in Africa and *campo cerrado* in Brazil). They are characterized by a lower tree density, massive trunks, and a concentration of the large crowns between 12 and 30 m. Most trees shed their leaves during the dry season. Since there is less competition for light, the undergrowth often consists of a dense shrub thicket with patches of herbs and tall grasses. Lianas and epiphytes are less abundant than in the rain forest. Information about ground litter is scarce. Anderson and Swift (1983) cite a value of 4.7 t ha^{-1} yr^{-1} of leaves for seasonal forests in Nigeria.

Under the same climatic conditions as the raingreen deciduous forests are found the 'savanna woodlands' (essentially synonymous with 'humid savanna', 'tall grass savanna', or 'tall-grass low-tree savanna'). They cover extended areas south of the Equator and especially in the Sudan region north of the tropical rain forest. Biogeographers hold the view that savanna grasslands owe their origin to:

(a) changing climate conditions that have chronically hindered the development of arboreous species but favoured the growth of grass species;

(b) edaphic drying conditions that relate to the physical character of soils and lead to the proliferation of grass varieties;

(c) anthropogenic circumstances that result from occasional or periodical forest burning for the purposes of rounding up game or clearing plots for episodical cultivation (Walter, 1977).

The plant formations of the savannas consist of widely spaced xerophytic trees of particularly fire-resistant species, embedded in a thick layer of tall grass (2 m and more). In locations where the soil-water regime is favourable, i.e. along river banks, in creek valleys, on natural river dams, or on abandoned termite hills, evergreen forest formations occur in the form of bands or islands.

4.   *The 'dry savanna' or 'thorntree–tall grass savanna'* ('acacia–tall grass savanna', 'Sudan savanna', *campo sujo* in Brazil). As the dry period lengthens to almost 6 months, the locations with evergreen forest disappear and the deciduous trees become more widely scattered. There is an abundance of flat-topped acacia; other species, such as the grotesquely shaped baobab in the Sudan savanna, have water-storing trunks and branches. The undergrowth of tussock grass (1–2 m in height) is found everywhere, though without covering the ground entirely.

5.   The dry savanna progresses into a belt that is dominated by relatively dense stands of low thorn trees, thorn brush woods, thorn shrubs, and various species of succulents with sparse undergrowth of raingreen herbs and grasses: *the thornbush–succulent savanna*, 'thorntree–short grass savanna' or 'acacia–short grass savanna', *caatinga* in the *noreste* of Brazil.

6.   Beyond the thornbush–succulent savanna begins the open *semi-desert* with its widespread small xerophytic shrub vegetation and poorly developed herbaceous layer.

Figure 1.9 offers an illustration of the regional distribution of natural vegetation in tropical Africa and South America coinciding with the hydrometeorological conditions on both continents.

With regard to the discussion of population distribution and agricultural activities within the tropics, it should be noted that the poleward boundary of rainfed agriculture is to be found in the outer fringes of the dry savanna. In these areas yearly mean precipitation oscillates between 400 and 500 mm distributed over 3–4 humid months. The thornbush–succulent savanna and the semi-desert serve only as periodic pastures for nomadic herding and in restricted areas for artificially irrigated agriculture.

1.4   Population distribution in the context of natural regions

In order to prove a presumptive spatial coincidence between natural regions and population distribution it is appropriate to use demographic data from the years subsequent to the Second World War. On one side, the data are reliable and regionally well differentiated; on the other, the influence of the post-war population explosion and of internal/external migrations did not have the masking effect that they have today.

Concerning the continental extension of the tropics, Nye and Greenland (1960) provided the following values:

Continent	Area	Inhabitants	
	(million km²)	Total (millions)	Density (per km²)
Africa	17	104	7
South America	15	65	5
Asia	9	530	70

The aggravating differences between the African and South American tropics, on one hand, and the Asian tropics, on the other, necessarily compel us to look for the reasons that can explain the differences. This will be done in Chapters 5 and 7. For now we concentrate our attention on the less intensive inhabited tropics of Africa and South America.

In the framework of the natural regions that were outlined above, the population distribution shows the following characteristics (Figs 1.10–1.12):

1. The highest agricultural population density over extended areas is found in the tropical highlands. In South America, the difference in human occupancy between the eastern tropical lowlands and the highlands of the Andes is so fundamental as to present two different worlds. This has been most dramatic during pre-Columbian times. In the Andes we find the high cultures of the Inca and Muisca with their relatively high population densities while the eastern lowlands were thinly populated by hunters and collectors or shifting cultivators (see Ch. 2). The difference continues until today in the contrasting occupancy of the Colombian, Ecuadorean and Peruvian highlands versus the respective 'Orientes' (eastern lowlands). Most interesting examples are Rwanda and Burundi, the smallest but most densely populated countries in Africa with 209 and 156 inhabitants per square kilometre respectively (Fig 1.12). This exception with regard to human

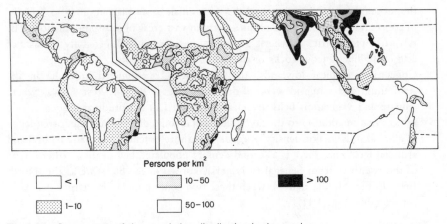

Persons per km²

< 1   10–50   > 100

1–10   50–100

Fig. 1.10   Survey map of the population distribution in the tropics.

23

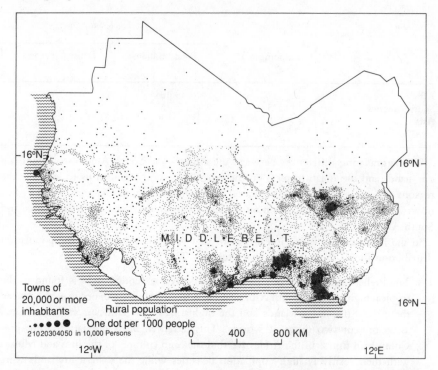

Fig. 1.11 The distribution of urban and rural population in West Africa. (Source: Morgan and Pugh, 1969)

occupancy coincides with a natural geographic feature. Almost the entire territory of both countries, located at the 'elbow' of the East African trench, extends over young basic volcanic rocks. Other more restricted areas which are built on young effusive materials, such as the volcanic district north of Lake Victoria, the mountains of Cameroon, and the Abyssinian basalt plateau, are also 'islands' of concentrated agrarian communities within vast areas of sparse population. The rest of tropical Africa overlies a geological basement which generally consists of acid crystalline igneous rocks or of non-volcanic sediments.

2. Concerning tropical lowlands, the most densely populated areas lie in the semihumid and semiarid wet-and-dry areas (Fig. 1.10). In West Africa, where climate and vegetation belts are clearly arranged in a latitudinal sequence, one notices that in addition to the favoured coastal zone, due to the greater frequency of alluvial soils, major density of roads, and more potential for many natural and artificial fertilizers, there is a second densely populated belt in the northern parts of the countries from Senegal to Nigeria, known in the literature as the 'North Belt'. In the 'Middle Belt' between these two zones population density decreases considerably (Fig. 1.11).

The natural environment of the 'Middle Belt' is the humid savanna and, at first

24

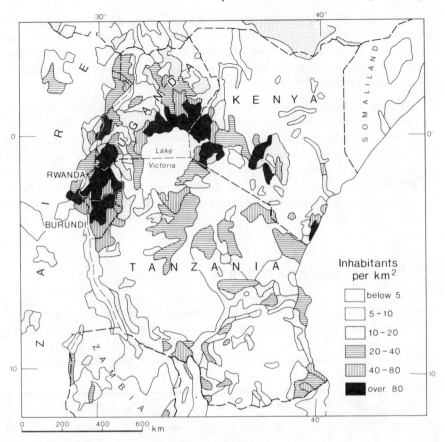

Fig. 1.12 The distribution of the population density in the East African highlands (after Schmidt and Mattingly, 1966, taken from Trewartha, 1972)

glance, it seems paradoxical that the more densely occupied areas are found in the drier savanna with severe hygroclimatic restrictions. The reasons for this seeming paradox will be taken up in Chapters 5 and 6.

A similar discrepancy occurs in the South American tropics between the *agreste* of north-eastern Brazil (*Noreste*), which has been overpopulated for generations, and the practically uninhabited Amazon basin (see Fig 1.10). The *agreste* is a type of dry savanna submitted to the catastrophic consequences of recurring *secas* (droughts). The contiguous areas towards the Amazon, the *sertão* and the *selva*, are climatically more favourable; still, human occupation has begun only recently and settlement is sporadic.

The number of people who depend on agrarian activities in the relatively well-populated regions of the wet-and-dry tropics amounts to 10–20, or, at best, 20–40 per square kilometre. These are remarkably low figures in comparison to rural population densities in subtropical regions, and even the higher mid-latitudes, where

25

1 km² supports 80–100 people on an agrarian basis. (For a detailed illustration, see Appendix 1, which presents the total area, cultivated land, pastures, forests and population to be fed for some West African states and India, as compared to Great Britain and West Germany. These data reveal that a similar agrarian area feeds four times as many people in Europe as in the Upper Guinean states.)

3. Progressing from the wet-and-dry climatic regions of the outer tropics to countries in the rain forest belt near the Equator such as Gabon, the West–Central African republics and Cameroon, or to Surinam, Guyana and French Guiana in South America, population densities decline drastically to only one-tenth or less than that in the cited countries of the wet-and-dry tropics. In South America the ever-humid rain forest region is only inhabited by a thin population in scattered settlements with the notable exception of the fringes on alluvial deposits along the white-water rivers. The latter will be discussed in Chapter 5.

1.5 Notorious nutritional problems

In view of the relatively low densities of agriculturally supported populations in the tropical lowlands of South America and Africa, it is especially remarkable that all these countries have had serious food problems, even when the ratio of people to be fed per hectare arable land has been essentially lower than that of comparable

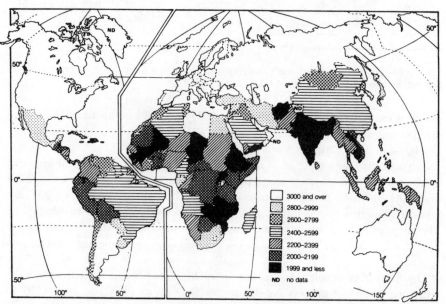

Fig. 1.13 Map of the food supply situation in countries of the tropics. Total Food supply, available calories per caput per day, average 1978–80. (Source: Grigg, 1985)

habitats in subtropical or extratropical regions. In this respect countries of the inner tropics, where conditions for plant growth are optimal and droughts unheard of, are no exceptions. Particularly critical was and is the situation in tropical Africa.

For the years 1978–80, Grigg (1985) has quantitatively documented the nutritional deficiencies. Figure 1.13 shows that most of the tropical countries lay below the minimal norm of 2600 calories per day, the problem being more acute among rural dwellers.

During recent years the food problem has grown worse. The Secretary-General of the FAO has repeated in successive general assemblies since 1982 that agrarian production in most of the tropical countries in Africa and South-America has not been keeping up with population growth and that the need for imported agrarian commodities has been on the rise even in countries not beset by climatic hazards (Table 1.4). It is undoubtedly difficult to discern in detail the reasons for the increase in food imports. How much is due to changes in consumption habits from domestic to imported foods (millet to rice or wheat for example) among the increasing urban population? How much depends on the international market situation and how much is the consequence of the inability to develop internal production? Nevertheless the figures in Table 1.4 clearly indicate that in the countries in question nutritional standards have declined during the last 15 years.

Table 1.4   Cereals (sum of wheat and wheat flour, rice, maize, barley, oat and rye) (100 units) (after *FAO Trade Yearbook*, Rome, 1971-81)

	Imports				Exports			
	1968	1972	1976	1980	1968	1972★	1976	1980
Indonesia	12 461	13 238	71 302	35 489	660	1 618	63	375
W. Malaysia	7 044	7 067	8 873	11 036	253	334	71	161
India	56 569	6 608	23 519	3 899	32	2 796	708	6 689
Brazil	27 204	18 998	35 282	67 404	14 158	1 813	15 210	285
Surinam	278	267	363	445	319	307	575	1 107
Guyana	484	488	521	512	957	773	718	812
Congo	236	371	319	869				1
Cameroon	630	902	745	1373	10	8	12	72
Ghana	1 145	1 128	1 165	2 110	—	—	—	—
Guinea	424	478	508	1 719				
Ivory Coast	1 117	1 539	1 223	4 428★	1	6	335	21
Niger	47	194	648	582		500†	400†	300†
Nigeria	1 080	3 555	8 516	17 756	—	—	7	—
Senegal	2 657	2 878	4 175	4 137	230	13	71	65
Togo	94	206	152	566			5	—
Zaïre	1 405	2 238	3 657	3 501		—	1	—

★ Unofficial figure.
† Estimated values.

Equally significant as these lamentable facts is the more promising fact that in particular countries of Asia, since the mid-1960s there has been a positive development through the implementation of the 'Green Revolution' package: intense management of soils, improved seeds of high-yielding varieties for rice, corn and wheat, and especially application of high rates of fertilizers, herbicides, insecticides and fungicides (for details see Ch. 7). Real progress was achieved, however, only under optimal soil conditions and where irrigation was practicable when needed. In this respect, India experienced a remarkable upsurge in food production during the good monsoon years of the 1960s, but during the monsoon failures in 1972, 1977 and 1982–84 – related to the global climatic anomaly of El Niño–Southern Oscillation (ENSO) – the positive trend was interrupted by drought occurrences. As climatological conditions returned to normal, the positive trend was re-established, India even being able to produce a surplus in wheat and rice.

A few other countries also managed to increase production, but of lesser degree, and there were finally some, expecially in tropical Africa, where no tangible results were gained during the expansion phase of the 'Green Revolution technological package'. The reasons for these diverse results will be discussed in Chapter 7.

In view of the mostly unsatisfactory developments during recent decades it is of interest to note the high expectations which some authoritative experts have placed on the potential productivity of tropical lands and to learn about the calculation methods on which the expectations are based.

1.6 High expectations of production capacity

Well into the 1930s, scientists were convinced that the tropical lowlands enclosed the most fertile regions with the greatest potential for agricultural production. In 1924, Albrecht Penck, a leading geographer of the time, addressed the problem in an academy paper entitled *Das Hauptproblem der physischen Anthropogeographie* (The main problem of physical anthropogeography). Penck advanced the opinion that, on an agricultural basis, the tropical rain forest areas might have a potential supporting capacity of nearly 200 people per square kilometre, as compared with only 100 in the humid parts of the mid-latitudes. Towards the end of the 1930s, Karl Sapper (1939), an expert in the geography of tropical South America, questioned Penck's assumption by drawing attention to the noticeable decrease in yields that occurred wherever former tropical forests or savannas had been converted to agricultural use. He cited soil conditions, not climate, as Penck had done, as the factor that determined agricultural productivity. However, at that time, and for years to come, the regional differentiation of soil properties and their ecological implications were insufficient to prove this hypothesis. Consequently, the idea of the enormous fertility of the tropics took a firm hold.

During the 1960s a very thorough and critical survey of the problem was performed in the United States on the basis of new data by the Panel on World Food Supply of the President's Advisory Committee (Revelle *et al.*, 1967). The chairman of the committee, Roger Revelle, summarized the findings in *Scientific American* (1976). Next to the subarctic, mountainous or arid lands, the largest geographic area of non-arable surfaces is that of the

> reddish or yellowish brown lateritic soils of the savannas and forests of tropical and subtropical regions. . . . Nonarable latosols cover 1.4 billion hectares, yet the latosols also include the largest area of arable land, more than a billion hectares. For high-yielding agriculture, however, these soils must be extensively treated with various supplements, including lime to reduce acidity and trace metals.

He excluded most of the humid tropics as a source of potentially high-yielding food production, because 'except for the island of Java and a few other areas with deep, recently weathered soils, no technology is currently available for high-yielding agriculture on a large scale'. For regions outside the humid tropics he stated: 'The highest concentrations of arable land lie in the semiarid, subhumid and humid grasslands.' The potential gross cropped area, that is, the sum of potentially arable areas multiplied by the number of crops with a 4-month growing season that could be raised in each area, totals 3.8 billion ha including areas that must be irrigated.

> Making the conservative assumption that the lower-quality soils and the uneven topography would limit the average yields to half those obtained in the U.S. Midwest, 11.4 billion tons of food grains, or their equivalent in food energy, could be grown on this potential gross-cropped area, enough for a minimum diet of 2,550 kilocalories per day for nearly 40 billion people (if pest and nonfood uses could be kept to 10 per cent of the harvest) (Revelle, 1976, p. 277).

A potential feeding capacity for 40 billion people on an agricultural basis in the tropics is an impressive figure indeed even with all the assumptions made by the author.

In 1973 appeared a posthumously published paper entitled 'The calculation of theoretical feeding capacity for tropical Africa' by the economic geographer Hans Carol. His calculations were based on 34 case studies of crop yields achieved from different agricultural resource systems in 4 agroclimatic zones. Systems and zones are specified in Table 1.5.

The calculation model for the 'theoretical feeding capacity' (TFC) is as follows:

$$\text{TFC (people per km}^2) = \frac{\text{yield} \times \text{caloric value} \times 100}{\text{food consumption} \times \text{rotation factor}}$$

Yield means staple food crops in kilograms per hectare, caloric value is expressed as calories per kilogram of the food crops yams, maize, groundnuts and millet respectively, and 100 is the multiplying factor for converting hectares into square kilometres. Food consumption is based on the standard nutrition unit of 1 million

29

Table 1.5 Theoretical feeding capacity (TFC) (millions) per agroclimatological zone of tropical Africa (Carol, 1973)

Agricultural resource system	Agroclimatological zone (after Bennett, 1962)			
	Single crops Dry savanna		Continuous crops	
	U (dry) 10% arable	X (wet) 20% arable	Humid savanna Y (moist) 30% arable	Tropical rain forest Z (wet) 40% arable
II₁ Shifting cultivation, land rotation	6.06	81.45	137.37	54.78
II₂ Improved permanent hoe cultivation	60.64	448.00	366.33	246.53
IV₁ Low scientific agriculture	94.60	868.86	815.09	536.88
IV₂ Medium scientific agriculture	203.75	1561.23	1318.80	865.59

calories per year, or 2460 calories per day, and the rotation factor is the number of years of a land rotation cycle divided by the number of years under cultivation.

Taking into account the total area of the different agroclimatic zones (Bennett, 1962) and the respective rate of arable land (*FAO Production Yearbook*), the TFC can be calculated with this model. Carol's values are listed in Table 1.5. He states: 'Under a Type IV₂ Resource System, Tropical Africa would potentially support a population of 4 billion' (Carol, 1973, p. 91).

The interesting inference of Carol's model is that a type IV₂ resource system based on medium scientific agriculture with continuous cropping and the application of fertilizers, is assumed to be applicable to 30 per cent of the humid savanna and to 40 per cent of the tropical rain forest region.

Another contribution pertinent to the problem stems from the Dutch project group 'Food for a Doubling World Population', published in 1979 as the *MOIRA Report* (*Model in International Relations in Agriculture*) (Linnemann *et al.*, 1979). Chapter 2 of that report presents the results of the computation of the 'Potential world food production' elaborated by Buringh and other scientists at the Agricultural University of Wageningen. The main findings of this study group have also been published in a separate form by Buringh (1977).

The purpose of *MOIRA* is 'to try and compute the absolute maximum food production for the world. . . . Moreover, an assessment is made of the land resources and productivity of more than 200 regions of the world' (Linnemann *et al.*, 1979,

30

p. 19). Buringh (1977, p. 482) qualified this study, with its division of the world in 222 broad soil regions, as 'the most detailed study made up to now'.

Taking Africa as example, the calculation procedure and the results of Linnemann *et al.* (1979) and Buringh (1977) are presented in Table 1.6 and Fig. 1.14. On the basis of the FAO/Unesco *Soil Map of the World*, the suitability and quality of the soils were studied and broad soil units (TA) distinguished. Depending on their geographic location, the soil units were classified as lowlands (A in Fig. 1.14), uplands (B), high mountains (C) or deserts (D). The potential agricultural land (PAL) is only a fraction of a broader soil region area. The reduction factor (FPAL) lies between 0.05 and 0.6. The values of the resulting PAL are given in column 6 of the table.

According to Buringh (1977), 'in many regions soil conditions may prove to be a limiting factor in crop production. Therefore, a reduction factor caused by soil condition (FSC) is introduced'. It is remarkable that the limiting factor for the soil conditions of the rain forest region is assumed to be very low.

By multiplying the PAL by the FSC one obtains the estimated area of potential agricultural land without irrigation (IPAL). After taking into account other factors for water deficiency (FWA) and for potentially irrigable land (PIAL) the total imaginary area of potential agricultural land including irrigation (IPALI, column 13) can be deduced for all broad soil regions.

For the calculation of the theoretical potential production on these imaginary areas the authors make the decisive assumption that 'crop production takes place by applying modern farm management practices best suited to the prevailing local environmental conditions' (Linnemann *et al.*, 1979, p. 21). Buringh defines the procedure as follows: 'The computation of this theoretical potential production is based on the assumption that the standard cereal crop is healthy, green, closed, well supplied with nutrients, oxygen, water and foothold and therefore production is only limited by the daily photosynthetic rate, that depends on the state of the sky, the latitude and the date' (Buringh, 1977, p. 480).

Based upon these premises, the maximum production of dry matter per hectare and year is calculated. In converting these values into maximum production of grain equivalent (MPGE, column 15), the following assumptions have been made: the dry production for cereals consists of 25 per cent roots and stubble, 75 per cent straw and grain at a 1 : 1 ratio, 2 per cent grain as harvest loss and 15 per cent grain moisture content.

The final results, expressed in amounts of grain equivalents per hectare and year for the different African broad soil regions, appear in Fig. 1.14. The authors of the study believe it possible in the lowlands of the inner Congo basin to have a maximum production of grain equivalents of more than 25 t or 25 000 kg ha^{-1} yr^{-1}. Since the optimal climatic growth conditions allow three harvests per year, the yields would be on the average 8300–8400 kg ha^{-1}. That is more than the average of the best yields produced by the high-technology agricultures of North America and Europe.

The maximum land productivity for the inner tropical lowlands of South America was calculated at 15–20 t ha^{-1} yr^{-1}, and for the savannas of Venezuela and the *campos* of Brazil at 20–25 t.

Table 1.6 Calculation table of the maximum potential production in grain equivalents (MPGE in million t yr⁻¹) for the different broad soil regions of Africa (Linnemann et al., 1979)

1	2 TA	3 TA%	4 PDM	5 PAL	6 PAL	7 DCC	8 FSC	9 FWD	10 IPAL	11 MPDM	12 PIAL	13 IPALI	14 MPDMI	15 MPGE
A1	40.3	1.33	74	0.3	12.1	2	0.7	0.2	2.4	179	0.4	2.7	197	85
A2	8.2	0.27	84	0.6	4.9	3	0.9	0	0	0	4.2	3.8	318	137
A3	40.3	1.33	76	0.4	16.1	2	0.5	0.6	8.1	613	0.6	8.2	626	271
A4	14.2	0.47	68	0.5	7.1	2	0.6	0.8	4.3	290	0	4.3	290	125
A5	30.9	1.02	74	0.3	9.3	5	0.8	0.9	7.4	549	0	7.4	549	237
A6	62.9	2.07	76	0.5	31.4	4	0.5	0.8	15.7	1195	0	15.7	1195	516
A7	15.4	0.51	84	0.1	1.5	4	0.4	0.1	0.2	13	0.1	0.2	19	8
A8	17.3	57	84	0.6	10.4	3	0.6	0.2	2.1	174	0	2.1	174	75
A9	15.4	0.51	84	0.5	7.7	2	0.7	0.4	3.1	259	0.1	3.1	262	113
A10	26.0	0.86	80	0.6	15.6	2	0.7	0.6	9.4	749	0	9.4	749	323
A11	8.6	0.28	76	0.4	3.4	2	0.5	0.1	0.3	26	0	0.3	26	11
B1	36.4	1.20	74	0.3	10.9	2	0.7	0.2	2.2	162	1.9	3.3	249	106
B2	99.7	3.29	74	0.2	19.9	3	0.4	0.1	2.0	148	0.1	2.1	153	66
B3	231.9	7.65	88	0.2	46.4	2	0.5	0	0	0	0.6	0	42	18
B4	212.5	7.02	82	0.4	85.0	2	0.6	0.3	25.5	2091	1.8	26.4	2165	935
B5	52.7	1.74	76	0.5	26.4	3	0.6	0.6	15.8	1202	0	15.8	1202	519
B6	73.1	2.41	86	0.6	43.9	2	0.5	0.1	4.4	377	3.1	6.6	564	244
B7	164.2	5.42	76	0.5	82.1	4	0.5	0.7	41.1	3120	0	41.1	3120	1348
B8	118.8	3.92	80	0.4	47.5	4	0.6	0.7	28.5	2281	0	28.5	2281	985
B9	157.1	5.18	82	0.4	62.8	3	0.6	0.6	37.7	3092	2.0	38.1	3125	1350
B10	155.7	5.14	80	0.6	93.4	1	0.5	0.5	46.7	3737	0.3	46.8	3744	1617
B11	119.2	3.94	80	0.3	35.8	3	0.6	0.5	17.9	1430	1.3	18.3	1462	631

Table 1.6 Continued.

1	2 TA	3 TA%	4 PDM	5 PAL	6 PAL	7 DCC	8 FSC	9 FWD	10 IPAL	11 MPDM	12 PIAL	13 IPALI	14 MPDMI	15 MPGE
B12	75.4	2.49	76	0.4	30.2	3	0.6	0.6	18.1	1375	0.5	18.2	1385	598
B13	8.5	0.28	76	0.3	2.5	2	0.6	0.4	1.0	78	0.3	1.1	87	37
B14	114.8	3.79	80	0.2	23.0	2	0.5	0.1	2.3	184	0.4	2.6	206	89
B15	31.2	1.03	80	0.4	12.5	3	0.5	0.7	6.2	499	0.5	6.4	511	221
B16	31.8	1.05	84	0.4	12.7	2	0.8	0.2	2.5	214	1.0	3.1	264	114
C1	63.9	2.11	78	0.05	3.2	3	0.7	0.5	1.6	125	0.1	1.6	127	55
C2	70.0	2.31	80	0.05	3.5	3	0.7	0	0	0	0.4	0.3	26	11
D1	803.7	26.52	82	0	—	—	—	—	—	—	—	—	—	—
D2	114.8	3.79	76	0	—	—	—	—	—	—	—	—	—	—
D3	15.1	0.50	80	0	—	—	—	—	—	—	—	—	—	—

Notes

1   Symbol of a broad soil region in a continent.
2 (TA)   Area of a broad soil region (10^6 ha).
3 (TA%)   Area (TA) as percentage of the total area of the continent.
4 (PDM)   Potential production of dry matter (10^3 kg ha^{-1} yr^{-1}).
5 (FPAL) Fraction of potential agricultural land.
6 (PAL)   Potential agricultural land (10^6 ha).
7 (DCC)   Development cost class.
8 (FSC)   Reduction factor caused by soil conditions.
9 (FWD)   Reduction factor caused by water deficiency.
10 (IPAL) Imaginary area of PAL with potential production, without irrigation (10^6 ha).
11 (MPDM) Maximum production of dry matter without irrigation (10^6 t yr^{-1}).
12 (PIAL) Potentially irrigable agricultural land (10^6 ha).
13 (IPALI) Imaginary area of PAL with potential production, including irrigation (10^6 ha).
14 (MPDMI) Maximum production of dry matter, including irrigation (10^6 t yr^{-1}).
15 (MPGE) Maximum production of grain equivalents, including irrigation (10^6 t yr^{-1}).

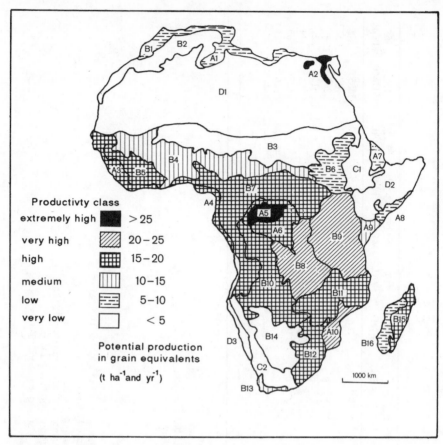

Fig. 1.14 Map of the maximum potential production in grain equivalent for different broad soil regions of Africa (after Linnemann *et al.*, 1979)

When looking at these figures it must be kept in mind that the purpose of the *MOIRA* study was 'to calculate the absolute upper boundary of food production (in grain equivalents) for the whole world, and that many other factors must be taken into account if one wants to get results that could have some meaning in practice' (Linnemann *et al.*, 1979, p. 41).

The factors that the authors had in mind emerge from the following statement: 'No attention was given to factors in the field of economics, and to social, cultural and political obstacles. The interrelationship of some of these factors and the complexity of the world food problem should not be overlooked' (Linnemann *et al.*, p. 53). Thus, all the factors specified by the authors proceed from socio-economic disciplines while natural factors are assumed to be encompassed in the calculation procedures.

Despite the restrictions, the authors are convinced that with suitable management optimal food production is possible in the tropics. Therefore, Buringh (1977, pp. 482–3) stresses emphatically: 'When Eckholm (1976) and some other authors try to convince us that the tropics never will be the food basket of the world, our only comment is: it is not true.'

The weak point of this deterministic model is the failure to discuss whether the premisses for modern agriculture are fulfilled for soil conditions in the humid tropics. Neither the *MOIRA* study nor Buringh deal with the problem of which agricultural technique would be able to keep the standard cereal crop 'well supplied with nutrients'.

A major study on the potential agricultural production and population-supporting capacity of the tropics – not to be taken lightly because of the authority of its sources: FAO, the International Institute of Applied System Analyses (ASA), and the United Nations Fund for Population Activities (UNFPA) – is the *Potential Population Supporting Capacities of Lands in the Developing World* published by FAO in 1982 (Higgins *et al.*, 1982).

The basic assumptions on alternative land uses, climatic adaptability of crops, soil requirements, net biomass production and crop yields for Africa and Latin America had been previously published in 1978 and 1979 in two *World Soil Resources* reports.

Table 1.7 contains the final results of the comprehensive study, this time using the South American tropics as an example. In the table, the alternative levels of input circumstances mean:

1. *Low level* of inputs assuming only hand labour, no fertilizer and pesticide applications, no soil conservation measures, and cultivation of the presently grown mixture of crops on all potentially cultivable rainfed lands. They represent, approximately, those found at present among farmers practising traditional land rotation systems (characterized in Ch. 2).
2. *Intermediate level of inputs*, implying the use of improved hand tools and/or draught implements, some fertilizer and pesticide application, some simple soil conservation measures destined to lessen productivity losses from land degradation, and cultivation of a combination of the presently grown mixture of crops and the most calorie (protein) productive crops on potentially cultivable rainfed lands,
3. *High level of inputs*, assuming complete mechanization, full use of optimum genetic material, necessary farm chemicals and soil conservation practices, and cultivation of only the most calorie (protein) productive crops on all potentially cultivable rainfed lands. They must be equated with systems utilizing modern agrarian technology.

The LGP values represent the length of the growing period, which is defined as the number of days when rainfall is half, or more than half, the potential evapotranspiration and when the mean daily temperatures are higher than 5° C to permit crop growth.

For the permanently humid inner tropics (1), with an LGP between 330 and 365 days, the potential population-supporting capacity, assuming low level of inputs, is 1.02–1.39 persons per hectare, or 102–139 persons per square kilometre. Given the

Table 1.7 South America as an example of present populations and potential population-supporting capacities of lands in the developing world by major climate and length of growing period (LGP) zone (density in persons per hectare) (extracted from Table 3.2, p. 109 of the FAO Technical Report, Higgins *et al.*, 1982)

	LGP zone (days)	Total land area* (million ha)	Year 1975 population† (millions)	Year 1975 population density	Potential-population supporting capacity		
					Low inputs	Intermediate inputs	High inputs
Major climate 1: warm tropics							
(1)	365+	110.4	2.2	0.02	1.02	2.91	8.31
	365–	150.9	7.6	0.05	1.39	4.08	9.85
	330–364	76.3	8.1	0.11	1.16	4.11	9.93
	300–329	269.0	11.8	0.04	0.79	5.00	9.21
	270–299	173.0	18.5	0.11	0.63	3.48	9.25
	240–269	126.8	16.0	0.13	0.71	3.74	10.07
	210–239	63.1	7.0	0.11	0.71	3.12	10.35
	180–209	24.2	2.9	0.12	0.55	2.39	7.88
(2)	150–179	31.1	8.8	0.28	0.53	2.12	6.29
	120–149	30.1	8.8	0.29	0.17‡	1.02	3.55
	90–119	3.0	2.2	0.71	1.13	1.50	4.21
	75–89	5.2	2.8	0.53	0.48‡	0.66	2.21
	1–74	14.4	5.0	0.35	0.20‡	0.21‡	0.22‡
	0	3.6	1.1	0.31	0.67	0.67	0.67
Major climate 2: moderately cool tropics							
(3)	365+	—	—	—	—	—	—
	365–	7.2	1.2	0.17	0.01‡	0.07‡	0.87
	330–364	14.3	4.4	0.31	0.02‡	0.08‡	0.85
	300–329	3.8	0.3	0.07	0.03‡	0.07‡	0.95
	270–299	4.5	0.3	0.07	0.07	0.10	0.79
	240–269	2.9	0.9	0.32	0.10‡	0.48	2.70
	210–239	0.5	0.1	0.26	0.01‡	0.05‡	0.56
	180–209	1.0	0.2	0.15	0.01‡	0.03‡	0.47
(4)	150–179	2.2	1.9	0.85	0.32‡	0.93	2.76
	120–149	3.7	0.7	0.19	0.15‡	1.60	2.01
	90–119	1.5	0.6	0.37	0.25‡	0.41	1.02
	75–89	1.3	0.4	0.32	0.02‡	0.14‡	0.41
	1–74	1.1	0.3	0.31	0.01‡	0.01‡	0.02‡
	0	8.4	3.2	0.38	0.01‡	0.01‡	0.01‡

* Land areas derived from FAO/Unesco *Soil Map* excluding areas mapped as water bodies.
† UN data for 1975, millions of persons, UN 1979.
‡ Indicates potential, less than present population.

utilization of modern agricultural techniques, the supporting capacity is estimated at 831–903 persons per square kilometre. For the wet-and-dry tropics (2) the calculated potential population densities are significantly inferior to those of the permanently humid tropics. Regarding the actual densities in these two LGP zones the relation is reversed; the real density in the wet-and-dry tropics is almost 10 times that of the inner tropical zone. For the tropical highlands (moderately cool tropics (3) and (4)) the potential population is assumed to be less than the number of people who actually live there.

Since these results are in strong contrast with reality, they are astonishing and prompt a close scrutiny of their underlying assumptions and calculation methods. While the climate inventory is relatively easy to establish, a particularly close look must be taken at the criteria used in the description of the soils.

Basic for the soil inventory is the 1:5 million FAO/Unesco *Soil Map of the World* (FAO, 1971 and 1977). The 106 different soil units comprised in this map are compiled in 26 major soil units. Table 1.8 lists the total number of hectares (in millions) and the percentage of the most extensive soil units for Africa, South America and South-east Asia (for the names and ecological significance of the soil units, see Ch. 3).

For each of the dominant soils of the respective major soil units, three textural classes and three slope categories are differentiated. These are coarse, medium and fine-textured soils. The slope level categories are: (a) level to gently undulating, (b) rolling to hilly, with slope values ranging between 8 and 30 per cent, (c) steeply dissected to mountainous, with slopes greater than 30 per cent.

As a general limiting property coarse soil texture is mentioned. Moreover the report

Table 1.8   Distribution and extent (million ha) of major soils in three regions of the developing world (FAO Technical Report, Higgins *et al.*, 1982)

Major soil units*	Africa		South America		South-east Asia	
	Area	%	Area	%	Area	%
Lithosols	372.5	12.9	185.8	10.5	90.4	10.1
Ferralsols	319.8	11.1	425.8	24.1	15.9	1.8
Yermosols	373.8	13.0	56.2	3.2	30.1	3.3
Acrisols	92.0	3.2	290.2	16.4	197.7	22.0
Arenosols	304.4	10.6	90.5	5.1	24.8	2.8
Luvisols	231.2	8.0	80.3	4.5	97.6	10.9
Regosols	263.1	9.2	41.9	2.4	27.5	3.0
Cambisols	99.7	3.5	51.6	2.9	122.3	13.6
Gleysols	130.4	4.5	117.9	6.7	49.5	5.5
Fluvisols	101.3	3.5	64.5	3.6	54.6	6.1
Vertisols	98.8	3.4	24.9	1.4	57.9	6.5
Xerosols	85.4	3.0	37.7	2.1	12.6	1.4
Nitosols	98.4	3.4	36.5	2.1	37.5	4.2
Solonchaks	50.1	1.8	23.9	1.4	17.3	1.9
Others	52.4	1.8	235.6	13.2	56.0	6.2
Miscellaneous†	204.8	7.1	6.9	0.4	5.9	0.7
Total	2878.1		1770.2		897.6	

* Includes all extents, whether occurring as dominant soil, or as associated soil or as inclusions.
† Excludes soil-designated miscellaneous land units.

recognizes 12 land characteristics to be of significance for soil management: stony, lithic, petric, petrogypsic, petroferric, phreatic, fragipan, duripan, saline, sodic and cerrado. Acrisols, ferralsols and podsols are mentioned as having severe fertility limitations. The consequences that severe fertility limitations have for the production potential of rainfed agriculture are not discussed in detail. Nevertheless, some insights can be drawn from the respective cultivation factors (see Table 1.9) or rest period requirements.

According to these values, at low-level inputs it is possible to cultivate the land on the ferralsols and acrisols of the tropics in only 15 per cent of the total length of the cultivation and fallow period, which means 3 years within a 20-year cycle. At high-level inputs the cultivating years are assumed to be 7 and 6.5 respectively out of 10.

The inventoried climatic and soil attributes of land must be compared with the crop requirements. For every crop that can be grown in an area, there is an optimum agroclimatic yield potential dictated by climatic conditions.

Soil conditions modify the agroclimatic potentials and determine attainable land productivity potentials. Limitations imposed by degradation hazards such as soil erosion, salinization and alkalinization, waterlogging, depletion of plant nutrients and organic matter, deterioration of soil structure and pollution are taken into account, as well as crop-specific allowance for seed and planting material requirements and wastage losses. Some crop-specific soil limitation ratings, used in the study, are shown for the main soils in Table 1.10.

Once deductions are made for land required for non-agricultural use, for irrigation and for fallow, each area is analysed separately to pinpoint the most productive crop(s) for specific soil and climate conditions.

When certain crops or crop mixtures have been chosen for specific areas, the maximum calorie (and associated protein) production for the given areas of rainfed agriculture and irrigation systems can be calculated.

The areal results are totalled to arrive at calorie (protein) potentials for each LGP zone. Once the maximum potential calorie (protein) combination is ascertained,

Table 1.9 Cultivation factors for some major soils according to level of inputs (%)

Major soil unit	Low inputs		Intermediate inputs		High inputs	
	Humid	Semiarid	Humid	Semiarid	Humid	Semiarid
Ferralsols	15	20	35	40	70	75
Acrisols	15	20	40	60	65	75
Luvisols	25	35	50	55	70	75
Cambisols	35	40	65	60	85	80
Nitosols	40	75	55	70	90	90
Vertisols	40	45	70	75	90	90
Gleysols	60	90	80	90	90	90

Table 1.10   Limitation ratings for growth of various crops

Major soil	Millet	Maize	Wheat	Cassava	Potato
Leached acid soils (ferralsols)	S2	S2	S2	S2	S2
Lateritic soils (acrisols)	S2	S2	S2	S2	S2
Base-saturated soils (luvisols)	S1	S1	S1	S1	S1
Weakly developed soils (cambisols)	S1	S1	S1	S1	S1
Shallow soils (lithosols)	NS	NS	NS	NS	NS

NS = major limitations making the soil not suitable for the specific crop.
S2 = moderate limitation.
S1 = no or slight limitation.

application of FAO/World Health Organization country-specific per capita calorie (protein) requirements allows the calculation of the potential population-supporting capacity for each country of the different LGP zones.

Given these elements of judgement, the relationship between population growth, food requirement and potential food production at different levels of input is discussed for South-west Asia, South-east Asia, Africa, Central and South America as a whole, and for individual countries.

In the context of our book, the details of the FAO study presented are not decisive. What is decisive is the fact that the permanently humid inner tropics with growing periods of 300–365 days are considered as potentially the most productive areas regardless of the level of inputs.

African countries like Cameroon, Zaïre, Equatorial Guinea, Congo, Ivory Coast, Liberia and Ghana – contrary to the present situation – are assumed to have the potence to support nearly twice as many people as countries like Rwanda, Burundi or Ethiopia, in the volcanic regions of East Africa. In South America the maximum potential population-supporting capacity per surface unit is calculated for Surinam, Guyana and French Guiana.

Drawing a balance of the FAO Report, we are confronted with these discrepancies:

1. The areas of the African and South American humid inner tropics, which are still scarcely populated and hold vast uninhabited tropical rain forests, are considered as potentially the most population-supporting regions of these continents.
2. The wet-and-dry outer tropics, on the other hand, especially the dry savannas, which presently have a considerably higher agricultural population density, are assessed as being potentially less productive and less population supporting.
3. The fact that there are densely populated, or even overpopulated areas amid thinly populated zones is only considered as a hopeful sign that the population-supporting capacity in the surrounding areas could also be successfully raised.

4. The discrepancies between reality and calculated scenarios remain as unexplained as the fact that the South Asian tropics exhibit the overall highest population densities.

Comparing the FAO and the other computations that were detailed above one perceives that all of them express high expectations of the potential productivity of the tropics. There are, however, basic differences so far, as Revelle and the Panel on World Food Supply exclude the humid inner tropics as a source of potentially high-yielding food production, while the other proposals consider this region as precisely the potentially most productive. This discrepancy arises from the different views on the suitability of the deep weathered inner tropical soils to the application of high-yielding modern agrarian technologies.

The discrepancy can only be solved after a thorough analysis is made of the decisive soil qualities and their regional differentiations (see Chs 3, 5 and 6). At any rate, the calculation of the perceived carrying capacities implies an environmental reasoning. This deterministic approach is, of course, not to everyone's liking. In particular, the socio-economic development experts hold a different view on the role played by natural factors on agricultural progress and population development. A short review should be made of the respective standpoints in order to highlight their decisive arguments and to see what inferences can be drawn for our approach.

1.7   Questioning environmental determinism? The role of interregional exchange

As representative for socio-economic reasoning, passages from Bronwell W. Hodder's *Economic Development in the Tropics* (1980) can be quoted:

> Even the most cursory glance at natural resources – water, forests, minerals, or power – reveals that there is indeed little correlation between the occurrence of these resources and the level of economic development in any particular tropical country. The presence or absence of naturally occurring material resources in no way immutably determines the economic development of a country (p. 8).

> Yet even though it is necessary to emphasize our ignorance rather than our knowledge about the physical factors and natural resources of the tropics, it is also necessary to emphasize that these are no grounds for assuming that natural resources are anywhere a limitation to economic development or an excuse for the present relatively low standard of living in any part of the tropics (pp. 12, 13).

'The early traditional viewpoint of the tropics as having exceptionally rich natural resources and the "modern" view of the tropics as poor or inferior in natural resources are equally false.' Specifically concerning soils, the statement is made: 'There is already sufficient evidence to make it possible to refute the general contention that tropical soils are a serious or permanent limitation to the agricultural development of the tropical world' (pp. 37, 38).

He continues:

> The standpoint adopted in the pages that follow is that any realistic analysis of the applied situation in tropical economic development must start from the assumption that the natural resource base is potentially adequate for substantial development, given sufficient knowledge about these resources and a ready adaption of planning to the opportunities and limitations set by them' (p. 13).

The same author also writes, however, the following: 'Furthermore, a good deal of the available empirical evidence supports the contention that natural resources are by no means an irrelevance, especially in the early stages of development' (p. 9).

Considering the arguments concerning natural resources, the tenor is that natural resources certainly have an impact on the economic situation of the tropical countries, but that they are in no way decisive, or apt to explain the present, relatively poor, economic development anywhere in the tropics.

The basis for this conviction is the economic thesis that the limitations imposed upon a particular society by the lack of certain natural resources can be overcome through exchanges between people and countries. 'Trade clearly opens up the possibility of breaking through any constraints imposed by natural resource deficiencies and links growth more intimately and directly to population and capital accumulation' (p. 8).

The arguments against natural resources being the causative factors for the production capacity and state of development of the tropical countries are summarized in the words of Bauer and Yamey (1957) also quoted by Hodder: 'The Creator has not divided the world into two sectors, developed and underdeveloped, the former being more richly blessed with natural resources than the latter.'

These ideas are, by no means, uncommon. It has become a widely accepted practise to consider viewpoints in favour of the influence of environmental conditions on the state of development or economic behaviour of social groups as 'determinism', the words 'determinism' or 'environmentalism' having acquired a negative connotation.

1.8 The Reasons for an actualistic approach

Instead of discussing natural resources cumulatively, a differentiation should be made, first, between natural conditions that are unalterable and those that can be compensated for by man. Second, proper consideration should be given to the reality that this differentiation applies to specific geographical regions which must be clearly defined. Political or administrative limitations are, in most cases, unsuitable for that purpose. Third, it should be taken into account that the significance of certain natural resources is subject to technical progress and attitude change. And fourth, natural influences that were strong in the past should be examined and divided into those that have vanished in the course of time and those that persist into the present.

Regarding the exchange factor as the decisive socio-economic argument, one must acknowledge that in recent centuries and for today its significance and effectiveness are certainly unquestionable. Yet what was the situation like before exchange became a forceful factor? In our context this question is crucial in connection with the North–South socio-economic decline.

Fortunately, we have ample and accurate information about the time when the tropics and the extratropics were two separate worlds without any trading going on between their societies. At the time of the European discoveries, in the fifteenth and sixteenth centuries, the intellectual impulses and the technical achievements to establish contacts and exchanges with different parts of the world had been developed by the nations of the subtropics and mid-latitudes of the northern hemisphere. They already had at this time the political systems, the social make-up, and the technical superiority which allowed them to install themselves as trading partners and, sooner or later, as colonial powers in the tropics. These indisputable facts show that a sharp socio-economic and technological gradient from North to South already existed before the age of colonialism. The so-called underdevelopment of the tropics today can, therefore, not be considered as a direct consequence of colonialism; at most, colonialism has contributed to reinforce the existing disparities.

A realistic approach to the problem of uneven growth must begin with the analysis of the two systems that were independent of each other and for which intercontinental exchanges were totally irrelevant.

The fact that the total separation of the two systems lies a few centuries back is not a handicap for scientific analysis, because physical conditions such as water budget, soil genesis, vegetation sequence, morphogenesis and all their ecological interactions, have not greatly changed since. The sparse population density of the tropics and the large tracts of land that are still in a natural state allow this type of environmental study to be conducted. Even in those locations where the landscape has been altered by man, the changes in many parts have been quantitatively and qualitatively so small that a reliable estimation of the early natural conditions is by no means impracticable.

Thus, with actualistic reasoning and with information yielded by modern environmental research the question of the ecological disadvantage of tropical realms can be adequately approached. The purpose of the following sections will be to explore whether, beyond the socio-economic factors that have been traditionally invoked, there are properties of the natural environment that can help us to understand the difficulties with which large areas of the tropics have been confronted in the course of their development.

2

Characteristics of traditional rainfed agriculture in the tropics

Summarizing the viewpoints pertaining to the characteristics of tropical environments presented in the introductory chapter one is confronted with these realities:

1. In contrast to the countries of the extratropics, many countries of the inner tropics have nutritional problems due to scarce agricultural production. Most affected by this situation are the countries of tropical Africa, while the situation is less dramatic in tropical South America and Asia.
2. Among the troubled regions some face the climatic risks of periodic dryness and episodic droughts, while others enjoy optimal climatic growth conditions.
3. A remarkable facet is, however, that in the risk areas of climatic wet-and-dry conditions of the outer tropics the population density is considerably greater than in the climatically favourable humid regions of the inner tropics, where temperature and precipitation regimes permit year-round plant growth. It is well documented that the natural vegetal formations of the tropical-lowland rain forests exhibit the largest net biomass production on earth (Table 2.1).

The conclusion that has been all too readily drawn from these facts is as follows: with the natural system of tropical rain forests exhibiting such a high vegetative production capacity, properly operated artificial agricultural production systems in that environment should result in similar production levels. If these levels are not reached by the actual land-use systems, then these systems are inadequate and must be changed!

Seduced by the evident superiority of their production systems in other important sectors of the economy, the representatives of advanced nations of the extratropics reached exactly that same conclusion about the agricultural systems of the tropics. However, 30 years of experience with the real difficulties and the scientific endeavours to solve them, have rendered very thoughtful and modest those who realize the intricacies of the natural circumstances, not to mention the man-made problems.

Before dealing with the difficulties of agricultural production in the humid tropics the following questions must be answered: which are the characteristic rainfed farming systems practised until recently by the native populations of the humid tropics; how

43

Table 2.1 Productivity of main types of vegetation (t ha^{-1}) (after Rodin and Basilevič, 1968)

Characteristics	Tropical rain forests	Savannas	Savannas dry	Sub tropical forests	Beech forests	Fir forests, middle taiga
Biomass	> 500	66.6	26.8	419	370	260
Of which:						
Green parts	40	8.3	2.9	12	5	16
Perennial overground parts	370	54.4	12.6	316	270	185
Roots	90	3.9	11.3	82	95	60
Net primary production (per year)	32.5	(12)*	7.3	24.5	13	7
Litter-fall (per year)	25	(11.5)	7.2	21	9	5
Real primary production (per year)	7.5	(0.5)	0.1	3.5	4	2

* The numbers in parentheses are approximate data calculated by the authors.

do they function; what are their advantages and disadvantages, and how are they to be assessed from an ecological viewpoint?

2.1 The areally dominant and representative form of tropical rainfed agriculture

W. B. Morgan (1969a) has published a very detailed cartographic representation of the peasant agriculture (staple food crops) of tropical Africa. For tropical America Sanchez and Cochrane (1980) offer an approximate overview of the major agricultural production systems. In the case of Asia one has to resort to the map compiled by Spencer (1966).

According to Morgan (Fig. 2.1), shifting cultivation and rotational bush fallow are the predominant land-use systems of peasant agriculture in tropical Africa. Permanent and semipermanent cultivation occupy only relatively small and isolated regions.

Shifting cultivation or shifting field agriculture is the characteristic agricultural technique of much of Central and East Africa, and some small positions of West Africa. Rotational bush fallow methods are characteristic of what may broadly be termed the Sudanic Zone stretching from Senegal in the west to Ethiopia in the east, and most of West Africa. They are uncommon in Central and East Africa (Morgan, 1969a, pp. 248 and 251 respectively).

'Shifting cultivation' in the words of Morgan (p. 248) – which are similar to Watters (1960) – is

44

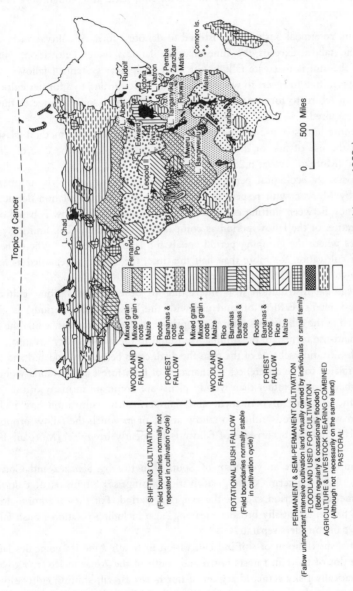

Fig. 2.1 Peasant agriculture in tropical Africa (staple food crops). (Source: Morgan, 1969a)

SHIFTING CULTIVATION
(Field boundaries normally not repeated in cultivation cycle)

WOODLAND FALLOW
Mixed grain
Mixed grain + roots
Maize

FOREST FALLOW
Roots
Bananas & roots
Rice

ROTATIONAL BUSH FALLOW
(Field boundaries normally stable in cultivation cycle)

WOODLAND FALLOW
Mixed grain
Mixed grain + roots
Maize

FOREST FALLOW
Rice
Bananas
Bananas & roots

PERMANENT & SEMI-PERMANENT CULTIVATION
(Fallow unimportant intensive cultivation land virtually owned by individuals or small family
Roots
Bananas & roots
Rice
Maize

FLOODLAND USED FOR CULTIVATION
(Both regularly & ocassionally flooded)

AGRICULTURE & LIVESTOCK REARING COMBINED
(Although not necessarily on the same land)

PASTORAL

45

characterized by a rotation of fields rather than crops, by short periods of cropping (one to three years) alternating with long fallow periods (up to twenty years or more, but often as short as six to eight years), by clearing by means of slash and burn, and by use of the hoe or digging stick, the plough only rarely being employed.

The term 'rotational bush fallow' is used to denote a form of cultivation in which the time in fallow exceeds the time that the land is under cultivation. For this technique the aim is to create fallows, or rather a regular system of fallows, which are never permitted to revert to woodland or forest. Returning to a former cultivation site means a return to former fields, except in cases where some change in the field pattern is required.

'It is a more intensive use of land (p. 255) although the clearing method (slash and burn) and the use of hoe or digging stick is as common as in the case of shifting cultivation' (Morgan, 1969a, p.257).

From a strictly ecological perspective, and without considering the implications indicated by Morgan with respect to settlement forms and population density, the differentiation between shifting cultivation and rotational bush fallow is based on the varied duration of the fallow period as compared to cultivation. The limiting period of time lies where the cropping period equals the land rest period. 'Where the land was under cultivation for more than half the time, the system is regarded as having broken down already' (Morgan, 1969a, p. 251).

In its regional distribution it is interesting to observe that shifting cultivation (i.e. the less intensive form of land use with the longer fallow periods) is clearly concentrated in the rain forests with their optimal climatic conditions, while the more intensive land-use forms are predominantly found in the wet-and-dry savanna.

In the classification scheme of the map the distinction between woodland and forest fallows is related to the differences in character of high forest and savanna woodland fallows, where grasses rather than woody plants are dominant in many of the fallow areas. In the category of permanent and semipermanent cultivation 'are included practices in which there is either no fallow at all, or in which the fallow proportion is very small, and the occurrence of fallow is frequently irregular' (Morgan, 1969a, p. 256).

In the approximative survey map of South America by Sanchez and Cochrane (1980, Fig. 2.2), the term 'shifting cultivation' comprises all forms of cultivation in which the fallow period exceeds the cropping period. For these authors, shifting cultivation has an essentially broader meaning, also including rotational bush fallow, which Morgan considers separately.

The spatial distribution of shifting cultivation in South America coincides largely with the realm of the rain forests north and south of the Amazon River. However, in the climatically risky semiarid region of north-east Brazil, shifting cultivation has been replaced by the multiple cropping systems of permanent rainfed agriculture. The north and south margins of the tropics are dominated by intensive annual crop

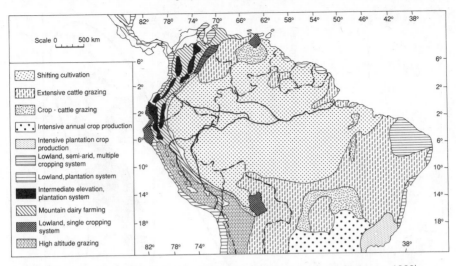

Fig. 2.2 Land-use forms in tropical South America (after Sanchez and Cochrane, 1980)

and intensive plantation crop production. Neither is there shifting cultivation in the intermediate and high elevation fringes of the Andes.

While in most of the African and South American tropics shifting cultivation is the absolutely dominant form of agriculture, 'in southeastern Asia it takes a minor position', an observation by K. J. Pelzer made as early as 1948 in his classic *Pioneer Settlements in the Asiatic Tropics*. An overview of the areas where the practice of shifting cultivation is dominant, frequently present, remnant only, or totally absent, is provided by Fig. 2.3, taken from J. E. Spencer's *Shifting Cultivation in Southeastern Asia* (1966). A characteristic for that part of the tropics is that shifting cultivation distribution is not only rather restricted, but also clearly confined to particular areas, which, at times, lie right next to others with no shifting cultivation at all. The ecological causes for this phenomenon will be detailed in later parts of the book (see Ch. 5 and Section 9.3).

2.2 Specific meanings of shifting cultivation and rotational bush fallow

From the presentations of Morgan, Sanchez and Cochrane or Spencer a certain inconsistency is observed in the use of the term 'shifting cultivation'. Other authors differ even more widely, so that: 'Whenever the term shifting cultivation was used, an explanation had to be given as to whether this collective term was being used in the narrower or broader sense' (Manshard, 1974, p. 53).

47

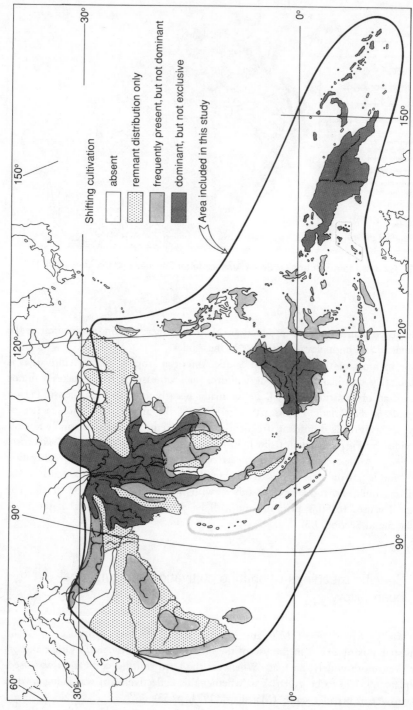

Fig. 2.3   Present practice of shifting cultivation in tropical Asia. (Source: Spencer, 1966)

As to the areal distribution of the various forms of agriculture in the tropics it is necessary to clarify the meaning of the different terms, especially when applied to southern Asia where circumstances are more complicated than in Africa or South America.

In its original and restricted sense shifting cultivation refers to a form of agriculture 'in which both, economic area and settlement, are moved at certain intervals' (Manshard, 1974. p. 52). It is an itinerant agriculture – *agricultura migratoria* in Spanish America, and *Wanderfeldbau* in German. With growing population density and improving infrastructure in many regions, settlements are being shifted less frequently, but even there the ecologically decisive shifting of fields after a short cropping period has remained. More precisely, we are dealing here with a field–forest (or woodland) economy, or field–forest (woodland) rotation or rotational crop-and-fallow economy. The German term *Landwechselwirtschaft* was adopted by both Unesco and the International Geographical Union (IGU) in the abbreviated form of land rotation (Manshard, 1974, p. 53), but this clear descriptive term has unfortunately not found general acceptance. Instead, shifting cultivation in its broader connotation appears almost everywhere in the specialized literature.

Shifting cultivation in its broader connotation

> may be briefly defined as an economy in which the main characteristics are rotation of fields and not of crops; short periods of soil occupancy alternating with long fallow periods; clearing by means of fire; absence of draft animals and of manuring; use of human labor only, employment of the dibble stick or hoe. . . . Many shifting cultivators no longer change their dwellings when they turn to new land; they have become sedentary at least as far as their houses are concerned. They may build a hut in the *ladang* if it is too far away from their village and may occasionally live in such a shelter . . ., but after the harvest they return to their permanent houses in the village (Pelzer, 1948 p. 17).

Especially in South-east Asia, land clearing by fire is described by the Old English word swidden, and the agricultural form is called *swidden cultivation*. A modern term for that procedure is the *slash-and-burn method*.

The generic contrast to shifting cultivation in this broader connotation is the permanent field, or sedentary, cultivation. In this type of tillage 'agriculture is practiced year after year on the same well tended plot, the fertility of which is maintained by manuring, crop rotation, intensive cultivation, terracing, and often the regular addition of mineral matter by silt deposition on the land by irrigation water. Only occasionally may the land be allowed a short fallow period' (Pelzer, 1948, p. 6).

Ruthenberg (1980) calls the opposite of shifting cultivation 'permanent rainfed cultivation systems'. It is characterized by the continuous annual cultivation of a plot within well-established crop field boundaries.

This generic differentiation must be clearly defined, and a detailed subdivision within the two larger groups made. First the meaning of 'short' and 'long' with regard to periods of cropping and fallow needs to be clarified. The attempt was made by

49

Ruthenberg with his cultivation frequency value and by Allan with his land–use factor (Ruthenberg, 1971, 1980; Allan, 1965). The cultivation frequency factor expresses in a percentage value the length of a cropping phase divided by the sum of crop years plus fallow years. A value of 30 per cent differentiates shifting cultivation from what Ruthenberg calls 'semipermanent rainfed cultivation'. This 30 per cent means 3 years of crops followed by 7 years of fallow. Norman (1979) adopted Ruthenberg's 30 per cent as the boundary value between shifting cultivation and, what he characterized as 'semi-intensive rainfed cropping'. Morgan's rotational bush fallow cultivation systems and Sanchez's version of shifting cultivation go beyond the limits recommended by Ruthenberg, since the condition of 'the time in fallow exceeds the time that the land is cultivated' is applied somewhat loosely. In the case of 3 years of cultivation, 4 years of fallow satisfy the condition, the cultivation frequency factor being 43 per cent.

Allan's land–use factor is the sum of the area in cultivation plus the area in fallow divided by the area in cultivation. In practice it is often difficult to determine the areas of cultivation and fallow.

Within the generic group of shifting cultivation in its broader connotation, Greenland (1974) offers a summarized overview of variants in which he distinguishes 'simple shifting cultivation' when dwelling and newly cultivated fields shift together; 'recurrent cultivation' when the farmed areas, comprising several field types, shift with more frequency than dwellings; 'recurrent cultivation with continuously cultivated plots' when cultivation spreads over an entire field complex; and 'continuous cultivation' when land use is alternated with cultivated pastures for husbandry and fallow crops.

Decisive in all these differentiations is the distinction between shifting cultivation in its broader connotation with the characteristic alternation of periods of land occupancy and longer periods of forest or bush or woodland fallows, and permanent rainfed cultivation systems with continuous annual cultivation of plots within a landscape of clearly defined crop field boundaries.

2.3 The areal extent of shifting cultivation in its broader connotation

As to the spread and importance of shifting cultivation within world agriculture, Hauck (1974) estimates the area occupied by this form of agriculture to be 36 million km^2, comprising 30 per cent of the global arable land and supporting somewhat more than 250 million people (7 per cent of the world's population). In Africa, conservative estimates of the areas farmed by subsistence agriculturalists under rotating field-fallow systems run close to 14.5 million ha (145 000 km^2) (Braun, 1974). According to Grenzebach (1984) rudimentary forms of shifting cultivation and pastoralism are practised today by less than 1 per cent of Africa's population. That is true. The problem is, however, that this small percentage occupies the enormous area of land mentioned above. According to Sanchez and Cochrane (1980, p. 127), 'shifting cultivation is practiced in localized areas over the largest expanse of tropical America – about 720 million hectares (7.2 million km^2) of forested regions'.

As an example from South Asia, Spencer (1966) presented data with the number of families, the acreage of land cleared annually, the acreage of land still being harvested, the acreage of fallow land, and the total acreage for different areas from Pakistan in the west to New Guinea in the east. According to his estimates at that time,

> land in crop, in the harvest-gleaning stage, and in various stages of fallow, including all land that is required to maintain the annual cropping sequence, may reach the figure of 250–275 million acres. The annual crop area may amount to about 55 000 square miles, whereas the total land needed to maintain the annual crop area may perhaps be as much as 400 000 square miles.

(Transforming these values to hectares means 250–275 million acres = 100–112 million ha, 55 000 square miles = 14.2 million ha, 400 000 square miles = 104 million ha.)

> The total share of population which lives by shifting cultivation reaches 50 million, against perhaps 675 million living by permanent agriculture. The latter must provide the great bulk of the food supply going to feed the nonagricultural population, possibly 200 million, making up the balance of perhaps 925 million people residing in the area covered by this study. The 675 million have at their disposal 560–580 million acres of permanent cropland (the best agricultural land available). The 50 million shifting cultivators have some 250–275 million acres as their total land resource (the poorest tracts for agriculture), but cultivate only some 45–52 million acres in any one year (Spencer, 1966).

These figures show the South Asian tropics to be in an exceptional situation.

The generalizations drawn from the data cited so far can be summed up as follows:

1. In the African and South American tropics, shifting cultivation in its broader connotation is the dominant form of rainfed agriculture.
2. Shifting cultivation in its restricted connotation and with an especially low cultivation frequency is the prominent land-use form in the climatically optimal humid inner tropics.
3. Towards the wet-and-dry tropical margins, as well as at higher elevations, the cultivation frequency increases and the ratio between cropping period and fallow period becomes more favourable. In some of these areas, shifting cultivation has already been replaced by semipermanent or permanent semi-intensive rainfed cultivation.
4. In the Asian tropics, shifting cultivation is of minor importance. The area occupied by that type of land use is only half as large as the area occupied by permanent field sedentary cultivation.
5. The areas where shifting cultivation is dominant are characterized by a very low population density.

In order to assess critically shifting cultivation (in its broader connotation) it is necessary to take a closer look at its functions, and its ecological and socio-economic consequences.

2.4 Characteristic procedures of shifting cultivation land-use systems in Africa

Impressive descriptions of the appearance of the agricultural landscape shaped by shifting cultivation are offered in the book of Allan (1965) *The African Husbandman* and in the work of Nye and Greenland (1960) *The Soil under Shifting Cultivation*. Only a few sentences can be cited here.

> The first sight of native subsistence farming in the semi-deciduous forest region, e.g. in Ghana, presents an appearance of bewildering confusion to anyone familiar only with the pattern of well ordered fields under single crops characteristic of more advanced systems of farming. There are no clear boundaries, individual fields can scarcely be discerned, and while some patches of land are definitely under crops, and others are under a thick regrowth of forest, there is a middle group in which perennial crops survive amidst a regrowth of forest which is gradually choking them. Some patches of land carry only one kind of crop, yet others appear to carry a mixture of up to half-a-dozen kinds in a seemingly haphazard arrangement (Nye and Greenland, 1960: *1*).

With reference to the South-east Asian agrarian landscapes, Pelzer (1948, p. 6) states:

> Can there be a greater contrast than that between the small, irregular, temporary clearings with their chaotic jumble of fallen and half-fallen tree trunks and stumps of the shifting cultivators on the one hand, and, on the other, an irrigated plain, subdivided into small fields, each surrounded by a dyke – or a steep slope – where a succession of beautifully built terraces leads up several hundreds or even thousands of feet?

A most insightful documentation with regard to the special organization, the work procedures and the socio-economic consequences of this special form of agriculture is still Pierre de Schlippe's *Shifting Cultivation in Africa* (1956) which represents years of personal experience among the shifting cultivators of the Azande group. The Azande district lies near the border between the north-east Congo and the south Sudan, inside the deciduous tropical forest belt which experiences a relatively short dry season of 3 months. The agricultural practices among the Azande can be regarded as prototypical for those areas in which rural settlements consist of isolated farmsteads.

The land use on two holdings over a period of 4 years is illustrated in Fig. 2.4. There are two main types of cultivated fields: home gardens and cropland in forest clearings. The inner parts of the gardens are composed of the courtyards surrounding the dwellings of the cultivator and his wives. These locations lack topsoil, but weeds, ash and household refuse serve as fertilizer. A variety of legumes and medicinal plants are grown here, interspersed occasionally with pineapples, maize, eggplants, or even fruit trees such as fig, mango, palm or citrus. Particularly humid and shady places are used as nurseries.

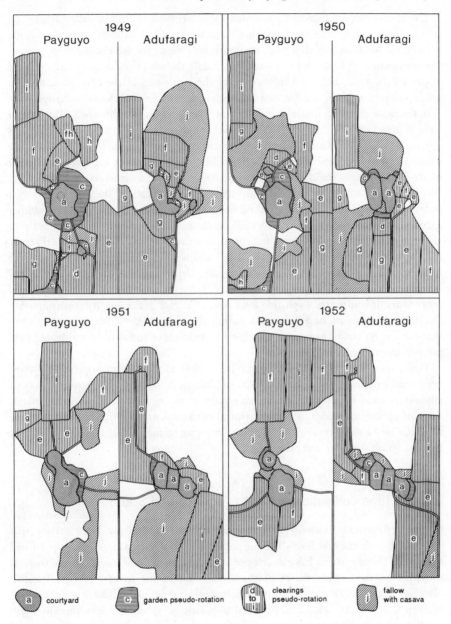

Fig. 2.4   Pseudo-rotation of shifting cultivators in the Azande district, Central Africa. (Source: De Schlippe, 1956)

The courtyards are usually surrounded by an artificial ridge and valley system on which, immediately after the first rains, a garden pseudo-rotation is started with the topsoil that has been removed from the courtyard. Along the crest of the ridges, maize and pumpkin are planted on the same day, while the sweet potatoes on the outward slopes follow later. Since the ridges receive the sweepings of the courtyard, kitchen refuse, crop processing residue, and ashes, they are the first to yield their crops.

In the more extended outfields the slash-and-burn system is established. These fields vary in size between 0.25 and 0.33 ha, each of them growing a different crop. As a general rule, immediately after burning and hand clearing, maize and cassava are planted in widely spaced holes; 15–20 days later, when the maize has reached a height of 20–25 cm, finger millet (*eleusine*) is broadcast into the maize–cassava field. Other crop sequences are the groundnut and finger millet association, the maize and sweet potato association and the maize and oilseed gourd association, which all have similar agronomic requirements. The cassava fallow, however, is different. In the second year after the initial burning, cassava is interplanted in the main associations. After that, the land reverts to fallow and the cassava is left unattended. Sometimes it develops into lush groves, sometimes it vegetates miserably, depending on how aggressively the weeds return. During the first 2 years, cassava is harvested as needed, then, for several more years, it constitutes a valuable food reserve in case of famine. By superimposing the different maps of land cultivated during the years 1949–52, de Schlippe shows that there is no rigid or planned succession of field types and that the cultivated field boundaries are extremely unstable.

The examples previously mentioned relate mainly to the African tropics. They have been used to illustrate the essential facts of shifting cultivation. Numerous works of ethnologists and geographers provide a picture of the many facets and details of this form of agriculture under the varied natural environments in different regions of the tropics. Only a short review can be given, referring to the two remaining parts of the tropics, South America and South Asia.

2.5 Shifting cultivation in South America and South Asia

In South America, rotational field farming systems comprise a large variety that reflects different cultural levels among populations and varied degrees of use of soil resources. These systems have developed mostly in the humid forests of the tropical lowlands on the eastern flanks of the Andes and in the *cerrados* of the Brazilian *planaltos*. In areas of denser forests, in which aboriginal populations have sought refuge or keep little contact with Europeanized populations, aboriginal agriculturalists still practise a variety of farming systems on forest clearings that are commonly referred to as swidden fallow. Research conducted in the tropical lowlands by Denevan (1971), Harris (1971), Basso (1973), Posey (1985), Stocks (1983), Beckerman (1983), Vickers (1983b) and Denevan *et al.* (1985) coincides in pointing out that after as little as 2 or 3 years of intensive planting, a cleared plot declines in soil fertility and an aggressive onslaught of noxious vegetation leads to 'phased field abandonment' during which vegetables are progressively replaced by productive trees. In this way, gardening is gradually replaced by a form of subtle agroforestry, abandoned fields being reclaimed for cultivation only after 20–35 years of fallow.

In Beckerman's work on swidden gardening among the Bari Indians, a fact that emerges conspicuously in all varieties of farming is that shifting cultivation is explained by pedologic conditions, rather than by human volition or increasing population density pressures (Beckerman, 1983). Among the Candoshi and Cocamilla Indians of the Huallaga River in Amazonian Peru, the use of concentric rings in crop zoning obeys ecological conditions. Shade reduction, isolation of plague-sensitive vegetals in the centre of the fields and away from the insects and weeds that populate the periphery of a newly opened field, and location of species that are nitrogen demanding, such as maize and plantains, on the edge of a plot to take advantage of litter supply from the adjacent undisturbed forest, are indications of the weighting that native cultivators give to ecological constraints (Stocks, 1983).

In general, these works reveal that shifting cultivation as practised by the native populations of South America is to be considered as an ecological strategy intended to mimic forest structure and to foster natural recovery after natural or human destruction in order to maximize soil use without total effacement of its natural cover.

Shifting cultivation has also been the prime form of farming among the mestizo colonists of Andean countries who descended into the tropical lands of eastern Peru, Ecuador and Colombia, as well as the Brazilian squatters who progressively pushed into the heart of Amazonia. Subsistence farmers – *conuqueros* in Spanish-speaking countries, or *caboclos* in Brazil – practise itinerant agriculture referred to as *conucos* or (*roças* in Brazil) through which selected places of the virgin forest are cleared by machete and fire and, subsequently, farmed (Watters, 1971).

The main points of shifting cultivation in the tropics of South America can be summarized as follows:

1.   The movements of cultivators who farm crops such as maize and rice, which need better care and more fertile soils than traditional cultivars such as manioc, yam or arrowroot, are determined by the progressive empoverishment of the soils a few harvests after a forest has been cleared;

2.   Polycultures, i.e. the planting of several staples in one plot, are predominant in fields that have been recently opened, while, as soil fertility decreases and the growth of weeds increases, the cultivator is forced into monocultures, until finally, 3–5 years after the initial clearing, he abandons the cultivated plot and leaves it fallow – *en rastrojo* – for an undetermined number of years.

In the Asian tropics shifting cultivation is not the dominant form of agriculture. In the rainfed agriculture of the Deccan Plateau of India, permanent field cultivation prevails. As far as alluvial lowlands are concerned, the intensification of agricultural practices caused the forcible translation of shifting cultivation from the lowlands into the forested areas of the uplands, where remnants of Asian shifting cultivation can still be observed (Grigg, 1974).

In the South-east Asian tropics different forms of agriculture show strong variations in regional as well as in local scale. The areal overlapping between shifting cultivation with its subtypes and permanent agriculture, the differing environmental and socio-economic circumstances and the newest developments have been discussed by Uhlig and collaborators (1984) in *Spontaneous and Planned Settlement in Southeast Asia*. We shall return later to the regional differentiation and its reasons.

Today swidden practices and bush fallowing are still found in hilly areas of northern Thailand, on Sarawak (Borneo), in the Philippines, and in isolated pockets of north-eastern Cambodia and southern Laos. In northern Thailand, the survey by Kundstadter *et al.* (1978) established that swidden was applied to clear forests in order to cultivate paddy rice, peanuts, mung beans and, on some occasions, maize. On Kalimantan, Indonesia, dry shifting agriculture and the use of tree shading to foil the growth of perennial weeds have been detailed by Seavoy (1973). In the hilly areas of Sarawak, which are occupied by the Iban, Freeman (1955) reports the use of shifting cultivation as a farming modality to grow dry hill rice. Shifting cultivation practices in the forest areas of the Philippines have been documented by Eder (1977) and by the now classic study of Conklin (1957) on shifting cultivation among the Hanunoo.

Rotational fields cleared by fire among aboriginal populations of New Guinea have been described by Brookfield (1962), Clarke (1976), Rappaport (1967), Waddell (1972), Seavoy (1973) and Manner (1981).

Among the swidden cultivators of northern Thailand, the Hill Lua and Karen populations apply a 'short cultivation–long fallow' system. One or two years of cultivation are followed by fallow periods of 6–7 years, 12–15 years as a maximum. The dwellings of these people are permanent, but the cultivators are highly mobile in the forested hills that surround the villages. Among the Hmong group, long cultivation periods are followed by 'very long fallow' periods. Once the land fertility is exhausted, the temporary settlements are abandoned and the search for new upland forests to be opened begins. Kunstadter *et al.* (1978) acknowledge that ecological constraints cause the mobility of these cultivators and that soil impoverishment lies at the root of these itinerant farming practices.

The documentation offered by Freeman (1955) for the Iban of Sarawak reveals a very high mobility of the cultivators. The penetration of the rain forest is made from a long farmhouse, the *dampa*, which acts as the spearhead into a territory of 0.8–1.6 km in radius. From here the forest is felled and burnt for cultivation of dry rice. Seldom does the cultivation period of newly opened plots exceed 5 years, and the return to a field that was formerly opened may not happen before 15 or 25 years. If crop production declines before the first 5 years or when population pressure demands, a *dampa* is moved to a new location. Freeman (1955) calls the practices of the Iban 'land squandering', but he also hastens to point out that the soils taken under cultivation by the Iban are among the poorest of Sarawak.

In the forests of western Kalimantan, slash-and-burn agriculture and dry rice cultivation (*ladang*) on hilly terrain are conducted, and no field appears to be used for more than 5 years. After that the plots are left fallow for periods of 5–8 years (Seavoy, 1973). This author points out that the ever-changing and rather short-term use of cleared land is determined by pedological conditions: the lateritic soils are poor in nutrients because of intense outwash and leaching related to high rainfall.

2.6   Critical assessment of shifting cultivation in its broader connotation

2.6.1   *Why field shifting?*

A critical point of shifting cultivation revolves around the reasons for abandoning a cultivated area after a relatively short period of use. Has the abandonment been forced by the environmental circumstances? Is it simply advantageous for the cultivator? Or could it even be avoided through adoption of particular measures?

Consensus exists among all scholars who have dealt with shifting cultivation that the general cause for abandonment of a plot after only a few harvests (from two to five, or at most six) is a decline in yields combined with an increase of weeds.

Both are obviously related, but do they weigh equally in the decision to abandon the plots? Are both processes interrelated in the sense that the same events that lead to a lowering of the yields are also responsible for the increases in weed associations? With respect to the preponderance of both causes, an attempt must be made to prove their importance.

Figure 2.5 illustrates the yield decreases in slash–and–burn plots in different shifting cultivation areas with varied soil and hygro-climatic conditions, as published by Sanchez (1976). The original data proceed from different experimental stations that maintain adequate weed control; the numbers refer to consecutive crops. The decline is extreme on oxisols and ultisols under tropical rain forests. In the wet-and-dry savanna of Yambia, Sudan, it is much less pronounced, even on oxisols. The most stable yields come from alfisols and entisols. (Properties of the mentioned soils are given in Ch. 3.)

Since all data were obtained under conditions of adequate weed control, the initial indication is that the yield decline occurs independently of the growth of weeds. On the other hand, the data testify to a gradual dependency on pedological and rainfall conditions: while, in the humid tropics, dry rice cultivated on oxisols and ultisols renders unacceptable yields after only two cropping periods, in the wet-and-dry savanna it takes more than twice that time before the rice yields become unprofitable. Under equal climatic conditions, in Yurimaguas, the yields on relatively nutrient-rich alfisols are significantly higher than on ultisols. On immature entisols abundant in residual minerals, long-lasting fertility is maintained even under the rain-forest conditions of the Pacific islands (New Britain).

The gravity of the weed problem for a shifting cultivator cannot be appreciated by people accustomed to the conditions of permanent cultivation on well-tended plots in extratropical regions. The magnitude of the problem can be quantified – to a certain extent – by the length of time and human effort dedicated to weeding. As an example

57

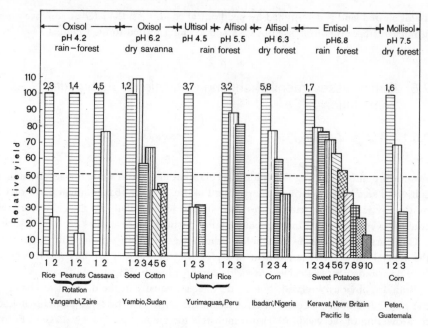

Fig. 2.5 Yield decline in slash-and-burn agriculture. (Source: Sanchez, 1976)

be noted that, among the Zaria cultivators of the dry savanna of northern Nigeria, a great part of every work day between May and September (the rainy season!) is spent weeding. In June, for instance, weeding takes nearly 7 hours, and in September – at the beginning of the harvest season – it still requires at least 4 hours (Norman *et al.*, 1982, p. 120).

The very instructive documentation of de Schlippe (1956) on this subject shows which amount of time in relation to the entire agricultural work the Azande cultivators spend weeding in the different field types. In Fig. 2.10 the time occupied for weeding is represented in the hatched area under curve 3. In the courtyards and on the plots with pseudo-rotation (ridge cultivation and maize to sweet potatoes) the application of labour for weeding purposes is minimal. It is, however, significant in the groundnut–eleusine succession and in the maize–eleusine association. In the cotton fields it takes up most of the cultivation time. These differences indicate that in those plots that have been kept under permanent cultivation, the weed problem is minimized.

It follows then that, also in tropical environments, the weed problem could be significantly reduced – even without the use of herbicides – if continuous cropping was practised and the land permanently tended. Continued tending of a piece of land will, in the long run, systematically reduce the weed potential.

We conclude that the field changes are not necessarily caused by an overabundance of weeds; rather, the weed problem may appear as a side-effect of the field change.

The primary reason for moving on to a newly fire-opened field is the decline in yields on the old fields. This decline must be caused by circumstances different from the weed problem.

2.6.2 The effects of slash-and-burn

Theoretically, one might expect the slash-and-burn procedure to have a generalized scorching effect on the upper soil layer. But upon examination of a clearing, shortly after burning, only small patches of brick-red loam or clay are discovered. Scorching is relatively mild because the material closest to the surface consists of fragmented twigs and leaves which burn quickly without developing much heat, while the heavier branches and trunks, which smoulder for hours and even days, are not really in contact with the soil but tangled above the ground. The fauna and microfauna in the upper soil are destroyed, of course; only worms and ants survive by retreating into the deeper soil. With their subsequent return to the surface the microfauna will be restored.

Several comparative studies conducted in the Amazon confirm that hand-clearing methods that involve burning are more advantageous than mechanical clearing by bulldozer which causes soil compaction, topsoil displacement and lacks the fertilizer value of the ash. Table 2.2 illustrates clearly the advantages of slash-and-burn over mechanical soil clearing.

Noteworthy is the fact that, even if a full fertilizer treatment is given to a

Table 2.2   Yields of different crops planted after the use of slash-and-burn or mechanical soil clearing means (t ha^{-1})

Crops	Fertility levels	Crop yields	
		Slash-and-burn	Bulldozed
Upland rice	Without fertilizer	1.3	0.7
	With fertilizer*	3.0	1.5
	Fertilizer + liming	2.9	2.3
Corn	Without fertilizer†	0.1	0.0
	With fertilizer*	0.4	0.04
	Fertilizer + liming†	3.1	2.4
Soyabeans	Without fertilizer	0.7	0.2
	With fertilizer*	1.0	0.3
	Fertilizer + liming†	2.7	1.8

* 50 kg ha^{-1} N, 172 kg ha^{-1} P, 40 kg ha^{-1} K.
† Plus 4 t ha^{-1} of lime.

Source: Seubert *et al.* (1977).

mechanically opened field, crop yields are always superior in slash-and-burn plots. Using African experiences, Lal (1979b) favours the slash-and-burn practice and later mulch-farming combined with a no-tillage system (see section 8.3.3). He states that

> the dramatic failure of many large-scale schemes for mechanized arable farming in the tropics of Africa is an indication of factors that strongly limit crop production other than just the availability of commercial fertilizer. An important factor in continuous productivity of tropical soils, therefore, is the maintenance of soil physical characteristics at optimum level. (Lal 1979b:397)

After studying the positive effects of shifting cultivation on soil properties such as bulk density, nitrogen and carbon content, soil acidity, and calcium content in a rain-forest area of coastal Guatemala, Popenoe concluded as early as 1957 that careful consideration should be given to new land-clearing alternatives before replacing the shifting cultivation practices used by indigenous peoples. Slash-and-burn is obviously a good guarantee for keeping optimal physical soil conditions.

An even more beneficial aspect of the burning practice is the nutrient supply to the soil that the plants require for growth. The relevant question is: what is the actual nutrient content of the vegetal ash and how much of it is ultimately used by the cultivated plants?

2.6.3 *Nutrient content of forest biomass and ashes after burning*

The first comprehensive quantitative analysis of the nutrient content of forest biomass and its conversion into soil nutrients in the slash-and-burn cycle was undertaken by Nye and Greenland (1960). The diagrams in Figs 2.6 and 2.8 are based on their findings.

The distribution of nutrients in two different vegetation–soil systems is illustrated in Fig. 2.6. One is a 40-year-old secondary evergreen forest on ferrallitic forest oxisol with phyllites as parent rocks located in Kade, southern Ghana; the other an 18-year-old secondary forest in Yangambi, Zaïre, on oxisol (details about these soils are given in Ch. 3). The different contents of nitrogen (N), phosphorus (P), potassium (K), calcium (C) and magnesium (Mg) found in the leaves of the canopy, in lianas, branches, trunks, dead wood, and litter, appear in the upper part of the vertical scale, the contents for the roots and soil layers down to a depth of 30 cm in the lower part of the scale (both values in kilograms per hectare). In the case of Yangambi, Nye and Greenland give the calcium and magnesium content in a single value. We have separated the components, which is indicated by the broken line in the corresponding columns. The calcium values in the experimental system of Kade are extraordinarily high, almost twice as high as those given by Nye and Greenland for other tropical sites (1960, Table 2). It may be assumed that the mineral content of the parent rock plays a major role in the calcium proportion.

The conclusion drawn from these empirically obtained values is that, in the forests

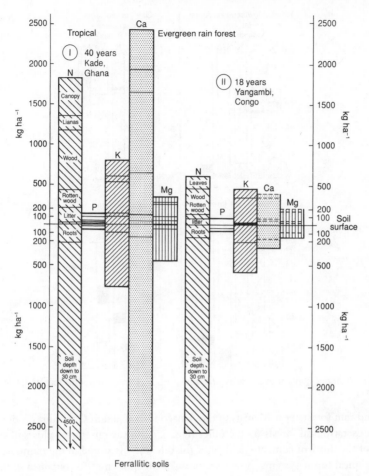

Fig. 2.6 Nutrient content in biomass and soil in tropical evergreen forests (after Nye and Greenland, 1960)

of the humid tropics, most of the nutrients, except for the nitrogen, are actually stored in the phytomass and not in the soil itself.

Sanchez (1976) summarized a vast array of published data on dry matter and nutrient content of the biomass from evergreen forests in Zaïre, Ghana, Panama and Puerto Rico as follows: dry matter 200–400 t ha^{-1}, nitrogen 700–2050 kg ha^{-1}, phosphorus 33–137 kg ha^{-1}, potassium 600–1000 kg ha^{-1}, calcium 650–2750 kg ha^{-1} and magnesium 400–3900 kg ha^{-1}. The wide range of values is probably due to the differences in parent rocks; the higher values correspond almost exclusively to the forests of Panama and Puerto Rico. Lacking here is, however, the ecologically important information about the corresponding values of the nutrient contents of the upper soil layers.

Klinge (1976b) compared nutrient distribution in the biomass and the soil in tropical

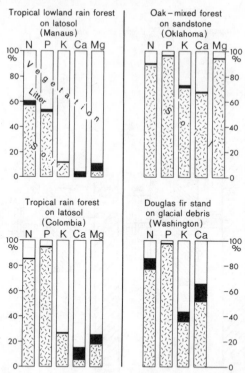

Fig. 2.7   Nutrient content in biomass and soil in tropical and temperate forests (after Klinge, 1976b)

lowland rain forests near Manaus in the Amazon basin and in Colombia on one hand, and in extratropical forests on the other (Fig. 2.7). In the rain forest-plus-soil system, most of the nutrient content – and all of the calcium – is stored in the biomass. In the extratropical forest systems, however, most of the nutrients are contained in the soil. This is an ecologically decisive difference with far-reaching consequences, to which we have to refer in various contexts.

By burning forest biomass, a large percentage of nitrogen and sulphur escapes into the air through volatilization. Most of the nutrients, however, are concentrated in the remaining ash. The amounts can be found in Seubert *et al.* (1977). The ash from a 17-year-old forest near Yurimaguas (in the Peruvian segment of the Amazon basin) contained 67 kg ha^{-1} nitrogen, 6 kg ha^{-1} phosphorus, 38 kg ha^{-1} potassium, 75 kg ha^{-1} calcium, 16 kg ha^{-1} magnesium, 7.6 kg ha^{-1} iron, 7.3 kg ha^{-1} manganese, 0.3 kg ha^{-1} zinc and 0.3 kg ha^{-1} copper.

In Fig. 2.8, to the values of the different nutrient contents given for the roots and the soil by Nye and Greenland (Fig. 2.6), the gains in phosphorus, potassium, calcium and magnesium yielded from the ash after burning the forest have been added (indicated in the diagram by a plus sign). The resulting amounts represent the nutrient potential at the beginning of agricultural use.

As to the nutrient demand by different crops, the respective values are indicated on

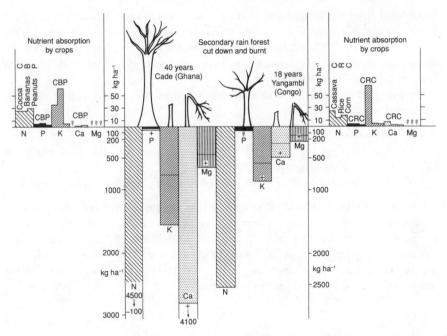

Fig. 2.8 Nutrient supply in ash and uptake of cultivated plants (after Nye and Greenland, 1960)

both sides of Fig. 2.8. (It should be noted that the scale of that particular section was amplified to accommodate low values!) The calculations of the nutrient demands are based on yields of 1100 kg of maize or rice, 770 kg of peanuts, 11 200 kg of plantains or cassava, and 1100 kg of cocoa per hectare.

When comparing the nutrients supplied by the burning of forest biomass with actual nutrient extraction by annual or biennial crops, the latter is so small that it is difficult to believe that the yields of second and third harvests decline so drastically.

However, it must be borne in mind that the duration of possible cultivation periods is by no means determined by nutrient consumption alone. The values cited by Nye and Greenland were obtained right after the forest burnings. Sowing or planting takes place a few days later. Until the plant cover has reached a certain height and density, the ash layer on the bare ground is fully exposed to the effects of erosion and leaching in connection with rainfall events of tropical intensity. Since, during the first 2 weeks of growth, the root system of the new plants is still weak and shallow and the demand for nutrients is minimal, it may well happen that, by the time the need for nutrients increases, a considerable portion of them has been leached out of the reach of the roots. (The insufficient cation exchange capacity (CEC) of tropical humid soils has negative consequences in this context, as will be explained later.) Annuals are at a particular disadvantage in comparison with deeper-rooting perennials.

The exact nutrient losses are difficult to trace. That they must be considerable can

63

be deduced from the comparison of the nutrient contribution by burning and their uptake by plants. Chemical analyses of stream waters point in the same direction. In the Amazon, Sioli (1967) and Brinkmann and Dos Santos (1970) detected a sudden rise in the nutrient content of otherwise very poor clear- or black-waters far downstream from a slash-and-burn clearing site.

Regarding the effects of slash-and-burn practices we may conclude that the advantages are as follows:

1. For a few crops nutrition is provided without artificial fertilizing;
2. There is no need for heavy mechanical tools; planting stick, hoe, cutlass or machete being the essential farming tools;
3. Weeds are suppressed for a while and pests are kept at bay to a certain degree.

However, it must be said at the same time that the quantitative analysis of nutrient content in the biomass, in the ash after burning and then after uptake in the cultivated plants, makes it evident that under the particular conditions of leaching and erosion, slash-and-burn practices are, in fact, nutrient squandering. This is true especially with the previously described precipitation patterns of the tropics (Section 1.2) and with the soil properties that will be described in Chapter 3.

2.6.4 Impact of shifting fields and dwellings on tropical forests

The impact of shifting cultivation on tropical forests depends mainly on two variables: population density and duration of the rotations.

For many centuries only small social units populated the humid tropical forests. Under this condition long periods of fallow could re-establish the original climax vegetation. This explains why, until the middle of this century, about 60 per cent of the world's total tropical moist forest area had been spared from destruction or severe degradation (Table 2.3, after Sommer, 1976). In this context it is interesting to note that reductions were least where the evergreen tropical rain forests dominate. In contrast, human impact was extreme on the drier margins of the forested areas, as in East and West Africa, or in South Asia with losses of 72 and 63.5 per cent respectively.

Meanwhile, tropical rain forests are being disturbed and degraded at an increasing rate by human activities. The FAO made several attempts to quantify objectively the extent of the losses. In the first, a reduction quotient of 5–10 million ha yr^{-1} was calculated for the total area of humid forests in 1969, i.e. some 0.5–0.6 per cent of the still existing 935 million ha.

In the second, Sommer (1976) calculated from 1970 onwards a minimal reduction rate of 1.2 per cent to all humid forest areas, resulting in an annual deforestation of approximately 11 million ha. However, these estimates were questioned by Grainger (1983b).

In 1978 FAO started a new 'Tropical Forest Resources Assessment Project' which completely covers the 75 larger countries of tropical America, Africa and Asia.

Table 2.3 Summary of the status of humid tropical forests (million ha) (after Sommer, 1976)

Region	Climax area				Actual area		
	Tropical rain forest	Semi-deciduous forest	Deciduous forest	Total moist forest	Total moist forest	Regression of climax	% of climax
East Africa	13		12	25	7	18	72.0
Central Africa	197		72	269	149	120	44.6
West Africa	50		18	68	19	49	72.0
Total Africa	260		102	362	175	187	51.6
South America	600		150	750	472	278	37.1
Central America*	27	26		53	34	19	35.8
Total Latin America	627	26	150	803	506	297	37.0
South-east Asia	237	15	50	302	187	115	38.1
South Asia	12	15	58	85	31	54	63.5
Pacific region	48			48	36	12	25.0
Total Asia	297	30	108	435	254	181	41.6
Total humid tropics	1184	56	360	1600	935	665	41.6

* Including Mexico and Caribbean.

The findings of this important survey are presented in four technical reports. The first three consist of a regional synthesis for each of the continents mentioned. The fourth (Lanly, 1982) summarizes the overall results for 1980 and gives some considerations on the evolution of tropical forest resources up to 1985. The study contains a large number of different 'natural tropical woody vegetation' classes. The classification differs from that used in Sommer's 'Assessment of the world's tropical moist forests', so that the figures of different classes are not directly comparable. For our purposes the main interest is directed to the class of 'closed broad-leaved forest stands' and the fallows of closed broad-leaved forests. The first have not been cleared in the recent past (during the last 20–30 years) and cover with their various storeys and undergrowth a high proportion of the ground. They may be evergreen, semi-deciduous, wet, moist or dry. The latter represent the complex of areas of shifting cultivation and of the various types of secondary forest regrowth. Table 2.4 presents the respective values given by Lanly (1982).

The area of closed broad-leaved forests cleared annually is around 7 million ha, i.e. a rate of reduction of 0.6 per cent per year. This rate is more or less the same for

Table 2.4 Areas of closed broad-leaved forests, average annual deforestation and forest fallow areas, 1980 and 1985 (thousand ha)

		Total broad-leaved forest	Undisturbed broad-leaved forest	Forest fallow
Tropical America				
Annual deforestation	1976–80	3 807	1 135	
Total area	1980	653 926	492 976	99 338
Annual deforestation	1981–85	4 006	1 196	
Total area	1985	633 893	437 196	106 431
Tropical Africa				
Annual deforestation	1976–80	1 319	220	
Total area	1980	214 403	118 180	61 631
Annual deforestation	1981–85	1 318	225	
Total area	1985	207 805	113 889	66 685
Tropical Asia				
Annual deforestation	1976–80	1 767	483	
Total area	1980	291 951	97 259	67 246
Annual deforestation	1981–85	1 782	379	
Total area	1985	283 038	85 139	71 573
Tropical world				
Annual deforestation	1976–80	6 893	1 838	
Total area	1980	1 160 280	668 415	228 215
Annual deforestation	1981–85	7 106	1 800	
Total area	1985	1 124 737	636 224	244 689

Source of values: Lanly (1982).

the three regions. Since tropical America has over 56 per cent of this type of forest it accounts for an almost equal proportion of the total deforestation, i.e. around 4 million ha yr^{-1}. In South America, Brazil alone loses 1.36 million ha each year, although the corresponding deforestation rate is relatively low (0.38 per cent) due to the considerable size of the closed forests in this country (356 million ha). The highest rates are found for Costa Rica, El Salvador and Paraguay with 3.5 per cent.

In Africa more than half (55 per cent) of the total deforestation in the 37 countries studied, occurs each year in the 9 countries of West Africa. Ivory Coast and Nigeria alone, with 310 000 and 285 000 ha deforested annually during the period 1976–80, are responsible for 45 per cent of the nearly 1.8 million ha of Africa in total. In Central Africa the situation does not give reason for concern. The Cameroon–Congolese forest is only being reduced by 50 000 ha yr^{-1} (0.2 per cent of the total area). In East Africa clearing affects 0.8 per cent of the total area per year. The levels are 2.9 and 2.4 per cent for Burundi and Rwanda respectively, where the last remaining areas of forest are gradually being eaten away by farmers looking for new land.

In tropical Asia the annual deforestation was greatest in insular South-east Asia (890 000 ha) and continental South-east Asia (Burma and Thailand 428 000 ha). The countries where areas deforested annually are largest are Indonesia (550 000 ha), Thailand (330 000 ha) and Malaysia (230 000 ha). The highest rate of deforestation is in Thailand with 3.6 per cent. Also in this region, shifting cultivation, particularly that which follows the logging front, appears to be the main reason for deforestation (Lanly, 1982; Scholz, 1986).

In our context the following findings are interesting: comparing the figures on the reduction of forests during the period 1981–85 with those of the increase in the same period of forest fallow areas; the conclusion is that shifting cultivation (with the complete cycle) is responsible for around 35 per cent of the total deforestation in tropical America, more than 70 per cent in tropical Africa and 49 per cent in tropical Asia (Lanly, 1982).

This supports and underlines quantitatively the statement of Ruddle and Manshard (1981, p. 65) who see the following relation between the different deforestation activities:

> The main causes for this large scale disruption of tropical vegetation lie in the practice of shifting cultivation and small-scale sedentary agriculture, mostly by peasant groups, whereas forest clearance for the establishment of large plantations, pasture lands and the commercial exploitation of timber resources has been of lesser although still significant importance.

The preliminary results of the FAO's 1990 reassessment of tropical forest cover (Lanly *et al.*, 1991) reveal a dramatic increase in the rate of deforestation from 9.2 to 16.8 million ha annually in the periods 1976–80 and 1981–90 respectively. This significant difference can be attributed to the following main factors:

(a)   an actual increase in the annual rate;

67

(b)   an underestimation of the rate in the 1980 assessment; and

(c)   an overestimation in the 1990 assessment.

More accurate values will be provided by the results of the uniform remote sensing observation in the second phase of the global assessment. These values will not be available until 1992. However, the preliminary results already indicate that

> the reduction and degradation of the world's tropical forests have continued and even been aggravated in most countries during the 1980s as the main causes have remained largely unattended: increasing numbers of rural poor in desperate search of land for their immediate survival, decreasing yields, exhaustion of agricultural lands, inadequate landownership and tenure systems, lack of sound land-use planning, unsustainable forest management practices, lack of alternatives to fuel wood and charcoal for energy supplies etc. (Lanly *et al.*, 1991, p. 21).

To conclude we can say that, besides the above-mentioned disadvantages, shifting cultivation is also a forest-squandering form of land use.

2.6.5   *Does rotational bush fallow practice constitute real progress?*

So far, the functioning and implications of shifting cultivation have been discussed for areas where natural or secondary forest is still preserved. Now we must deal with the transition to 'rotational bush fallow' systems in regions where cultivation is practised from nucleated sedentary settlements (shifting field agriculture or shifting cultivation in its more restricted connotation) and where the natural vegetation has already been degraded to bush or open forest as it occurs, for example, in vast regions of West Africa (see Fig. 2.1).

While it is obviously advantageous for rural populations to live in well-demarcated villages with ready access to highways, railroads and other infrastructural facilities, the step to permanent settlements does not necessarily entail a change as far as agricultural land use is concerned. The examples of concentric rings of cultivation around some Nupe villages in central Nigeria, published by Morgan (1969b, Fig. 2.9), are certainly somewhat exceptional. But they demonstrate more clearly than others the considerable area in waste, woodland, forest, grassland and fallow, which is normally obscured by the pattern of dispersed distribution due to land tenure differences among the farmers. In the example of the Mokwa village, the cultivated fields were found beyond a wide zone of fallow at a distance of 4–7 km (3–5 miles) from the village in the year of survey. In subsequent years the area under cultivation moved closer to the village, but then again there were times when the farmer had to travel great distances to get to his fields. Would he be doing this without necessity?

The surface area which is demarcated by the outer ring of fields comprises almost 130 km^2 (13 000 ha). Morgan (1969b: 316) does not mention specifically the number

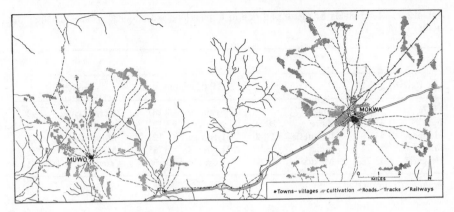

Fig. 2.9 Areal extent of field and fallow around some Nupe villages in central Nigeria (from Morgan, 1969b)

of inhabitants for Mokwa, but he states: 'The apparent optimum development of land-use rings, with reference to a single settlement, takes place where the settlements are of the order of 2,000–3,000 population, cultivating over an area some 2–5 miles in diameter allowing considerable area in waste, woodland, forest and fallow.'

Taking these facts into consideration, it turns out that in the humid savanna the transition from real shifting cultivation to crop–fallow land-use systems without displacement of homesteads brings gradual change only in as far as the fallow percentage of the total area decreases somewhat. Basically, the land-demanding character remains unchanged. The inhabitants of permanent settlements enjoy the advantages of better developed infrastructures, for which they are paying a price, however, with shorter rotation periods, smaller amounts of fertilizing ash and longer travelling distances to their fields.

In the case of dry savanna the conditions are more favourable in that the land use is more differentiated, that there is a manured and permanently cultivated ring outside the village wall and that the crop–fallow ratio is greater.

2.6.6 *The yields obtained from shifting cultivation*

The shortcomings associated with this land-extensive system of agriculture would not be that critical if, at least, the yields could be kept in a reasonable relation to the total area exploited.

Given the scarce data available, it is difficult to assess statistically the yields of crop associations from slash-and-burn plots and to compare their feeding potential with that of extratropical crops. In general, experts agree that in the tropics less food is produced per unit of arable land than in the extratropics, not to mention the larger areas required by tropical agriculture.

Table 2.5   Mean yields of maize and rice in tropical and extratropical regions (bushels per acre)

Maize				Rice			
Extratropics		Tropics		Extratropics		Tropics	
USA	50.66	Brazil	29.80	Spain	187	Sierra Leone	62
Italy	47.68	Zaïre	29.80	Italy	122	Thailand	50
Hungary	44.70	Indonesia	29.80	Japan	107	Indonesia	47
France	38.74	India	25.82	USA	65	Malaysia	35
Bulgaria	32.78	Mexico	17.88	Korea	56	Vietnam	32

Source: Gourou (1966).

Maize is perhaps the crop that lends itself best to a comparison. For the first decades of this century Gourou (1966) published these values shown in Table 2.5. In all cases the proportion is 1 : 1.5 or 1 : 2 in favour of the extratropical countries. As to the nutrient value of paddy rice – which is the most intensive crop grown by tropical cultivators – it still lies far below the value of European and North American wheat crops. Considering the dry rice cultivated in crop–fallow rotation, which yields about a quarter of wet rice or less, one again arrives at the conclusion that the average yield of crops per hectare in the tropical non-irrigated lands lies between 50 and 100 per cent below that of the extratropics.

Information collected from more recent publications indicates the following yields: maize in Ghana, 1200 kg ha^{-1} (Nye and Greenland, 1960); average maize yield (from 11 small farmers) near Yurimaguas in the Amazon basin of Peru 2.44 t ha^{-1} the first harvest, 1.77 t ha^{-1} the second harvest; average upland rice yield 1.91 t ha^{-1} in the second harvest (Bandy *et al.*, 1980). These are very discouraging statistics indeed, when compared to yields of more than 5 t ha^{-1} in the fields of extratropical countries. The values are in line with statements of world food production experts in the sense that today the discrepancies between the yields in tropical and extratropical agriculture are even greater than they were a few decades ago.

2.7   Pros and cons of shifting cultivation

Given the evidence of low agricultural output per individual and acreage involved, the small proportion of cultivated land in relation to the total areas taken up by swidden practices, and the devastating consequences for the forests in areas where shifting cultivation is practised, it is not surprising that the persistence of this form of land use is upsetting to those concerned about the development of tropical regions. 'A shifting cultivation system can at the maximum feed only 40 to 50 people per square kilometer, and is the cause of undernourishment in many overpopulated developing countries' (Andreae, 1981, p. 301). It is 'becoming increasingly evident that, owing to the explosive growth of world population, such systems are not longer rational (and that) shifting cultivation may become maladaptive' (Ruddle and Manshard, 1981, p. 76).

However, there are also other opinions worth mentioning:

The historical success of systems of shifting cultivation as rational systems of resource use is unquestionable, for until the last century and, indeed, until the last two or three decades in many cases, they have sustained the bulk of the human population in most tropical regions. . . . Integral systems of shifting cultivation represent remarkably perceptive and sophisticated ways of using what, to most westerners, are difficult regions to sustain societies. . . . The great triumph of shifting cultivation has been its capacity to support large numbers of people through many centuries not by an extraordinary and laborious reworking of the natural landscape, but rather through a more subtle and ingenious imitation of the natural order by a miniature tropical forest of cultivated plants (Ruddle and Manshard, 1981, p. 74).

It all sounds reasonable, but in order to uncover the whole truth one must also ask what price has been paid. What we are sure of at this point is that these positive attributes could only be maintained under strict conditions of low population density.

Maintenance of a low population density over many generations is not something that can be taken for granted. If anything, a sustained population growth would be a more natural thing. At a growth rate of only 2 per cent, it takes 500 years for a group of 1000 individuals to reach the 1 million mark. Since the past centuries obviously experienced no such population increases in areas of shifting cultivation, it becomes necessary to pose questions about the 'limitations to natural growth' and the associated 'limitations to cultural development' such as Frank (1987) did in his empirical study of the Indian tribes of the eastern Amazon. He arrives at the conclusion that, in pre-Columbian times, the demographical stagnation cannot be explained as the result of high mortality caused by illness and starvation, but rather by the circumstance that the Indians 'invented/adapted certain cultural practises in order to curb the birth rate and to raise the mortality rate'. Such practises are still alive in the form of 'contraceptives, tabus against sexual intercourse, abstinence for long periods of time, extended periods of breast feeding, abortions, and infanticide' (Frank, 1987).

According to Frank the most plausible explanation for the adoption of such strict population control devices is the 'protein hypothesis'. The agrarian-based diet does not supply the essential amino acids which must be incorporated into the diet by ingesting meat obtained through hunting and fishing. Both sources are relatively scarce in the tropical forest and decline even faster around the dwellings of shifting cultivators, so that the settlement needs to be moved. If the number of people surpasses a certain upper supporting limit, the threat of an ecological protein crisis becomes real.

71

Taking into account the intricate web of socio-economic consequences and cultural implications of low population density and permanent movement as prerequisites of shifting cultivation, it looks as though de Schlippe (1956:192) was right when after spending many years among shifting cultivators, he stated: 'The periodical dying of homesteads is one of the most important features of shifting cultivation, and as a traditional limitation of general character it is the greatest obstacle in the way of Africa's progress.'

Numerous case studies and also very general rules demonstrate that in the whole tropical region, where shifting cultivation (in its broader connotation) ceased to be the integral system of land use centuries ago, the socio-cultural matrix of these societies denotes a higher level than that found among shifting cultivators. Illustrations of this widely accepted fact are the contrast between Java and Bali on one side, and Borneo and Zulawesi on the other; the high level of socio-political development of the peoples of the tropical Andes when compared with the majority of the lesser developed Amazonian cultures (Frank, 1987, p. 110); and also the cultural differences between the societies of Rwanda and Burundi versus the shifting cultivators of the neighbouring Azande district in Zaïre.

Countries like Malaysia, Thailand and Sri Lanka, which experience especially grave losses of natural forests, have taken a radical approach to the problem by declaring shifting cultivation illegal. Whether this administrative measure will help is doubtful because what alternatives are there for subsistence cultivators?

In view of the ecological, economic and cultural shortcomings attributed to shifting cultivation the overriding question is why rotational crop–fallow land usage with slash-and-burn methods of land clearing is so widespread in the tropics, and why does it still prevail after more than 30 years of agrotechnical assistance from abroad? The statement of C. S. Ofori at the FAO Seminar, *Shifting Cultivation and Soil Conservation in Africa*, held in Ibadan, 1973, is still valid: 'Despite the fact that the system increasingly fails to satisfy the requirements of higher production, it has not been possible to change successfully or even modify considerably this system of agriculture in the humid tropics' (Ofori, 1974:14).

The gist of the arguments repeatedly given by rural economists and social scientists to justify the persistence of shifting cultivation into the present time may be summarized in the following statements:

1. In its broadest sense, shifting cultivation is a pioneer form of agriculture which was also practised by most of today's high-ranking agricultural countries of the extratropics at a certain stage in their development.

The fact that, after more than 30 years of scientific research and technical assistance, tropical agriculture in vast regions could still not make the transition to more intensive forms of rainfed agriculture suggests that the step from rotational to continuous cropping systems must be intrinsically much more complicated under the environmental conditions of the tropics than under those of the extratropical regions.

72

2. Rotational crop–fallow is a proven method for controlling the rapid proliferation of weeds in newly opened fields.

Having shown already in Section 2.6 that yield reduction is the primary cause for field shifting it makes one wonder why, in the course of generations, it has not occurred to the cultivators that reserving a given tract of land for permanent cultivation might in the long run reduce the Sisyphean task of weeding (see Fig. 2.10). Farmers who cultivate permanent fields in tropical regions have apparently learned that it is much easier to keep a field well cleared in a deforested environment than in an area that has been fire-cleared in the middle of a weed-prone area.

3. Shifting cultivation and rotational bush fallow allow the most favourable relationship between labour input and food procurement. They require less average labour input per man-year than more intensive systems of rainfed cropping (e.g. see Boserup, 1965).

Firstly there are numerous objections on an empirical, basis (e.g. Morgan, 1969b; Grigg, 1979; Norman, 1979). The data provided by de Schlippe (1956) about the various activities of shifting cultivation allow us to formulate the general objections more concretely. Figure 2.10 shows the distribution of agricultural work according to field types through the 1950 season in the Blue Belt of the Azande district. The various fields of a shifting cultivator, from the courtyard to the cottonfield, are arranged in such a way that the sum of the work that is done in different plots at a certain time adds up to the total workload. For purposes of clarification, we have highlighted the labour connected with field preparation such as opening, hoeing, tree cutting, hoeing and hand clearing, or burning with dots, and weeding with hatches. As it turns out, the rotational land-use system entails considerable effort for preparing and clearing the fields, taking up almost one-third of the total labour time.

Apart from the work actually required, there is an essential objection against this thesis in that the clearing work constitutes a true Sisyphean task. This labour must be performed in order to prepare a piece of land that will provide food for only a short period of time. Afterwards the land returns to its original state, only to require the same effort years later. There is no improvement of the soil as a production basis, as happens with sedentary field agriculture where the work invested by one generation benefits the next.

4. Since the transitions from forest fallow to bush fallow and from this to short fallow periods mean always higher proportions of labour input they have only been adopted when pressures made them necessary (Boserup, 1981).

As an example, the case of the Cabrais of northern Togo is often cited. After migrating into the thinly populated Middle Belt, instead of holding on to the traditional semi-intensive land-use systems to which they had been accustomed for generations in the densely populated areas of their homeland, they readily adopted the practises of shifting cultivation hitherto unknown to them (Hodder, 1980).

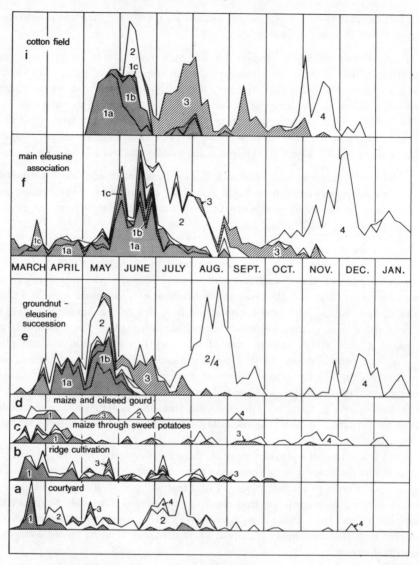

Fig. 2.10 Distribution of agricultural work according to field types in 1950 in the Azande district, north-east Congo. (Source: De Schlippe, 1956)

Another example are the Kofyar of the Jos Plateau in central Nigeria, who, at altitudes between 1500 and 1700 m, practise continuous farming of site-specific crop associations, but, at the same time, in the foreland of the mountains in the humid savanna, shifting cultivation is their preferred farming modality (Netting, 1968).

These examples are not conclusive proof of a more favourable relation between yield and labour input. The dry savanna and the mountains above 1500 m exhibit, from an ecological viewpoint, some attributes which are quite different from those of the humid savanna in the lowlands. A deeper understanding of these differences will make clear that the Cabrais and the Kofyar are prompted in their behaviour by certain natural geographic conditions that deserve closer scrutiny, as will be proved in Chapter 5.

As to Boserup's contention that labour-extensive shifting cultivation is practised for as long as population density remains low and the incentive for progression towards a more labour-intensive system of higher productivity is lacking, the principal question must be posed about the densely populated or even overcrowded pockets which exist either adjacent to, or in the middle of, vast areas of very low population densities? Normally, these pockets have advanced agrotechnical standards and also a higher socio-cultural matrix. If Boserup's thesis is correct, the shifting cultivators of the Middle Belt in West Africa or the Congo basin are to be considered the 'smart folks' who, taking advantage of their low population density, obtain reasonably good harvests with a minimum of labour investment. The people of the North Belt, on the other hand, who have reached a higher socio-cultural level in spite of less favourable climatic conditions, are to be considered foolish for working excessively for their subsistence. What about the people that crowd Java, Bali or Lombock as compared to the scattered cultivators of Borneo, New Guinea and large areas of Sumatra? (For our view see Ch. 5.)

If rising population densities trigger the purported transition from subsistence agriculture based on shifting cultivation to permanent cultivation and higher agricultural density, then shifting cultivation (in its broader sense) should not be found in certain densely populated areas of the tropics. The data compiled by Turner *et al.* in their article 'Population pressure and agricultural intensity' (1977) suggest that shifting cultivation is not indicative of low-density populations since it occurs across a wide range of densities, from less than 1 to more than 500 inhabitants per square kilometre (Table 2.6).

Condensing the facts and findings expounded in this chapter, the following realities emerge:

1. In the tropics of Africa and South America, shifting cultivation in its broader connotation is spatially the absolutely dominating form of agriculture. In the Asian tropics it plays an overall lesser role, but in certain regions it is just as prevalent as in Africa and South America.
2. Shifting cultivation is land demanding, forest squandering, low yielding and last but not least, labour intensive.
3. The population-supporting capacity of shifting cultivation in comparison with permanent field systems is low.
4. The socio-cultural matrix of shifting cultivator societies is relatively poor in comparison with those which have progressed to permanent field cultivation.
5. Shifting cultivation is viewed by agricultural experts as the decisive obstacle in the endeavour to improve the food supply in food-deficient countries.
6. It is true that in the countries of the extratropics shifting field systems were

Table 2.6 Population density and agricultural cycle among subsistence agriculturalists of the tropics (after Turner *et al.*, 1977)

Country	People	Population density (km²)	Ratio crop–fallow
Africa			
South-east Nigeria	Okwe	150	1:4
South-east Nigeria	Omokile	425	1:7
South-east Nigeria	Owere-Ebeiri	750	1:5
Central Nigeria	Yako	90	1:5
South-west Mali	Ba Dugu Djobila	40	1:1.8
Uganda	Amba	141	1:0.5
East Nigeria	Tiv	103	1:1
Zambia	Gwemba Tonga	150	1:0
East Tanzania	Dodoma	15	1:9
South America			
East Peru	Campa	1	1:10
South-east Venezuela	Yaruro-Cano	27	1:4
Central Brazil	Kuikuru	1.1	1:17
North Venezuela	Karinya	12	3.5:1.5
East Peru	Bora	0.3	1:10
South Venezuela	Waika	0.1	1:5
Indonesia			
Sarawak	Iban	3	1:13
Kalimantan	Dyaks	?	1:3
Melanesia			
New Guinea	Tsembaga (Madang)	26	1:14
New Guinea	Bomagai-Angoiang	15	1:17
New Guinea	Kuma	38.6	1:3.8
New Guinea	Ipili	25	1:8.7
New Guinea	Abelam (Sepik)	498	1:8.5
Southeast Asia			
North Thailand	Lua	33	1:7
North Thailand	Karen	35	1:10

common in their historic past, but for the last 800 years or more they have been replaced by agriculturally more effective land-use systems.

In view of these undeniable facts one must search for the reasons why the field–fallow systems are still so widely used in the tropics.

The critical review of the main postulates of those theories which ignore environmental causes, leads to the realization that the arguments provided are at least disputable. It is necessary, therefore, to look for other explanations. Signals of the direction in which the search should proceed are hinted at by some remarkable coincidences between population and land-use distribution, and natural regions.

In West African countries north of the Gulf of Guinea a strong demographic concentration occurs in the rather unfavourable environment of the dry savanna. That this is not mainly a consequence of non-agrarian activities but the result of intensive agrarian colonization shows the fact that the maximum agricultural land occupied by native populations in sub-Saharan Africa is not found in the humid savannah, but in the area of uncertain rains of the dry savanna.

In South America, where the climatic/vegetational belts of the tropics are not as explicit as in Africa, a similar contrast exists – and, according to Parsons (1980), it appears to have been there since pre-European times – between the populated areas of the climatically unstable '*cerrado savannas*' or '*caatingas*' of north-eastern Brazil or the 'savannas' of Venezuela, on one side, and the Amazon basin with its optimal

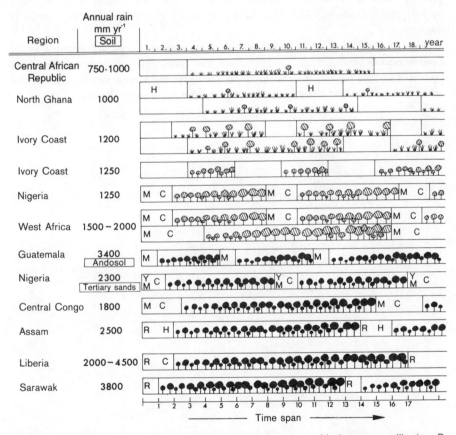

Fig. 2.11 Field–fallow alternation in shifting cultivation under stable long-term utilization. R = upland rice; C = cassava; H = millet; M = maize; Y = yams, (values after Nye and Greenland, 1960)

77

climatic conditions for plant growth on the other. Pertaining to Asia, the relationship between population density and agricultural intensity in the thorn forests and savannas of India have prompted Sopher (1980) to state that 'the savanna environment occupies the larger, central part of the Indian cultural space'.

Conspicuous in all cases is the concentration of rural populations close to the boundary of rainfed agriculture, an indication that the native cultivators, instead of flocking into areas of secured hygro-meteorological conditions, tend to push towards the outer limits of acceptable climatic risk.

Basically the same conclusion is arrived at if the data presented by Nye and Greenland (1960) concerning the duration of the crop–fallow period are placed in the context of the African natural geographic belts of the evergreen tropical rain forest, the deciduous forest, the tree and tall-grass humid savanna, and the short-grass dry savanna (Fig. 2.11). The rain forest, in spite of the large biomass it produces, allows only very short cropping periods: 1–2 years of cultivation followed by 12–15 years of fallow. By contrast, in the dry savanna 3–4 years of cropping can be followed by 7–10 years of fallow, which demonstrates the much greater agricultural potential of the latter.

In total, the fact that the most intensive agrarian activities are pursued in an environment where climatic conditions are less favourable and recurrent droughts present a serious hazard, directs the quest for causative natural factors towards the other element that influences agriculture beside climate, namely the soil.

3

The soils of the tropics and their key role in an ecological context

3.1 The problem of classification and taxonomy

A cursory consultation of atlases and soil maps concerning the different soil groups in a meridional sequence from the rain forest over the humid savanna to the dry savanna results in the classification and zonal arrangement shown in Table 3.1. This table attempts to approximate the correlation between the most important classification systems with special reference to the tropics. But it must be stressed that there is never total correspondence between the different classes and orders. In view of this and the additional existence of a Brazilian, a Belgian and a German classification, each with its own nomenclature, many interested persons might be discouraged at this point from probing deeper into the knowledge of the varied attributes of tropical soils and their ecological implications. Indeed, the attempt to find one's way through the maze of classification norms and artificially constructed nomenclatures may prove frustrating to the non-specialist who tries to obtain a coherent picture of the decisive and ecologically determining properties of the different soil types.

The soil types mentioned represent systematized natural systems whose scientific names evoke among specialists certain associations related to their ecological significance. In contrast, the layman does not grasp the meaning of the artificial terms, therefore he may not be overly enlightened by learning that the oxisols and ultisols dominate the humid inner tropics, while the eutric cambisols and regosols occur in the arid outer parts of the tropics. On the other hand, a certain familiarity with some important soil properties that differ in tropical and extratropical soils is the key to understanding the problems in question. For this reason, an attempt must be made to introduce the reader to the most important pedological facts and processes so that, in the end, he will be able to draw his own conclusions concerning the implications of particular soil attributes for tropical agriculture.

In this endeavour we will try to strike a balance between keeping the problem simple enough for the interested reader to follow the argumentation and, at the same time, avoid oversimplification and consequential distortions. It can only be hoped that, at the end of the chapter, pedologists and non-pedologists alike will agree with the essentials of the following outline.

Table 3.1 The main tropical soil groups according to four major soil classifications

FAO (*World Map*)	US Soil Taxonomy	ORSTOM (French)	English classification
Orthic ferralsols, Xanthic ferralsols, Acric ferralsols	Oxisols, ustox or orthox	*Sols ferralitiques fortement désaturés*	Leached ferrallitic soils
Dystric nitosols, Ferric acrisols	Ustox, orthox	*Sols ferralitiques fortement désaturés*	Weathered ferrallitic soils
Dystric cambisols	Ultisols	*Sols ferralitiques moyennement désaturés*	Weak ferrallitic soils
Ferric luvisols, Eutric nitosols	Alfisols, ustalfs	*Sols ferrugineux tropicaux lessivés*	Leached ferruginous soils
Eutric cambisols	Alfisols	*Sols fersialitiques non lessivés*	
Eutric cambisols	Aridosols	*Sols bruns eutrophes tropicaux*	Eutrophic brown soils
Xerosols	Aridosols	*Sols peu évolués xeriques*	
Regosols	Entisols		
Fluvisols	Entisols, fluvents	*Sols minéraux peu évolués*	Alluvial soils

Sources: Sanchez (1976) and Young (1976).

3.2 Ecologically decisive soil properties

What we refer to as soil is the superficial layer of the earth's surface which, without sharp boundaries, passes gradually into the underlying rock or loose sediment. Because soil has developed at the parts of a 'parent rock' or sediment nearest to the surface, it shares, to a certain degree, the mineral and chemical attributes of the parent material.

The uppermost layer of soil which interfaces directly with the atmosphere and surface vegetation normally appears as a relatively darker horizon in different shades of brown or mixtures of yellow and black. This zone called *solum* is the place where organic matter of animal and vegetal origin is decomposed and where, at the end of a more or less intricate process, it is transformed into inorganic residues. A very important substance that develops in the course of the transformation process is *humus*, a semipermanent constituent of the solum. While *humification* is the process of alteration of organic matter, *mineralization* is the final breakdown into inorganic elements.

Under normal conditions, even the uppermost horizon of the soil contains fragments of the underlying parent rock. Particles larger than 2 mm constitute the *soil skeleton*, while the small-sized particles make up the *fine-earth fractions*.

Solum is considered a mixture of inorganic mineral substances and humified organic substances interspersed with pores, which may be filled with air and water as well as with small amounts of various acids and bases. The water content accounts for the *soil moisture*.

In a soil profile which has developed on top of a loose sediment, such as silty sands or loess, it is easy to conduct the following simple experiment to assess an important fact of soil formation: when a lump of topsoil is kneaded between the fingers, a certain degree of plasticity is noticed; the moister the soil, the more evident the effect. This plasticity is an indication that the soil contains certain components that permit the absorption of water and, thus, contributes to its flexibility – the *clays*. The same experiment with a probe from deeper horizons in the loose sediment, e.g. loess, reveals no plasticity in the parent material. This observation leads to the conclusion that clays must have developed from non-clayey components within the parent material when these came into contact with generating agents that operate only in horizons near the surface. The actions of these external agents are summarized under the term *weathering*. *Physical weathering* causes the disaggregation of rock into smaller particles without chemical modification, while *chemical weathering* leads to the decay and transformation of the mineral components of the parent rock.

From a genetic viewpoint, soil is primarily the result of three well-known abiotic processes and almost a 'black box' (Odum, 1970) of biotic processes: first, of physical weathering and chemical transformation acting upon the mineral substances of the parent rock; second, of the humification and mineralization of organic substances that are drawn from the top of the soil; and third, of the interreaction of the chemical elements or compounds that have been added by the air and/or by the water, and of others which have developed as a consequence of the physical and chemical decay of minerals or organic matter.

This set of abiotic processes is complemented by the actions of the soil flora and fauna. The effects of these actions are relatively well known, but the complex of physiological details remains fairly unknown.

Water in the soil is the activator of the chemical processes and also the transmission agent between potentially reactive substances that may be separated in the solum. Together with dissolved chemical substances, water forms the *soil solution*, the reservoir from which the roots extract most of the nutrients for plant growth and the water needed for their physiological functions.

Soil moisture, or the water that fills the pores in the soil, normally stems from precipitation which enters the ground, turns into soil solution and, during periods of rainfall, percolates slowly into the ground (gravitational water). This obviously means that the nutrients contained in the soil solution also percolate and that, in the long run, they are washed out of the upper soil horizons. Assuming that all other physical and chemical soil properties are kept constant, the more rain that enters into the ground, the stronger the downward movement of soil solution, and the more significant the role played by this process in the ecology of plant growth. Only during periods of dryness and strong evaporation is the soil solution drawn upwards ('capillary rise'). In humid regions where yearly precipitation surpasses evapotranspiration, the dry

81

periods do not last long enough to prevent the ecologically deleterious process of soil empoverishment through *outwash* (*eluviation, leaching*).

Chemical and physical properties of the parent rock, varying external conditions for physical and chemical weathering, precipitation amounts that fluctuate in quantity and seasonal distribution, differences in biomass input and drainage conditions emerge, then, as the most important controlling factors for soil development, and since they occur in different combinations, the final result is an almost endless array of soil properties. The classifications listed at the beginning of this chapter try to systematize this variety by using different sets of key properties in the definition of soil classes, groups, orders and families.

The details of soil classification are only of minor importance for the following deductions. The major question we are trying to solve is whether there are certain soil properties which, at times without interregional contacts, hindered agrotechnical innovations in large parts of the tropics, while in other parts, especially in the extratropics, they posed no obstacles to innovations. Second, if such soil properties still prevent development even under conditions of modern technology transfer.

Consequently, the task is to identify *soil properties which act as limiting factors for plant growth and which, at the same time, cannot be markedly altered or manipulated by man*. These soil properties will be highlighted as the *ecologically decisive soil properties*.

Under the criteria of limiting factors and not being changeable by man, a well-defined selection can be established among the numerous soil attributes. As a final result after many years of deliberation, the following three properties can be designated as ecologically decisive (Fig. 3.1):

1. The content of weatherable primary minerals;
2. The amount of organic matter;
3. The total cation exchange capacity (CEC).

A few other soil attributes are also very important although, for different reasons, they do not have the same significance as determining factors. Important among these

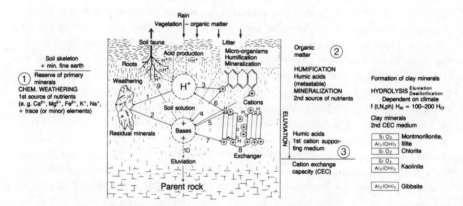

Fig. 3.1 The ecologically decisive soil properties in a schematic overview.

secondary properties are the soil pH as indicator of soil acidity, the anion exchange capacity, the aluminium toxicity, and the ability to fix phosphorus.

All the other soil properties, especially the physical soil attributes, are temporarily set aside because they can be manipulated by man. Principally, those physical soil properties that have a hampering or limiting effect on plant growth can be neutralized by appropriate measures. Moreover, this does not matter in our context, because there is general consensus among the experts that the soils in the tropics on which rotational bush-fallow land use is practised commonly display good to very good physical properties (Sanchez, 1973a; Young, 1976). The vertisols are exceptions; they are, however, restricted to certain areas and their physical problems can be overcome by the use of heavy machinery.

The *content of primary minerals* refers to the quantity of minerals and rock fragments remaining in the soil after the physical and chemical weathering of the parent rock, or to those newly provided by fluvial or aeolian sedimentation. These primary minerals constitute the soil's mineral reserve and form one of two sources of nutrients for the plants. From the 16 elements known to be essential for plant growth, 13 come from the soil, and all of them, except nitrogen, originate in the mineral reserve. They are set free in the course of weathering processes, especially during chemical weathering, and then enter the soil solution mostly as cations. From there they can be absorbed by plant roots. *Cations* are elements with different positive charges, such as potassium (K^+), phosphorus (P^{3+}), calcium (Ca^{2+}), magnesium (Mg^{2+}), sodium (Na^+), iron $(Fe^{2+}$ or $Fe^{3+})$, manganese (Mn^{2+}) and zinc (Zn^{2+}).

The *organic substances* in the soil perform three ecologically decisive functions: they form the second nutrient source for plants, are metastable and relatively resistant against outwash, and serve as one of two carriers of the soil's CEC.

When the organic materials decompose on the surface and in the upper horizons of the soil, the nutrients are set free by the processes of humification and mineralization and become inorganic chemical compounds such as carbon dioxide, nitrates and phosphates, or elements such as potassium, sodium, calcium and magnesium. They are incorporated into the soil solution and can be restored to the nutrient cycle in living plants (recycling). In conjunction with humification, the biochemical substance alterations induced by the soil flora and fauna play a key role, especially in the rapid changes that are caused by aerobic soil bacteria (biological humification).

During humification, fulvic and humic organic acids are formed. Their ecological significance lies foremost in the fact that they are relatively resistant against further decay and mineralization as well as against outwash. The acids have a high polymer structure and are partially bound to reactive inorganic soil constituents so that they do not percolate with the soil solution but remain concentrated in the upper soil horizon until a certain level of steady state is reached between current formation on one side and decay by mineralization on the other.

The organic matter content of a soil in equilibrium with the vegetation cover depends on the relationship between the annual input and the *mineralization rate of organic matter*, such as litter from leaves and twigs and final residuals. Assuming that this input is constant in a given soil, then, with the rise of the mineralization rate, the

83

humus content will drop, or, as the mineralization rate decreases, the humus content will rise. If the rate of mineralization remains constant, the humus content will decline as the input of fresh organic substances is reduced. The mineralization rate depends, among other factors such as soil acidity, chemical composition of organic matter, and soil aeration, particularly on the influence of climate, a fact which will be discussed in detail in Sections 3.6 and 3.7.

The function of organic matter as a carrier of CEC as the third ecologically determining soil property is expressed in that organic matter, especially the humic acids, exhibit the highest values among all soil substances. Small amounts of organic matter, therefore, can greatly affect the total CEC of the soil.

The *function of the CEC* can be illustrated in the following form: as discussed earlier, the major portion of the nutrients is contained in the soil solution in the form of cations which, under normal conditions of percolation, are subjected to downward movement along with the soil solution. This process is called 'leaching'. In humid climates where precipitation exceeds evapotranspiration, the leaching effect of gravitational water would result, if continued long enough, in the upper horizons of the soil being totally depleted of nutrients. Plant growth would no longer be possible. This effect of ecological self-destruction is prevented by the soil property of exchange capacity, which, roughly speaking, is the ability of the soil temporarily to offer lodging to a specified number of cations, thus keeping them from being leached out. For a certain time, the nutrients are adsorbed by locally stable soil substances, namely humic acids and clay minerals.

The process of cation exchange can be illustrated by means of the following schematic model (Fig. 3.1): the cations which have been incorporated into the soil solution in the course of chemical weathering of primary minerals or from decomposition of organic materials will be subjected to permanent molecular movement within the soil solution. The soil solution fills the capillaries within the soil and, under excessive rain, moves downward under the force of gravitation. The walls of the capillaries are built of local stable materials of the soil. Among these are the clay minerals and the macromolecules of the humic acids. Varying from soil type to soil type, these substances possess a specific number of sites of permanent negative electrostatic charge (negative poles of electricity) as adsorption points where positively charged cations can be attached, and the electrostatic attraction prevents further downward movement. However, the main nutrient cations have an avid competitor for the adsorption site in the small but very energetic hydrogen cation (H^+). The H^+ ions result from the dissociation of water (H^+OH^-) or inorganic acids or they are released by the plant roots; H^+ ions are strong enough to displace nutrient cations from the adsorption sites, returning them back to the soil solution.

A key question arises in regard to this permanent interplay: how many adsorption sites can be offered by clay minerals and humic acids per weight unit of fine soil? This concept is represented by the value of CEC measured in milliequivalents per 100 g fine soil (meq per 100 g).

84

1 meq (1 milli-equivalent) is defined as 1 milligram (mg) of H^+ ions or the equivalent of any other cation that will replace the 1 mg of H^+ ions. This value is related to 100 g of oven-dry substance, either fine earth, clay minerals or humus. For example, 10 meq per 100 g means that 100 g of oven-dry fine earth has the potential to adsorb 10 mg H^+ ions. If potassium (K^+) with its atomic weight 39 (compared to atomic weight 1 of H^+) is implied in the exchange, the same earth may adsorb $10 \times 39 = 390$ mg. Each of the two-valence cations, such as calcium (Ca^{2+}, atomic weight 40) substitutes, correspondingly, two of the H^+ ions. The same probe of fine earth with 10 meq CEC can adsorb $10/2 \times 40 = 200$ mg of calcium. The atomic weight of Na^+, Mg^{2+}, K^+, Ca^{2+} are 23, 24, 39 and 40 respectively. The milliequivalent weights are 0.023, $0.024/2 = 0.012$, 0.039 and $0.040/2 = 0.02$ respectively. The CEC value of a certain substance is determined by treating the probe with ammonium acetate (NH_4Hc). The NH_4 ions replace all the exchangeable cations from the adsorption sites in equivalent amounts. The soil–ammonium acetate solution is filtered and the soil is washed with alcohol in order to remove the excess solution. After this, the soil containing the adsorbed NH_4 ions is extracted with a solution of potassium chloride (KCl). The soil–potassium chloride solution is filtered and the ammonium contained in the filtrate is determined. The quantity of NH_4 ions extracted in the filtrate is a measure of the probe analysed.

A second condition in the interplay pertains to the relationship between the number of H^+ ions and the sum of the other cations such as Ca^{2+}, K^+, Na^+, Mg^{2+} and Al^{3+}. If H^+ ions prevail the soil solution has an *acidic* reaction, many of the adsorption sites being occupied by the H^+ ions, and most of their competitors remaining in the soil solution where they risk being leached away. If the Ca, K, Na, Mg cations prevail, the soil solution has an *alkaline* reaction; most sites are occupied by nutrients.

pH value, buffering capacity, and base saturation as soil characteristics.
All chemical processes in soils are decisively influenced by the relationship of hydrogen ions (H^+) and basic cations, such as Ca^{2+}, K^+, Mg^{2+}, and Na^+. The quantity of hydrogen ions per volume unit, i.e. the *hydrogen concentration*, is represented by the *pH values*. Even in the purest water there is always a small amount of water molecules dissociated into H^+ cations and (OH) anions. Under normal conditions (360 K = 0 °C), 1 l of pure water contains $0.000\,0001 = 10^{-7}$ g H^+ ions. This water has neither an acidic nor an alkaline reaction, it is *neutral*. If pure water is mixed with a small amount of mineral or organic acids (in the soil this may happen with carbonic, sulphuric or humic acid), the H^+ ion concentration increases. Assuming the increase to be one hundredfold, the concentration would be $0.000\,01 = 10^{-5}$ g l^{-1}. The solution now exhibits an acidic reaction. The same applies for all concentration values greater than 10^{-7}. If, inversely, basic cations, such as Ca^{2+}, K^+, Na^+ or Mg^{2+}, are added to pure water, some of the H^+ ions are bound in chemical compounds with the result that the quantity of free H^+ ions decreases. The concentration of H^+ ions drops to values below 10^{-7}, maybe 10^{-3}, and the solution exhibits an alkaline reaction.

In order to simplify the notation, there has been an agreement to use the pH value defined as the negative exponent of the prime number 10. An H^+ ions concentration of 10^{-7} g l^{-1} equals pH 7. *All pH values below 7 indicate a higher than 10^{-7} g l^{-1} H^+ concentration. The corresponding solution reacts acidic, 'the soil is acid'. Values of pH above 7 indicate alkaline soils.*

The actual pH in a given soil is determined primarily by the balance between hydrogen ions and basic cations. Involved in this balance are the negatively charged materials of the soil, i.e. those with CEC such as clay minerals and organic colloids of the humus fraction. The presence of the negative charge and the surface of these materials makes them attractive and adsorbing cations.

In acid soils, H^+ ions combine on the surface of the CEC carriers, while in base-rich alkaline milieux the main components adsorbed by clay minerals and organic colloids are the basic cations. These stages are, however, not permanent; what is permanent is the interchange of cations between the adsorbing surfaces and the soil solution. The hydrogen ion is more vigorous than all the others and therefore it is adsorbed more strongly so that, when an excess of H^+ is produced (during the formation of acids or as a consequence of heavy leaching of the basic cations), they will trade places with the basic cations on the surfaces and occupy the greatest part of the adsorbing sites. The quantity of H^+ ions held at the surfaces determines the *reserve acidity* in contrast to the *active acidity* in the soil solution, which is in equilibrium with the former. The percentage of the CEC occupied by the neutralizing bases, such as Ca^{2+}, Na^+ or Mg^{2+}, is determined as the *base saturation*.

The amount of H^+ in the reserve acidity of a soil is usually greatly in excess of the active acidity and thus acts as buffering agent for the soil. The greater the capacity for cation exchange of a given soil, the greater the buffering effect (*buffering capacity*). The lower the base saturation, i.e. the greater the excess of H^+ ions, the greater the reserve acidity and the lower the pH value (higher acidity) of the soil solution.

In a soil with low CEC the buffering capacity is also low with regard to the competitors for the adsorption sites among the basic cations themselves. This constitutes a problem especially when fertilization of nutrient- and CEC-poor tropical soils is attempted. Plant growth is limited by this nutrient which is minimally available (*minimum rule of plant growth*). Under low buffering capacity conditions it is very difficult to prepare, through fertilization, a soil solution with the adequate nutrient mixture that is needed by the plants at certain growing stages. It may well be that the greatest part of the nutrients, except those minimally available, remain unused and are leached away, leaving the plants to starve. This can only be avoided if, under condition of low CEC, the nutrient budget in the soil is closely monitored.

Under conditions of high CEC, the buffering capacity is higher too. High amounts of nutrients can be adsorbed, and in the case of shortage of certain cations in the soil solution it may easily be replaced by those from the exchanging surfaces.

A third determining aspect of cation exchange pertains to the relationships of the nutrient cations between themselves. For example, an abundance of aluminium cations and scarcity of calcium and potassium cations results in 'aluminium toxicity' of the soil solution, thus restricting plant growth. Good plant growth presupposes, then, a certain equilibrium among the different cations.

When the ecologically decisive soil properties were enumerated the CEC was mentioned with the prefix 'total'. 'Total' is added because the CEC is composed of a soil-acidity-independent, and a soil-acidity-dependent part; the *permanent CEC* and the *variable CEC*, respectively. The variable part can only be fully activated by lowering the soil acidity. The total CEC is the upper limit that cannot be surpassed. In our context the CEC is used in the sense of total CEC, even though the prefix is not mentioned.

In order to grasp the importance of the CEC as a soil property, let us imagine, for a moment, the following situation: there are two fields, both with equally nutrient-poor soils, but one with a considerably higher CEC than the other. Assuming that both fields are located in a climate with 1500 mm of annual rainfall and that the farmers have free access to all the machinery and fertilizers they need in order to achieve maximum yields – what will happen? The field with the high CEC can store a large amount of the plant nutrients which enter the soil as water-soluble cations from the fertilizers, while the other will soon run out of nutrient cations because they cannot find adsorption sites and are, consequently, out of reach of plant roots. The yields will differ accordingly. The example shows that the CEC works as the soil property which determines the amount of cation nutrients that can be adsorbed and stored in the soil after the application of fertilizers.

The other substances of the soil that act as bearers of CEC are the clay minerals. Before the application of electronic microscopy to clay analysis, in the 1950s, clays were simply defined as the smallest particles in the soil (less than 0.002 mm in diameter) and the ones that provide the soil with plasticity when water is absorbed. Since this is all that was known about clays at that time, decisive aspects of the soil's ecological role could not be explained.

The utilization of electronic microscopy offered the possibility to mineralogists and soil scientists to probe deeper into the secrets of clay mineralogy. From the many findings, well established by now, the following are of ecological importance:

1. Clay minerals consist of microcrystals which are synthesized as newly built materials during the alteration of primary minerals by hydrolysis. For this reason clays are considered as 'secondary minerals'.
2. There are several mineral groups of different crystallographic structure and varied physical and chemical properties. Among the latter the CEC is one of the most prominent.
3. Clay mineral synthesis, apart from its recessive dependence on the parent rock, is heavily influenced by climate-controlled exogenous factors.

An initial distinction must be made between the *clay fraction*, defined as particle size, and the proper clay minerals. The clay fraction includes all particle sizes below 0.002 mm = 2 μm diameter and consists of the finest residues of quartz and various primary minerals, iron compounds, aluminium sesquioxides and the clay minerals themselves. Clay minerals, in a strict sense, are layered silicates, aggregates

87

of submicroscopical lattices (phyllosilicates). Their length is less than 0.001 mm and their width is 50–500 times smaller than their length. These tiny dimensions are due to their formation under the low pressure and moderate temperatures that prevail near the earth's surface.

3.3 Crystallographic structure of different clay mineral groups

Clay minerals – like any other minerals – are built of electrically active atoms, divided into positively charged cations and negatively charged anions. Figure 3.2 lists the dominant building elements according to their electric charges, their relative size with respect to the large oxygen atom (O^{2-}), and their atomic weight or MOL mass.

The common clay minerals consist of a series of elementary layers. A crystallographically perfect layer is formed by two basic structural units: the silica tetrahedral sheet and the aluminium octahedral sheet (Fig. 3.3). The geometrical layout in these sheets is such that the relatively small Si^{4+} and Al^{3+} ions act as central atoms which are surrounded by the larger O^{2-} or $(OH)^-$ ions respectively. The variety of clay minerals results in part from the different arrangements of the tetrahedral and octahedral sheets. Three structural groupings of the two basic sheets are recognized. The first consists of a silica tetrahedral sheet attached to one side of an aluminium octahedral sheet, forming the *1 : 1 layer phyllosilicate*. The second group is built by a symmetrical arrangement of two tetrahedral sheets around a central octahedral sheet (*2 : 1 layer*

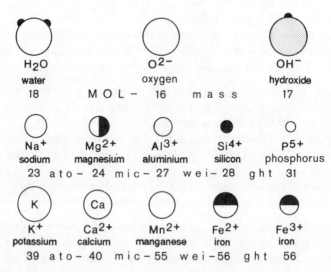

Fig. 3.2 The dominant building elements of clay minerals according to their electric charges and their size.

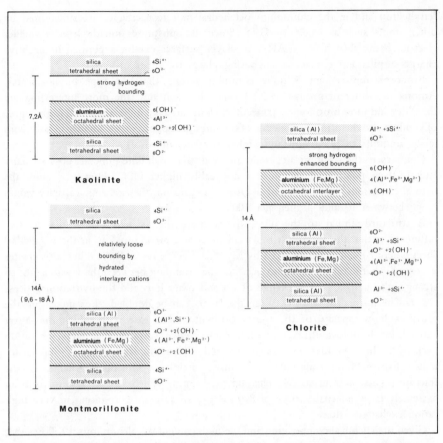

Fig. 3.3 The elementary layers of kaolinites, montmorillonites and chlorites as characteristic examples for 1 : 1, 2 : 1 and 1 : 1 : 1 clay minerals in a schematic structure (using figures from Laatsch, 1957; Scheffer and Schachtschabel, 1976; Birkeland, 1974).

clay minerals). The third arrangement is the presence of an octahedral sheet between adjacent 2 : 1 layers. These are known as *2 : 1 : 1 layer phyllosilicates*.

Examples of simply built 1 : 1 clay minerals are the *kaolinites*. Ideally, all the layers should be neutral since the sum of the negative charges equals the sum of the positive charges. Electrically neutral minerals cannot have adsorptive sites with permanent negative electrostatic charges to serve as lodging sites in the cation exchange process. However, real kaolinitic clay possesses a certain, though small, CEC due to its incomplete crystallographic structure, including broken edges. The less perfectly structured the kaolinitic clay minerals of a soil, the more effective they are with respect to cation exchange processes.

In more complicated clay minerals, there is a second path from perfect to imperfect structure and balancing of electrical charges within the crystals: *isomorph replacement*. This means that Si^{4+} and/or Al^{3+}, which serve as central atoms in the silicon tetrahedron and in the aluminium octahedral unit respectively, are substituted by other atoms such as Mg^{2+} or Fe^{2+}. Since the substitutes mostly have a smaller electric charge than Si^{4+} or Al^{3+}, the layer surfaces exhibit a permanent negative charge surplus and can act as anchoring places for other cations.

Isomorph replacement is quite common among the 2 : 1 type phyllosilicates. Among the different groups of 2 : 1 type phyllosilicates, the more interesting are the illites and montmorillonites (smectites). *Illites* are clay minerals with a high degree of similarity to the primary minerals of the mica group, but much smaller in size, with lesser amounts of potassium and higher contents of adsorbed water.

Contrary to the large mica crystals, many of the small illite minerals possess many lateral attack points and therefore a noticeably higher CEC. However, since the natural conditions create different degrees of expansion, the exchange capacity values range between 10 and 40 meq per 100 g.

A structure similar to the illites characterizes the *montmorillonites*; the decisive difference is that the isomorph replacement occurs predominantly in the octahedral sheet where Al^{3+} is often replaced by Mg^{2+} and Fe^{2+}. As with the illites, this creates a negative charge on the layer surface, but the bounding between the layers is not so strong, which makes the exchange of Ca and other ions and the hydration process easier. The normal distance between the layers can be doubled, thus increasing the great swelling capacity of the montmorillonites. The ease with which the layers expand turns almost all layer surfaces into active surfaces that facilitate cation exchanges. In sites that are poorly secured electrically, the entire crystal lattice may collapse. Montmorillonitic clay consists mainly of mineral fragments. At the fracture points, new negatively charged sites appear which all combine to form an extremely large internal surface of 800 m^2 g^{-1} of substance, resulting in very high cation exchange values.

Given the chemical composition and the high content of calcium ions it follows that montmorillonitic clays can be synthesized only in soil solutions rich in bases, especially calcium. Under acidic soil conditions, montmorillonites are easily reduced to a 1 : 1 type clay mineral.

Chlorites are the best representatives of the 2 : 1 : 1 clay minerals. They consist mainly of 2 : 1 type crystals, such as the illites, but a second octahedral sheet is inserted with alternating Al^{3+} and Mg^{2+} central ions. Physical and chemical properties of chlorites are similar to those of illites.

A sesquioxide of aluminium, the *gibbsite*, is left as a residual clay mineral, after extended periods of heavy weathering under acidic soil conditions. When kaolinites have lost their silica tetrahedral sheet through desilicification processes, only the aluminium octahedral sheet remains. The sesquioxide of iron (*goethite*) is normally linked with gibbsite.

Allophanes, which have been defined as amorphous clays, are commonly found in areas of volcanic ash accumulation.

3.4 The formation of clay minerals

The diagram in Fig. 3.4 schematizes qualitatively the general rules of the very complicated formation processes of different clay mineral groups based on original illustrations by Laatsch (1957), Scheffer and Schachtschabel (1976) and Birkeland (1974). For the English reader similar illustrations and further explanations can be found in the works by Brown *et al.* (1978) and Bonneau and Souchier (1982). Clay minerals originate during the hydrolysis of varied primary minerals of the parent rock, such as feldspar, mica, hornblende, augit, olivin and a great number of others which are not listed in Fig. 3.4. All are silicates of complicated chemical and crystallographic structure. Hydrolysis, the most intense modality of chemical weathering, begins with the release of potassium (K^+), calcium (Ca^{2+}), sodium (Na^+), iron (Fe^{2+}), and magnesium (Mg^{2+}) and their incorporation into the soil solution, where they are available for plant nutrition. The rest of the primary silicates synthesize the basic structural units of new secondary minerals, the clays.

Initially, when the content of basic cations, such as Ca^{2+}, Na^+, K^+ and Mg^{2+}, is still high, four-layer (2 : 1 : 1) clay minerals, such as chlorites, and three-layer (2 : 1) phyllosilicates, such as illites or montmorillonites as members of the smectite group, are formed. As hydrolysis progresses, the soil solution becomes poorer in basic cations and the clay minerals mentioned are destroyed through solution of the remaining basic cations, the removal of one silica tetrahedral sheet from the montmorillonite layer, the

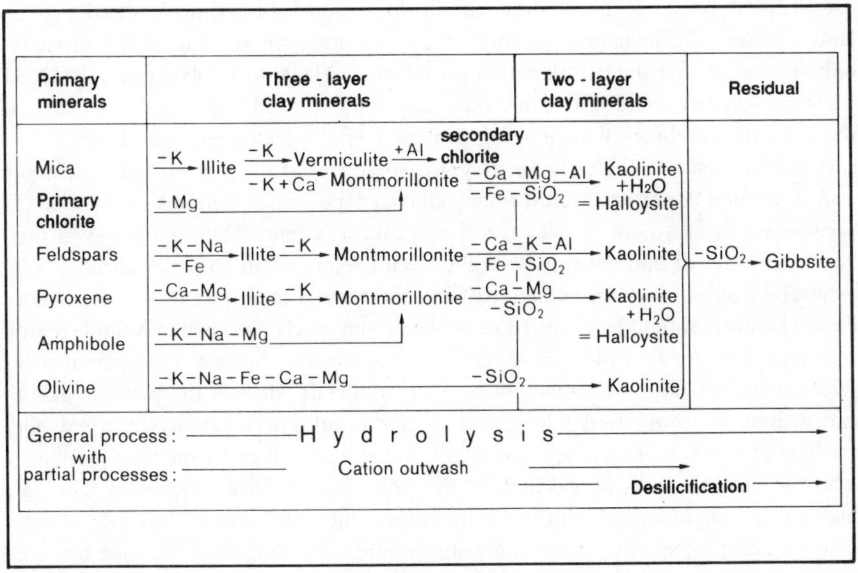

Fig. 3.4 Schematic flow chart of the clay mineral formation process by hydrolysis of primary minerals.

breakdown of the silicon tetrahedron, and the outwash of SiO_2 (*desilicification*). At this stage the kaolinites, the 1 : 1 type of clay minerals, are formed.

With advanced desilicification even the last silica tetrahedral sheet may be removed so that only gibbsite remains, a mineral that consists only of aluminium hydroxide, $Al(OH)_3$, sheets.

At a particular location in a well-drained soil over a given parent rock, the stage of evolution reached in the sequence of various clay minerals depends on the intensity of the hydrolysis and on the time-span during which hydrolysis has been in effect. In other words, clay mineral evolution in the soil is influenced by the intensity and length of time during which the weathering mantle of the parent rock has been exposed to specific external agents of decomposition.

Millot (1970), compared the intensity of hydrolysis on acid parent rocks (granite, sandstone or gneiss) in well-drained topographic locations in the permanently humid tropics and mid-latitudes. Under high tropical temperatures (30 °C annual mean) hydrolysis is four times stronger than in the mid-latitudes (10 °C annual mean). Where yearly precipitation ranges between 2000 and 4000 mm, as in the humid tropics, hydrolysis is about five times more effective than in areas with 1000 mm of annual precipitation, like the temperate regions. Increased soil acidity due to excessive rains and acidic residues from the decomposition of organic matter leads to a tenfold increase. With all these conditions working together, hydrolysis over an acidic parent rock may be 100–200 times stronger in the tropical rain forest than in humid forests of the temperate latitudes.

Considering the time-span during which hydrolysis has been effective, extended parts of the tropics are at a particular disadvantage. A large proportion of the extratropical soils, especially those in the high mid-latitudes, have developed on land surfaces which have been formed by accumulation or denudation processes only during or after the last glaciation period some 20 000 to 10 000 years ago. Other parts are somewhat older, but not exceeding 100 000 or 200 000 years. In contrast, the exposure to chemical weathering processes of the majority of the tropical land surfaces dates back to the early Tertiary, i.e. some tens of millions of years ago. Only a small portion is restricted to the young alluvial depositions along the rivers or to the accumulations of young or even recent volcanic eruptions. The implication of their youth, in terms of nutrient supply, will be examined in detail on p. 147, Section 5.3.

Several assumptions can be derived from the general principles of clay formation. First, kaolinites should be expected to be the dominant clay mineral in humid tropical soils that developed on old land surfaces over acid rocks. Second, over parent rocks which contain a large percentage of base-rich minerals, such as the plutonic gabbro and diabase in comparison with granites, or the volcanic basalt as compared with trachyte (Table 3.2), a rapid kaolinization is initially prevented by the high percentage of basic cations, especially calcium, in the soil solution. However, when there is a high weathering intensity, with severe hydrolysis, these elements are leached out faster than they can be supplied from the primary minerals, and when the rock has been subjected to this severe hydrolysis for such a long time that the mineral reserves have been depleted, kaolinitic clay minerals may also develop over basic parent rocks. Table

Table 3.2 Chemical composition of major rock groups, expressed in average percentages of their weight

| | Plutonites | | | Vulcanites | | Sedimentary |
	Granite (72)*	Diorite (50)	Gabbro (160)	Trachyte (24)	Basalt (137)	Shale
SiO_2	72.08	51.86	48.36	58.31	50.83	50.70
TiO_2	0.37	1.50	1.32	0.66	2.03	0.78
Al_2O_3	13.86	16.40	16.81	18.05	14.07	15.10
Fe_2O_3	0.86	2.73	2.55	2.54	2.88	4.40
FeO	1.67	6.97	7.92	2.02	9.00	2.10
MnO	0.06	0.18	0.18	0.14	0.18	0.08
MgO	0.52 }2%	6.12 }14%	8.06 }19%	2.07 }6%	6.34 }17%	3.30 }10%
CaO	1.33	8.40	11.07	4.25	10.42	7.20
Na_2O	3.08	3.36	2.26	3.85	2.23	0.80
K_2O	5.46	1.33	0.56	7.38	0.82	3.50
H_2O	0.53	0.80	0.64	0.53	0.91	5.00
P_2O_5	0.18	0.35	0.24	0.20	0.23	0.10

* The number of samples included appears in parentheses.
Source: Values from Wedepohl (1969).

3.5 gives an example of this action for the weathering cover of basalts in the Western Ghats of the Malabar coast.

Table 3.2 illustrates the chemical composition of the most common rock groups. The importance of the various groups of parent rocks for soil formation in different parts of the world depends on the frequency of their occurrence. Table 3.3 shows that, within the group of plutonites, the acid rocks, particularly granites, are the most abundant on the earth's surface, while the basic rocks are restricted to small areas. Within the group of volcanites, basalts, as the representatives of the basic rocks, are relatively widespread as compared with the acid volcanites.

Table 3.3 Frequency of occurrence of various groups of parent rocks (% of continental areas)

	Plutonites		Volcanites
Granite	45	Andesite	16
Diorite	2	Basalt	80
Gabbro	13	Alkali basalt	2

Source: Values from Wedepohl (1969).

Third, the differences in soil quality due to parent rock diversity should be of lesser importance in lowland soils of the humid tropics than in the mid-latitudes where lower hydrolytic activity prevails and shorter weathering time is the rule. Nevertheless, in restricted areas of the tropics where rock-induced soil quality differences do occur, they are of great importance.

Fourth, since the intensity of hydrolysis depends heavily on the water supply, in the wet-and-dry outer tropics desilicification and kaolinization processes should not be so strong and the rock-dependent soil differentiation more pronounced. For Hawaii, Hay and Jones (1972) have shown the overall trend of silica loss with precipitation in 10 000–17 000-year-old volcanic ash. Sherman (1952) reports that, in the continuously wet climate of Hawaii, montmorillonite predominates below about 100 cm annual precipitation, kaolinite between about 100 and 200 cm, and the aluminium and iron compounds above 200 cm. And Birkeland (1984, p. 293) summarizes the findings of various authors in a schematic global overview by Millot (1979) (Fig. 3.5). Regarding the tropics, he states: 'commonly, montmorillonite forms at low amounts of precipitation, kaolinite at higher amounts, and oxides and hydroxides of iron and aluminium at still higher amounts' (p. 293).

Aside from these general aspects of clay formation, more details about regional topographically induced differences (catenas) in soil clay properties will be discussed on p. 112, Section 3.11. For now, we will proceed to review the physical and chemical properties of the different clay mineral groups.

94

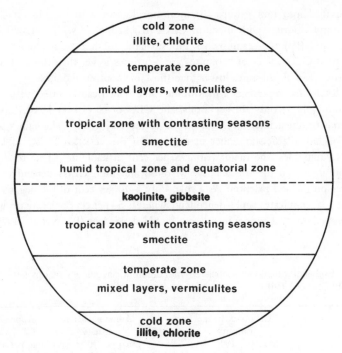

Fig. 3.5 Schematic global overview of clay mineral zones in relation to climatic zones (after Birkeland, 1984)

3.5 Properties of clay mineral groups

Table 3.4 synthesizes the most important chemical properties associated with the different clay mineral groups, such as their CEC, and anion exchange capacity (AEC).

Regarding the exchange capacity, the following ecologically relevant properties can be inferred:

1. The anion exchange capacity (AEC) is relatively low in all mineral groups and differs only slightly from one group to another. This has unfavourable consequences for the supply of nutritional anions to plants, especially such as nitrates (NO_3^-) and phosphorous ($H_2PO_4^-$ and $H_2PO_4^-$) compounds. This principle, however, applies to all soils, regardless of which clay group they contain.

2. The total CEC is comparatively higher and, apart from that, differs widely among the various groups. The commonly dominating clays, i.e. those which are not restricted to particular topographic or petrological conditions, can be

95

subdivided into four groups: (I) montmorillonites with the highest CEC; (II) illites and chlorites with medium CEC; (III) kaolinites with a remarkably low CEC; and (IV) the sesquioxides gibbsite and goethite with almost no CEC. The total CEC values of kaolinites are 3–5 times lower than those of illites and chlorites, and 10–20 times lower than those of montmorillonites.

3. The total CEC represents a maximum value which can be effective in a neutral medium (pH 7) only. In acid soils, the CEC of clays is reduced; how much, depends on the acidity level and on the special clay group. The values presented by Sanchez (1976) show that in the case of the very low CEC of kaolinites and sesquioxides, the major part (75 per cent in kaolinite) of its total capacity is contingent on the soil acidity. Medium and high CECs depend to a lesser degree on the soil reaction. As a rule, only 5–10 per cent of the negative charge in 2 : 1 layer silicates is pH-dependent, whereas 50 per cent or more of the charge developed on 1 : 1 type minerals can be pH-dependent (Bohn *et al.*, 1979).

Table 3.4 Exchange properties associated with different clay mineral groups (after Sanchez, 1976 and Birkeland, 1974)

Material	Cation exchange capacity (meq per 100g clay)			Anion exchange capacity* (meq per 100g clay)	Approximate CEC values† (meq per 100g clay)
	Permanent	Variable	Total	Total	Total
Montmorillonite	112	6	118	1	80–150
Vermiculite	85	0	85	0	100–150
Illite	11	8	19	3	10–40
Halloysite	6	12	18	15	3–15
Kaolinite	1	3	4	2	3–15
Gibbsite	0	5	5	5	4
Goethite	0	4	4	4	
Allophanic colloid	10	41	51	17	25–100
Peat	38	98	136	6	150–500

* Sanchez (1976) for Kenyan soils.
† After Birkeland (1974).

In acid soils with pH values around 4, the available CEC of kaolinite drops to one-tenth of the values of illites and to nearly one-hundredth of those of montmorillonites found in an equally acid medium. Lowering soil reaction from pH 4 to pH 6 through liming may improve the actual CEC of kaolinites, as proven by the experiments of Khomvilai and Blue (1976), but the improvement is always restricted to the total CEC as the upper limit which cannot be manipulated.

3.6 Climatic dependency of clay mineral formation and soil types

For over half a century textbooks have taught that the same parent rock will render decomposition materials of quite different chemical character, depending on whether rock and regolith decay occurs in the humid climates of the temperate extratropics or in the warm tropics. Scheffer and Schachtschabel (1976) offer the values of a quantitative geochemical analysis of the same rock type under different climatic conditions (Table 3.5).

Under the climatic conditions of the UK or the Mediterranean, the percentages of the main chemical elements contained in the weathering mantle of a dolerite basalt are essentially the same as in the parent rock itself, only the Mediterranean decomposition cover is somewhat richer in Ca, Na, K, and P components. By contrast, under the humid tropical conditions of the Malabar coast, India, not only are all the chemical substances which serve as plant nutrients (Mg, Ca, Na and K) lost, but also most of the silicic acid (SiO_2). The loss of nutrients is through 'leaching', the solution and outwash of silicic acid are caused by 'desilicification'. The material left as residue after tropical weathering consists almost entirely of aluminium and iron compounds. The completely different forms of weathering are referred to as *allitic weathering* and *siallitic weathering* respectively, denoting as the most important elements aluminium (Al) in the one, and silica (Si) and aluminium (Al) in the other.

Leaching under humid tropical conditions has led to the characterization of tropical soils as 'poor', 'nutrient-poor', 'infertile' or 'leached', terms usually found in older

Table 3.5 Chemical composition of rocks and weathering covers developed under allitic and siallitic weathering processes (values in percentage of the total weight) (after Scheffer and Schachtschabel, 1976)

Chemical compound	Great Britain Humid climate of the temperate high mid-latitudes		Mediterranean Wet and dry winter-rain climate of the temperate subtropics		Western Ghats Wet tropical climate of the outer tropical monsoon region	
	Rock	Weathering cover	Rock	Weathering cover	Rock	Weathering cover
SiO_2	49.3	47.0	44.7	35.7	50.4	0.7
Al_2O_3	17.4	18.5	15.5	34.9	22.2	50.5
Fe_2O_3	2.7	14.6	7.5	7.9	9.9	23.4
FeO	8.3	—	3.7	0.7	3.6	—
MgO	4.7	5.2	7.9	3.6	1.5	—
CaO	8.7	1.5	15.3	4.9	8.4	—
Na_2O	4.0	0.3	1.1	0.9	0.9	—
K_2O	1.8	2.5	1.4	3.1	1.8	—
P_2O_5			1.7	2.8		
H_2O	2.9	7.2	0.9	5.8	0.9	25.0
		Siallitic weathering			Allitic weathering	

treatises. From an ecological viewpoint these are unfavourable attributes, indeed. But they are not the worst. Much graver are the consequences of disilicification. The depletion of SiO_2 from the weathering materials is a strong indication that the clay dominating minerals in the soil belong to the group of kaolinites or even to gibbsite, and that, therefore, the soil possesses an extremely low mineral-dependent CEC.

Nutrient-poor soils can be artificially fertilized, provided they have sufficient CEC. Without the necessary CEC this easy way of improving the soil fails. This harsh consequence of desilicification is a discovery that became apparent only with the application of electronic microscopy to the structural analysis of soil particles. Unfortunately, even today this reality is not sufficiently considered by those who claim that it is possible to transfer extratropical agrotechnological methods, especially the application of fertilizing practices, to soils of the humid tropics.

In view of the importance of the problem, we must return to the classification and regional order of soils.

Figure 3.6 depicts the regional distribution of zonal soil types in tropical Central

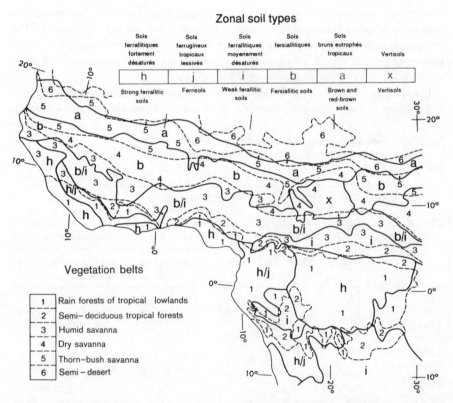

Fig. 3.6 Regional distribution of zonal soils in relation to vegetation belt in West and Central Africa (zonal soil types according to Ganssen and Hädrich, 1965; vegetation belts according to Schmithüsen, 1976)

and West Africa which has been superimposed upon a map of the natural vegetation zones according to Schmithüsen (1976). The soil map stems from Ganssen and Hädrich (1965), who employ the French classification. One may wonder why Central and West Africa and why the French, instead of the FAO/Unesco classification are used. The reasons are threefold: first, Central and West Africa is the most thoroughly surveyed part of the tropics; second, the original soil map of Africa by d'Hoore (1964) uses the classification and nomenclature particularly developed for this tropical region; third, the French classification seems to be the most meaningful and informative for the scientifically interested layman because its terminology refers to the soil-forming processes and chemical attributes instead of resorting to artificially devised codes. (The US taxonomy originates mostly from experiences gained in extratropical regions, and the FAO/Unesco classification has emerged as a composite of different nomenclatures.)

The abbreviations *fer* and *al* stand for the respectively dominating compounds of iron (L. *ferrum*) and aluminium in the case of *sols ferrallitiques* (ferrallitic soils) with the ever-present oxygen. In the case of the *sols fersiallitiques* (fersiallitic soils), silicium (*si*) is associated with the elements iron and aluminium. *Fortement désaturés* means heavily leached, *moyennement désaturés* moderately leached. The term *ferrugineux* indicates the extraordinarily high content of iron compounds. *Lessivés* is used internationally to refer to the displacement of clay minerals from the upper soil horizons to a lower illuvial horizon. The translation of *sols bruns euthrophés tropicaux* is brown nutrient-rich tropical soils, and that of *sols minéraux peu évolués*, is poorly developed mineral soils. In the case of vertisols, the term is the same as in other classifications. The name proceeds from the Latin *verto* (to turn) and refers to the fact that in these clay-rich soils horizon development is impeded by the churning effects of extreme expansion and contraction due to water absorption and drying of montmorillonites.

For those more familiar with the FAO classification or the US Soil Taxonomy, though, it is fortunate that several experts have repeatedly undertaken the task of assembling comparable taxonomic soil units (Sanchez, 1976; National Research Council, 1982; Bridges, 1978; Dudal, 1976). Table 3.1 offers a generalized overview. Once again it must be emphasized that the taxonomic units from various classification systems are not completely in agreement, which sets certain limitations to random interchange.

The example of Central and West Africa demonstrates, except for the vertisols, the close relationship between soil units and natural vegetation zones. The heavily leached ferrallitic soils with kaolinites and gibbsites as clay minerals dominate in the rain forest areas of the inner Congo basin as well as along the humid Guinea coast. The soils of the dry savanna are fersiallitics, and in the thorn bush savanna of the Sahel zone the brown and red-brown eutrophic soils prevail. Both groups are relatively rich in nutrients and are not desilicified, i.e. they possess three-layer clay minerals with a high CEC. The humid savanna is characterized on the map by a contiguity of weak ferrallitic and fersiallitic soils.

The influence of the ever-humid climate of the rain forest on soil formation is so strong that all soils are ferrallitic regardless of the chemical constitution of

99

the parent rocks and the drainage conditions. Under the periodically wet-and-dry conditions of the savanna, the chemical properties of the parent rocks and the topographically induced drainage conditions influence the soil formation to such a degree that ferrallitics only result under unfavourable circumstances. The appearance of contiguity in the field is a matter to be discussed further in Chapters 5 and 6.

The same procedure can be carried out for South America by using the maps of natural vegetation and soil orders published by Sanchez (1976). In South America the association of broad-leaf evergreen forests with oxisols (orthox) occurs in the interior of the Amazon basin and with ultisols (udults) along the eastern coast of Brazil. The association of deciduous forests and shrubs with alfisols (ustalfs) is found in the *noreste* (north-east of Brazil). The savanna zone and the grasslands of the *campos*, south of the Amazon basin, are characterized by a contiguity of oxisols (ustox), entisols (psamments) and ultisols (ustults). A comparison of the mentioned soil orders with the corresponding soil types of the French classification and their regional–climatic arrangement shows the principle of climate–soil relationship to be the same as in Central and West Africa.

A broad-scale association of clay mineral groups (families) with soil orders is expressed in Table 3.6 taken from Birkeland (1984). The figures are based on a large number of soil analyses from the United States, Puerto Rico and the Virgin Islands (Allen, 1977). In order to qualify a soil probe as belonging to the indicated groups of kaolinite, illite and so on, the mineral family of that probe must have over one-half of the indicated mineral or, in some cases, that clay mineral must be dominant. Mixed means a mixture of several clay groups.

The closest relationship exists between kaolinite and oxisols and, to a lesser degree, between kaolinite and ultisols. Smectites (montmorillonites) are quite common in aridisols and entisols, and absolutely dominant in vertisols, which are found, however, in different climatic regions under certain environmental circumstances. Mixed mineralogies appear to prevail in the alfisols.

Numerous and very detailed mineral analyses for a wide range of soils of the wet, the wet-and-dry, and the semiarid tropics as well as for the subtropics are contained in the classic work of Mohr *et al.*, *Tropical Soils* (1972).

Table 3.6   Dominant clay mineral families (%) in different soils of the tropics (after Birkeland, 1984)

Soil order	Kaolinite	Illite	Smectite	Vermiculite	Mixed	Number of families represented
Oxisols	100					6
Ultisols	32	4			64	81
Alfisols	5	7	35	1	52	291
Aridisols	1	2	68		29	131
Entisols	1	2	62		35	116
Vertisols	3		84		13	68

The clay fraction of soils on relatively young sediments from the temperate climatic region of Europe is normally dominated by illites and chlorites (Millot, 1970). As a summary of the relationships between climatic zone and dominating clay mineral groups, Millot (1979) offers the sketch presented in Fig. 3.5.

3.7 Organic matter in tropical soils

Soil classifications mentioned so far do not refer to the organic matter content in the various soil types, although organic matter, especially in tropical soils, is one of the determining ecological factors.

The value of organic matter as a nutrient source is evident in natural systems because plants and animals return to the soil the nutrients they have taken out during their lifespan. Its importance as a carrier of CEC can be best demonstrated by the results of analyses which Sombroek (1966) obtained from a large number of soil samples from the Amazon basin. The diagrams (Fig. 3.7 and 3.8) compare first the relationship between CEC values (indicated as T) and organic carbon content, and second between CEC values and clay content for well-drained kaolinitic forest and savanna soils. (The rate of organic carbon in biomass is commonly considered to be 58 per cent. Thus the organic matter content of soils can be calculated by multiplying the organic carbon values by 1.7. The data for the relationship between CEC and clay content refer, of course, to the B horizons.)

While the CEC rises very little with increasing clay content and remains between 3 and 5 meq per 100 g soil at a clay content of 90 per cent, it does rise significantly

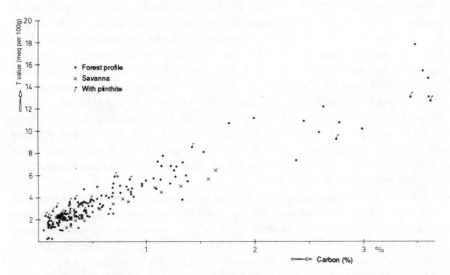

Fig. 3.7 Relation between CEC and organic carbon content for well-drained soils in the Amazon basin (after Sombroek, 1966).

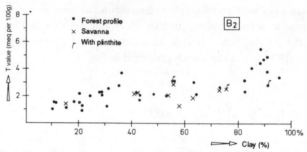

Fig. 3.8 Relation between CEC and clay content for well-drained kaolinitic soils in the Amazon basin (after Sombroek, 1966).

with higher organic carbon contents. With 2.5–3 per cent of organic carbon CEC values between 10 and 12 meq are reached for 100 g soil (dry matter), with more than 3 per cent CEC values range from 14 to even 18 meq. These values are still relatively modest as compared to those which are characteristic for soils of the temperate extratropics. This is due to the fact that in the soils of the extratropics with three-layer clay minerals the total CEC is the sum of both organic-matter CEC and clay-mineral CEC. In humid tropical soils the CEC is supported almost exclusively by organic matter only.

The diagram in Fig. 3.7 also reveals that the bulk of the forest soils is provided with organic carbon contents only between 0.2 and 1 per cent and that the organic matter content in savanna soils is generally somewhat lower than in forest soils.

As far as the organic matter content of tropical forest soils is concerned, there exists little consensus among soil scientists as to whether the humus contents of tropical soils should be qualified as 'poor' or 'normal'. This is especially true for the soils of the tropical rain forest. After consulting a vast amount of data, Sanchez (1976) came to the conclusion that 'organic matter contents in the tropics are similar to those of the temperate regions. Highly weathered oxisols have higher organic matter contents than their reddish colors would indicate'. For the upper 15 cm of 16 randomly chosen oxisols, ultisols and alfisols from Zaïre and Brazil, Sanchez reports organic carbon values between 2.01 and 2.13 per cent for the oxisols, 1.61 and 0.98 per cent for the ultisols, and 1.06 and 1.30 per cent for the alfisols. Considering the soil column of 0–100 cm depth, the values are 1.07 and 1.3, 0.88 and 0.35, 0.53 and 0.55 per cent, respectively. Obviously, there is a clear discrepancy between the average values reported by Sanchez and the great number of actual measured data diagrammatically presented in Fig. 3.7 by Sombroek.

However, for our purposes, we are not dependent on a definite solution to that problem. Dealing with the issue of the suitability of tropical soils for intensive agricultural exploitation, the absolute value of the original organic matter content under natural conditions is only of secondary importance. It is more pertinent to gain a proper understanding of the processes which determine the organic carbon content in soils and of the changes that affect former forest soils after clearance and after being submitted to agricultural land use.

3.8 Decomposition of organic matter and the formation of humus

The functioning of all processes in the biotic part of ecosystems can be recognized as occurring within three subsystems: the plant subsystem, the herbivore–carnivore subsystem and the decomposition subsystem. These three components are integrated by the transfer of matter and energy (Fig. 3.9, Table 3.7).

The annual gain of matter (and energy) by the plant subsystem, the net primary productivity (NPP), is distributed in three ways. A small part is stored in perennial matter and contributes to the biomass increase called 'net growth'. A minor part (in forest systems probably never more than 10 per cent) is consumed by animals. The bulk of the NPP in ecosystems is shed as plant litter or secreted as soluble organic matter. This part thus enters the decomposition subsystems as dead organic matter, or detritus. The faeces and carcasses of the herbivores and carnivores also contribute to the detrital input into the decomposition subsystem. In Table 3.7 the most accurate estimates of a number of parameters are listed in order to provide some comparisons of production and decomposition in different tropical and extratropical natural vegetation systems.

In comparison with all other natural vegetation systems, the tropical rain forest presents maximum values of net production (on the average 30 t ha^{-1} yr^{-1}) and biomass (500 t ha^{-1}). The tropical savannas carry a much smaller biomass (45 t)

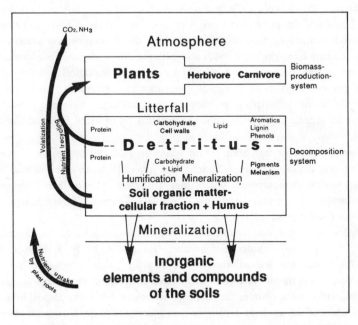

Fig. 3.9 Schematic flow chart of the decomposition processes of organic matter.

Table 3.7   Production and decomposition in six ecosystem types

	NPP (t ha^{-1} yr^{-1})		Biomass (t ha^{-1})		Litter input	Litter turnover ratio
	Mean	Range	Mean	Range	(t ha^{-1}yr^{-1})	t(yr^{-1})
Tundra	1.5	0.1–4	10	1–30	1.5	0.03
Boreal forest	7.5	4–20	200	60–400	7.5	0.21
Temperate deciduous forest	1.5	6–25	350	60–600	11.5	0.77
Temperate grassland	7.5	2–15	18	2–50	7.5	1.5
Savanna	9.5	2–20	45	2–150	9.5	3.2
Tropical forest	30	10–35	500	60–800	30	6.0

Source: Swift *et al.* (1979).

than the humid forests, while the NPP is proportionately not that much smaller. This is because of the relatively high percentage of the production which is diverted to perennial storage in woody tissue in the forest. The litter input in mature systems is equal to the value of NPP. According to the values indicated, the average litter input in the tropical forest is almost three times greater than in temperate forests.

Beside the NPP, the second determining process in all ecosystems is the decomposition of organic matter (Fig. 3.9). In general terms, 'decomposition essentially results in a change of state of a resource under the influence of a number of biological and abiotic factors' (Swift *et al.*, 1979, p. 50). Applied to the vegetation–soil ecosystem, the source consists of the litter as the major component and the corpses and faeces from the herbivore–carnivore food chain. They enter the decomposition system together with the leachate from the plant cover and with the rain-water and its chemicals. After the change of state under the influence of a number of biological and abiotic factors, the final product consists of inorganic elements and compounds. These are released into the soil or the atmosphere or they are recycled in the biomass, provided there is no loss of organic material to other ecosystems by artificial removal, by leaching or outblowing.

All the material that has been put in serves to feed the decomposer community which consists mainly of fungi, bacteria and invertebrate animals, whose carcasses enter the detritus compartment where they are acted upon by other decomposers. The decomposer community performs two major functions, mineralization and the formation of soil organic matter.

Mineralization is the conversion of a chemical element from its organic to its inorganic form. This includes, for instance, the release of P, S, K, Ca, Mg and nitrate (NO_3) into the soil solution where they are either recycled in living plants or form new soil material. The gases CO_2, from the respiration of carbohydrates, and ammonia (NH_3), from protein degradation, are released into the atmosphere.

The residues of this stage of decomposition contribute to the *soil organic matter* which consists partially of microscopically distinguishable components and is recognizable as digested plant residuals or faeces and as microbial cells. The ecologically more important part is the finer substance of organic matter which has no recognizable microscopic structure; that is, humus in its chemical definition. It is a mixture of complex polymeric molecules (fulvic and humic acids) which form negatively charged colloidal particles, which means that they may attract a swarm of cations which can be exchanged for others in the soil solution. This CEC has been qualified before as one of the three decisive ecological properties of soils. The humus colloids often associate physically with the equally negatively charged clay minerals and form the *humad complex*.

It is common practice that the two parts of organic matter distinguished above are summarized under the term 'humus' in its broadest meaning.

Related to the effectiveness of the decomposition processes there are two general rules. First, all organisms involved in biochemical processes conform to the relationship established in van't Hoff's rule that there is a change in activity of approximately 2–2.5 with every 10 °C rise or fall in temperature, provided there is sufficient moisture.

In textbooks of pedology, such as Duchaufour (1982:88), one finds in respect to the temperature dependency of litter decomposition: 'The speed of litter decomposition is roughly proportional to the mean temperature provided there is enough moisture.' In the specific case of the tropical zone it makes no great difference which of the rules is applied.

In any case, the relationships are complex due to the interplay between the composition and size of the decomposer community, the quality ('decomposability') of the resource material to be decomposed, and the physico-chemical environment on and in the soil. The decomposability depends on a variety of intrinsic factors, such as hardness, lignin content, nutrient content, particle size of the litter and secondary compounds from the leachate. Thus, leaves are of higher resource quality and are decomposed more readily. Governing factors of the physico-chemical environment are, besides the dominating temperature, the amount, intensity and periodicity of rainfall, the chemical nature of parent rocks and soils, especially the soil pH, the drainage conditions and the physical properties of the soil. All these variables influence the effectiveness of the decomposition process and thus the decomposition rate. But the interactions 'may be considered to have a hierarchical structure in the way that they influence one another in the decreasing rank order: macroclimate > microclimate > resource quality > organisms' (Anderson and Swift, 1983, p. 293).

The second rule (Jenny *et al.*, 1949; Olson, 1963) establishes that under steady-state conditions of mature ecosystems the ratio of dead organic matter production to the total decomposition approximates unity over a given period of time. In other words, since under steady-state conditions of an actual mature forest litter cannot be accumulated progressively on the forest floor, the decomposition of litter must be equal to its production in the course of years.

What is valid for the total decomposition system must also be valid for the subsystems of humification and mineralization, because under steady-state conditions, accumulation of soil organic matter, especially humus, should not occur either.

Applying these two rules to the comparison of decomposition conditions in the humid climates of the inner tropics and the temperate extratropics, some ecologically important consequences result for similar material (e.g. broad leaves, twigs and fruits) and similar chemical circumstances (neutral milieu).

As the mean annual temperature on the floor and in the upper soil horizons ranges around 30 °C, the biochemical processes of decomposition of organic material must be seven times more effective than under temperate forests, considering a mean annual temperature around 10 °C and taking 2 as the lower factor of van't Hoff's rule (three times for the first 10° span until 20 °C and four times for the second 10° span from 20 to 30 °C). According to Duchaufour, the decomposition is six times faster.

When the rule of Jenny and Olson is applied to the corresponding values of Table 3.8 (30 t ha^{-1} yr^{-1} litter input in the tropical forest, 11.5 t in the temperate forest), the decomposition in the humid tropical forest should be three times more effective than in the temperate zone.

The general rules are useful for the comparison of tropical and extratropical conditions. Missing, however, is the quantification of the decomposition rates. For this purpose, the litter turnover coefficient (*KL*) is widely used. Under steady-state conditions it can be calculated as the ratio of annual rates of litter fall (g m^{-2} yr^{-1}) and litter standing crops (= litter accumulation g m^{-2}) on the forest floor. The reverse relationship, i.e. litter accumulation (g m^{-2}) divided by the annual rate of litter fall (g m^{-2} yr^{-1}) shows the decomposing time.

Another method is the direct measurement of weight losses from litter bags. The results, however, are hardly comparable due to dissimilar experimental conditions.

Table 3.8 Litter production and litter turnover in tropical forests ((Anderson and Swift, 1983)

Forest type Locality	Altitude (m)	Litter-fall (t ha yr^{-1})	Litter standing crop (t ha^{-1})	*Turnover coefficient KL* Leaves	Small litter
Tropical lowland moist forests					
Dipterocarpaceae	10	10.6	3.2		3.3
Malaya		6.3 leaves	1.7	3.6	
Ivory Coast	50–100	8.1 leaves	2.5	3.3	
Alluvial forest Sarawak	50	9.4	5.5		1.7
Colombia	30	8.5	5.0		1.7
Rain forest Manaus (Brazil)	45	7.6 6.1 leaves	7.2 4.0	1.5	
Dipterocarpaceae Malaya		7.5 5.4 leaves	7.7 5.1	1.1	
Temperate deciduous forests					
Netherlands	100	3.1	3.6		0.9
UK		5.4 3.2 leaves	7.1 2.0	1.6	0.8
Minnesota		4.6	12.6		0.4

Anderson and Swift (1983) have critically reviewed the corresponding calculations of turnover coefficients. From the whole list of 26 different locations, the values of 2 maximum, 2 medium and 2 minimal turnover coefficients were extracted (Table 3.8). The listed values of litter turnover coefficient demonstrate, in a general overview, that

1. Tropical rain forests have *KL* values greater or equal to 1, suggesting that the turnover of soil organic matter occurs in a year or less;
2. The lowest values for the decomposition of small litter in tropical lowland forests exceeds the highest values of temperate deciduous forests;
3. The highest coefficients of the tropics are three times larger than the highest of the extratropics;
4. There is considerable variation within the tropics;
5. The fastest turnover takes less than 4 months;
6. 'There is some evidence of regionality in litter turnover coefficients, with values greater than 2 for most African forests and between 1 and 2 for forests in S.E. Asia and the Neotropics' (Anderson and Swift, 1983, p. 300).

A critical review of the results of litter bag studies by Anderson and Swift (1983) reveals that they generally support this pattern, although they are less comparable than the calculations of turnover coefficients cited above.

The litter coefficient *KL* describes the decomposition of only a component part of the total organic matter in the system soil plus litter cover, and thus relates only to the earlier stages of the change of state from dead organic matter to inorganic residuals. No account is taken of stages beyond a point at which detritus is no longer recognizable as plant residual. However, within a steady-state system, in which, by definition, accumulation of humus (in its broader sense) may not occur, the rate of decomposition of humus to inorganic substances must be strongly related to the formerly defined litter decomposition rate.

With respect to litter decomposition and organic carbon content in different natural environments, Sanchez (1976) summarized annual addition, decomposition and equilibrium levels of organic matter as shown in Table 3.9. For the areas covered by tropical rain forests, Sanchez and Buol (1975:599) concluded: 'These forests produce about five times as much biomass and soil organic matter per year as comparable temperate forests. The rate of organic matter decomposition, however, is also about five times greater than in the temperate forests. Thus, the equilibrium contents are similar.'

Apart from all the differences in the details of modelling and of experimental results, two facts stand out:

1. Organic matter decomposition under tropical moist climate conditions is, in the areal average, considerably more effective than in comparable temperate climates. The turnover time ranges between 1 year as a maximum, and 4 months as a minimum.

Table 3.9 Estimates of annual additions, decomposition rates, and equilibrium levels of topsoil organic carbon in some tropical and temperate locations

Location	b Addition of undecomposed organic matter (t ha⁻¹)	m Decomposition rate of fresh organic matter into soil organic carbon (%)	a Soil organic carbon addition (t ha⁻¹)	k Soil organic carbon decompostion rate (%)	C Soil organic carbon at equilibrium (t ha⁻¹)	(%)
Tropical forests						
Ghana (ustic)	5.28	50	2.64	2.5	106	2.4
Zaïre (udic)	6.05	47	2.86	5.2	55	1.2
Colombia (udic, andept)	3.85	51	1.97	0.5	394	9.0
Temperate forests						
California (oak)	0.75	47	0.35	0.4	88	2.0
California (pine)	1.65	52	0.86	1.0	86	1.9
Tropical savannas						
Ghana (1250 mm rain)	1.43	50	0.71	1.3	55	1.2
Ghana (850 mm rain)	0.44	43	0.19	1.2	16	0.4
Temperate prairie						
Minnesota (870 mm rain)	1.42	37	0.53	0.4	134	3.0

Source: adapted from Sanchez (1976).

2.   Aside from other influencing factors, such as the nature of the decomposer community, the characteristics of the organic matter which determine its degradability and the physico-chemical conditions which operate at microclimatic and edaphic scales, macroclimate, especially its temperature and moisture regime, is the governing variable factor.

3.9   Effects of forest clearing

Based on the knowledge of the processes, relationships and dependencies explained above, it should not be difficult to imagine what will happen when humid tropical forests are cleared and the soil is submitted to agricultural use, especially when the experiences of the change in soil climate are taken into consideration.

Prior to clearing, a great part of the solar radiant energy was reflected and absorbed by the trees and shrubs and a large proportion was sent back into the atmosphere. Bare soils, by contrast, are exposed to the entire solar beam.

The change from forested to bare soil results in a cessation in the input of fresh organic litter into the decomposition complex, while, at the same time, the decomposition process itself accelerates because of the increase of temperature on and in the soil. Without intervention by man, the organic matter content of the soil declines drastically and, within a relatively short time, the soil loses a decisive source of nutrients and its most important carrier of CEC. The preparation of land for cultivation involves some other modifications of the soil environment. The process of hoeing or, worse, ploughing alters the soil structure and favours aeration. The burial of litter results in extreme modification of the environment for the decomposing biota. In connection with the increase in temperature, organic matter decomposition is accelerated as is nutrient release.

The litter input from most arable crops lacks the lower resource quality components, particularly woody resources, which tend to stabilize the decomposition rate and nutrient release in forest ecosystems. Thus, the major part of the crop litter will decompose relatively soon with very little 'buffering' from slow-release components (Anderson and Swift, 1983). Under favourable conditions during the rainy season, 'relatively soon' actually means a few to several weeks. And since the initiation of decomposition will occur before planting, the maximum nutrient release may occur before the crop can benefit from the availability of nutrients. Clearly, under the leaching conditions of heavy tropical rains (see Section 1.2) lack of integration could result in the loss from the system of a significant component of the badly needed nutrient return.

Lack of integration between the release of nutrients by decomposition and their uptake by annual crops contrasts strongly with the 'tight cycling' described for forest ecosystems. It may well be one of the major features leading to nutrient loss and it is one that could be avoided by the correct timing of litter inputs, e.g. by mulching practices (Anderson and Swift, 1983:305).

109

Since every system of agricultural land use is necessarily tied to an export of biomass from the producing surface and since the quality in nutrients of the exported biomass is superior to the crop residual, the contribution of crop residuals to the soil account for only a single fraction of the natural input of the original forest. A commonly estimated ratio is one-fifth.

After clearing, soil organic matter will decompose at a rate approximately proportional to the existing amount in the soil, i.e, at first relatively fast, later more slowly so that, if calculation is continued for a given period, a new, but considerably lower level of equilibrium will asymptotically be reached. Conversely, the attempt to raise a low level of equilibrium requires an enormous quantity of fresh organic substance in order to achieve a noticeable effect.

The content of organic matter in the soil may be manipulated to a certain degree in specific locations, but not over large areas, since the quantity that is added at a particular place must have been produced elsewhere prior to its redeposition. This is valid everywhere. But since the rate of decomposition is much faster in the tropics than in the temperate regions, the application of organic matter in the tropics must be repeated more frequently.

3.10   Manipulating soil chemical properties

Aside from primary mineral content, humus matter and CEC – which defy manipulation – there are other soil properties which result from the combination of those three and which can be controlled by man to a certain degree.

Important physical soil attributes such as soil texture (distribution of aggregates that determine soil coarseness or fineness), structure (combination or arrangement of primary soil products into secondary aggregates), porosity (percentage of total pore space), bulk density (mass of dry soil per unit of bulk volume) and soil permeability (the ease of water infiltration and percolation) are strongly related to the mineral content, soil humus and the specific availability of clay minerals of varying crystallographic structures. Man can alter the physical properties by working the soil mechanically, but this is not necessary in the tropics – except in the case of vertisols and hardened plinthites (laterites) – since the physical properties of the tropical soils are generally good, and mostly excellent in the case of ferrallitics.

3.10.1   *The effects of phosphorous fixation and aluminium toxicity*

Essential chemical soil properties that affect agrarian productivity in the tropics are phosphorous fixation and aluminium toxicity.

Phosphorous deficiency is a very common ecological feature on strongly weathered ferrallitic soils of the tropics. This is a consequence of the fact that the total phosphorous content decreases with increasing weathering intensity. A further implication is the considerable phosphate-fixing capacity of the ferrallitics. While the total quantity of phosphorus in representative soils from the US Midwest lies around 3000 parts per million (ppm), oxisols generally exhibit less than 200 ppm. Even tropical ultisols and alfisols are relatively low in total phosphorus with values mostly below 200 ppm (Sanchez, 1976).

Part of the soil's total phosphorus is held in inorganic form as adsorbed anions on the soil colloids and part as organic phosphorus contained in the humus. (The latter is reported to account for 20–30 per cent of the total phosphorus in the tropical soils.) From both sources, phosphorus is slowly released to maintain the equilibrium level in the soil solution.

In the inorganic form the anions are held as compounds of iron, aluminium or calcium, forming the iron-bonded phosphates (Fe–P), the aluminium-bonded phosphates (Al–P), or the calcium-bonded phosphates (Ca–P). They are divided into active and inactive fractions. The active fraction is the Ca–P which is simply adsorbed at electrically charged surfaces and represents the soluble form of inorganic phosphorus. The inactive fraction consists of Fe–P and Al–P compounds which are covered by a coat of inert material, a fact that prevents the reaction of these phosphates with the soil solution. They are 'fixed', i.e. immobilized (*phosphorous fixation*).

In acid tropical ferrallitics, aluminium and iron are most abundant and react preferably with the phosphoric acid to form Al–P and Fe–P phosphates, the insoluble forms of phosphates.

The general rule is: the higher the content of aluminium and iron compounds in the soil, the higher the phosphorous fixation capacity. Consequently, tropical ferrallitics tend to retain considerable amounts of the already low total phosphorous content because the clay fraction is dominated by kaolinites, gibbsite and goethite; calcium is almost totally missing. Less acidic soils of the tropics with three-layer clay minerals, such as montmorillonites and illites, have much lower retention capacities (for more details see Sanchez, 1976 or Webster and Wilson, 1980).

A great number of cultivated plants are very susceptible to the aluminium concentration in the soil solution. Values above 1 ppm often cause direct growing reduction, but there is a wide difference in the ability to tolerate relatively high levels of aluminium concentrations. Coffee, rubber and pineapple are very tolerant; sorghum and cotton are intolerant. In between exists an important varietal difference in rice, corn, beans and soyabeans.

If soils have an aluminium concentration higher than a specific plant can tolerate one speaks of *aluminium toxicity*. The plants' development is restricted, their roots become thickened and stubby. Aluminium tends to accumulate in the roots and impede the uptake and translation of calcium and phosphorus.

Exchangeable aluminium cations (Al^{3+}) are kept close to the adsorption sites of CEC carriers. In the sequence of decreasing tenacity they rank after H^+ and before Ca^{2+}, Mg^{2+}, K^+ and Na^+. As demonstrated in Fig. 3.4, under high weathering conditions most cations are already leached when aluminium is still present and the soil consists of silica-poor clay minerals, such as the kaolinites, or of aluminium-

111

and iron hydroxides, such as gibbsite and goethite. Consequently, the danger of aluminium toxicity is always high in kaolinitic soils and even higher in the presence of gibbsite and goethite.

A certain influence is exerted by the soil organic matter content. Aluminium concentration in the soil solution decreases with increasing organic matter because the two form very stable complexes.

3.10.2 *Liming*

The growing conditions in acid and aluminium-rich soils can be corrected by applying lime ($CaCO_3$ granules, liming). In the soil solution there is always carbon dioxide (CO_2) which reacts with water molecules (H_2O) to create carbonic acid (H_2CO_3). This carbonic acid dissolves the lime according to the following formula: $CaCO_3 + H_2CO_3 = Ca^{2+} + 2(HCO_3)$. In this way the soil solution becomes charged with calcium cations Ca^{2+} which change places with H^+ and Al^{3+} ions in the exchange complex. The H^+ or Al^{3+} ions are released into the soil solution. The H^+ ions combine with OH^- ions to form water molecules.

When many of the H^+ and Al^{3+} ions are replaced by Ca^{2+} ions, the acidity of the soil is partially neutralized and the effective cation exchange capacity (ECEC) is increased. This results, first, from freeing of exchange sites on the clay mineral surfaces through a displacement of adsorbed triple-charged Al^{3+} by double-charged Ca^{2+}, and second, from the activation of additional charges on organic matter through the lowering of the soil acidity.

Because of the continuous generation of H^+ ions which themselves displace the Ca^{2+} ions from the exchange complex, the growing quantity of H^+ ions eventually makes the soil acid again. Moreover, carbon dioxide draws Ca^{2+} ions from the exchange complex and forms calcium bicarbonate $Ca(HCO_3)_2$, which is easily leached by percolation.

When excessive quantities of lime are added to soils poor in CEC the Ca^{2+} ions compete with other nutrient cations, such as Mg^{2+} or K^+, and displace them from the adsorption sites with the final consequence that they are washed out. The result is deficiency of one or the other nutrient cation. This triggering effect is referred to as *overliming*.

3.11 The role of topography on the formation of regular microscale soil sequences (toposequences, catenas)

Until now, the deduction of soil properties and soil regionalization has been made under idealization and presumption that all soils have been formed on well-drained sites over nearly horizontal surfaces. In reality, however, every terrain consists of sequences of summits, slopes and depressions. When the slopes are steep enough solid particles of varied dimensions are transported downhill. Soil erosion is the consequence. On the valley bottoms the eroded materials accumulate. Under these terrain conditions the formation of mature soil profiles is not possible. But these are special situations regionally restricted to mountainuous areas.

The dominating topography of tropical lowlands and uplands is that of the so-called peneplain, characterized by relatively gentle slopes, extended valleys and small altitudinal differences. Under natural conditions, i.e. when the natural vegetation is not disturbed, the complex of processes responsible for pedogenetic differences, in accordance with site differences on slopes or in depressions, can be classified in three groups:

1.   Selective transport of finer particles (silt and clay);
2.   Transport in form of solutions;
3.   Local changes in the moisture status and in the redox potential.

Thus, sites on the top of slopes tend to be impoverished, to the advantage of sites located on lower slopes, and particularly on depressions, that are enriched with materials from the higher positions. Such a regular succession of soils, identical along a given contour line but varying in continuous fashion downslope, has been called *catena* (a chain of soils) (Duchaufour, 1982, p. 132). In the English literature the term 'catena' is replaced by that of *toposequence*. Impoverishment results in that fine silt and clays, nutrient elements, especially the bases, soluble silica and pseudo-soluble complexes are gradually moved downslope. In the depressions with deficient drainage there occur an enrichment of bases, a lowering of acidity and in extreme cases, the resilicification of 1 : 1 layer clay minerals.

Moormann (1981) has published detailed data of a representative number of topo-sequence analyses for the typical soils included in a climatic profile from the humid forests to the savannas in Nigeria. Juo (1981) summarized the qualities in diagram form (Fig. 3.10). What are essential are the clay mineral outfit, the CEC and the base saturation, especially in deeper soil horizons (B horizons).

In the humid forest zone (around 2000 mm annual precipitation and 1–2 dry months) there is almost no difference in CEC and base saturation between the highly weathered upland soils and the hydromorphic soil of topographic depressions. The dominating clay minerals are kaolinites both in upslope and downslope sites. Hills, slopes and depressions in the terrain are equally carpeted by deep ferrallitic soils with low base status and extremely low effective CEC. Sanchez (1976), based on Brazilian experiences, referred to this type of soil associations as 'oxisol landscape'.

Within the drier forest zone near Ibadan (1200–1300 mm annual precipitation, 2 dry months), the weak ferrallitic, shallow sandy soils over gneiss of the higher parts of the terrain differ only slightly from the soils of the humid forest zone, as far as CEC and base saturation are concerned. However, the fine loamy hydromorphic soils in the lower parts of the slopes and in the depression floors exhibit a relatively high CEC and high exchangeable calcium and magnesium. The clay minerals consist of a mixture of

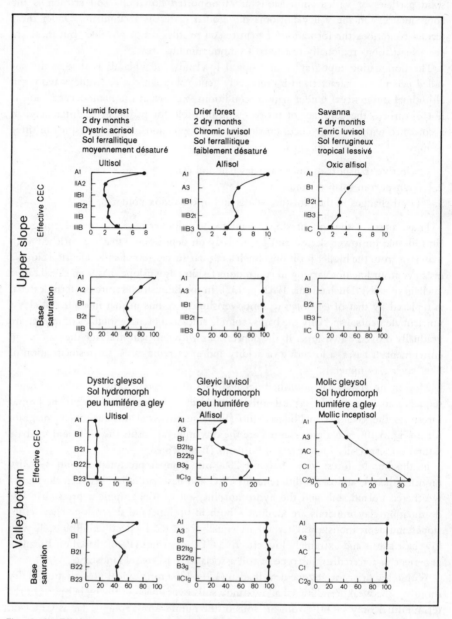

Fig. 3.10 Effective cation exchange and base saturation of soils in toposequences on intermediate crystalline rocks in different climatic zones of Nigeria (after Juo, 1981).

different mineral groups with smectites (montmorillonites) as the most characteristic. This toposequence association is called 'alfisol landscape' by Sanchez (1976).

The savanna zone of Guinea, with nearly the same annual precipitation as the drier forest zone, but with 4 dry months, still presents in the well-drained higher parts of the terrain highly weathered soils, characterized by kaolinites in the clay fraction and corresponding low CEC values. In contrast, the hydromorphic soils of the valley bottoms are less weathered, reveal a high smectite content and fair CEC values.

Even more pronounced is the change from strongly weathered kaolinitic soils with relatively low CEC of the summits and upper slopes to montmorillonitic soils of the depressions in the domain of the thornbush savannas with more than

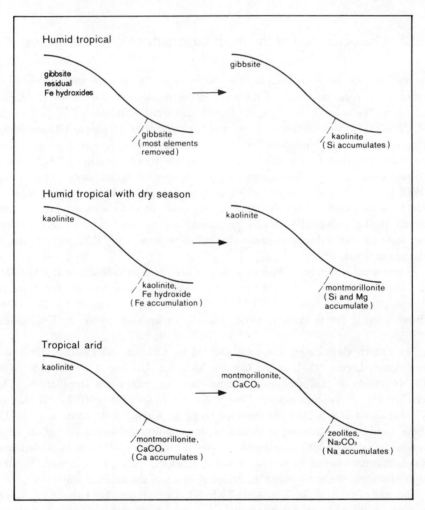

Fig. 3.11 Schematic overview of the range of soil associations in toposequences in different climatic regions (after Birkeland, 1984).

6 dry months. Bocquier (1973) has investigated such sequences in Chad. The lower parts of extended peneplains are occupied by dark, cracking vertisols with high montmorillonite content, even when the altitudinal differences between hills and depressions do not exceed 5 m. Sanchez (1976) has characterized regions with this soil association as 'Alfisol–Vertisol Landscapes'.

The foregoing detailed description deals with the agriculturally most relevant belts of the tropics, located between humid forest and dry savanna. A more extensive combination of climatic regions and rock-induced variations result in a wider range of different soil associations in toposequences. Birkeland (1984) has offered a scheme in this respect (Fig. 3.11).

3.12   Characteristics of the most important soils of the tropics

At the end of this introduction into the ecologically decisive soil properties, a summarized characterization of the most important soils of the tropics is deemed necessary. We will use as a basis the French classification, developed by pedologists of ORSTOM (Organisation de Recherche Scientifique et Technique d'Outre Mer). This classification is preferable for these advantages: first, it was developed with particular reference to the African tropics, while the American or FAO/Unesco classifications have been subsequently adapted to tropical circumstances. Duchaufour (1982, p. 173) remarks critically with respect to them: 'Soils being classified together that have no genetic similarity.' Second, the terms of soil classes are explicit with respect to the ecologically decisive properties, and not artificial constructs. Third, the scale of the soil regionalization is commensurate with those of climate or vegetation zones.

The original mapping of African soils was presented by d'Hoore in the *Soil Map of Africa 1 : 5 Million*, published by the Commission de Coopération Technique en Afrique au Sud du Sahara (CCTA) in 1964. The explanation of the ORSTOM classification is due to Aubert (1965). There is an updated version by Duchaufour (1982).

The French classification has been adopted by Ganssen and Hädrich (1965) and by Schmidt-Lorenz (1971) in the German literature. Bunting (1969) reproduced in *The Geography of Soil* the most essential lines of the French classification. The terminology in Young's book, *Tropical Soils and Soil Survey* (1976), is that of d'Hoore's *Soil Map of Africa* in modified form. In *Tropical Soils*, Mohr et al. (1972) chose another classification; nevertheless, it is indispensable because of its large number of tables on quantitative soil analyses. Fitzpatrick (1971) utilizes unusual and somewhat cumbersome nomenclature and systematics, but the book is useful because of its numerous illustrations and quantified sketches of the different soil types.

The map of soil types in West Africa, which has been combined with the vegetation formations according to Schmithüsen in Fig. 3.6, is a reproduction from Ganssen and Hädrich's map of Africa 1 : 25 million in *Atlas zur Bodenkunde* (1965). Their

116

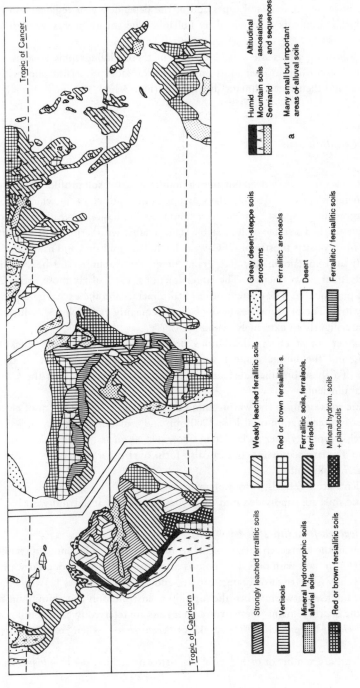

Fig. 3.12 Sketch map of the most important soil types of the tropics (adapted from Ganssen and Hädrich, 1965).

117

source was the *Soil Map of Africa* by d'Hoore (1964). An instructive summary of the distribution of climatic soil types and of the dependency of regional soil groups on geological basements is provided by Pullan's 'The soil resources of West Africa' (1969).

In the following pages the soils presented in these bibliographic sources will be characterized according to their most important ecological properties, similarities and differences, and their zonal climatic distribution (Fig. 3.12).

3.12.1 Ferrallitic soils

Ferrallitic soils offer a very deep but almost undifferentiated soil profile. The very fine-grained material consists of sandy clays or fine sands with clays, in which secondary quartz fragments are sparse. The mineral substances are basically weathering products which have reached a stable or metastable stage against weathering. The clay fraction (below 2 μm) consists mostly of CEC-weak kaolinites, gibbsites and iron oxides. The silts (2–20 μm) contain mainly fine grains of quartz, kaolinite and heavy minerals that are resistant to weathering. The larger fractions, particularly those above 2 mm, consist of pure quartz. The content of residual minerals and the exchange capacity are extremely low. The pH value lies below 6, and frequently even below 4, which means the soils are slightly to extremely acidic.

Even under forest cover, the humus horizons have a maximum thickness of 25–30 cm, and the humic content varies between 1 and 2 per cent of the total soil mass. To these organic substances is tied a good proportion of the CEC and the natural nutrients.

Ferrallitic soils occur in the warm–humid tropics, below 1000–1300 m of elevation, with a yearly precipitation of at least 1200 mm and a maximum aridity of 5 months. They develop typically over rocks rich in silica, such as gneiss, granite, sandstones, old schists and Tertiary sediments, but can also form over other rock types, except basic volcanites and plutonites.

Equivalents are ferralsols, or oxisols, *sols ferrallitiques*.

Ferrallitic soils are subdivided into two types, strongly leached and weak.

Strongly leached ferrallitic soils
They occur in the evergreen forests of the tropical lowlands with annual precipitation above 1500 mm, and with only 1–2 months of aridity: therefore strong outwash. The colour is yellowish-brown or orange-brown, and the pH value lies between 4 and 5. In cases of intense outwash, when the topsoil has lost the iron and aluminium oxides and only the quartz remains, these overly poor soils are referred to as 'tropical podzols' (Klinge, 1976a). Fitzpatrick (1971) speaks of 'krasnozems' when referring to strong ferrallitic soils.

Equivalents are xanthic or orthic ferralsols, ustox or orthox, *sols ferralitiques fortement désaturés*.

118

Weak ferrallitic soils

These are commonly found in the areas of deciduous tropical rain forests, with yearly precipitation between 1200 and 1500 mm and 3–5 months of aridity. The colour is red or reddish-brown, and the pH value ranges between 5 and 6 in the topsoil, or 4.5 and 5 in the lower horizons. In wooded areas, humus content and exchange capacity are slightly higher than in strongly leached ferrallitic soils. In the humid savannas the values decrease again. These soils develop over less deep weathering mantles; nevertheless, the weathering depth still comprises several metres.

The usefulness of ferrallitic soils is qualified by Pullan (1969:532) as follows:

> In well-drained sites they are excellent media for plant growth because of the high soil moisture and room for root development, despite their lack of nutrients. The application of chemical fertilizers for annual crops proves uneconomical owing to superficial roots and the high rate of outwash. Cations are not kept in the clay–humus complex because of the humus scarcity and the low exchange capacity of the kaolinitic clays. However, higher efficiency from added nutrients is assured in the case of deeply-rooted tree cultivars.

Comparable soils of other classifications are ultisols, ferric acrisols, dystric nitosols, *sols ferralitiques moyennement désaturés*.

3.12.2 *Ferrisols*

Although these soils occur in the same tropical humid zone as the strong ferrallitic soils, the ferrisols occupy an intermediate position between ferrallitic and fersiallitic soils. This is due to the concurrence of two favourable conditions: the existence of a parent rock with a high content of nutritive basic minerals and the occurrence of a relatively rain-poor region within the subequatorial tropics.

Ferrisols differ from ferrallitic soils not only because of their strong red colouring but also because of the somewhat higher – though still poor – content of weatherable residual minerals and because they have in addition to the dominating kaolinites a scarce proportion of three-layered clay minerals. Concerning their fertility, Pullan (1969:532) states that 'there are no considerable differences between ferrallitic soils and ferrisols with respect to nutrients, only a slight tendency towards larger nutrient reserves'.

As a consequence of the conditions in which they originate, these soils only develop in relatively reduced areas. Corresponding names in other classifications are ferric luvisols, eutric nitosols, *sols ferrugineux tropicaux lessivés*, ultisols, ustalf.

3.12.3 *Fersiallitic soils*

These soils occur in areas where yearly precipitation lies around 1000 mm and the dry period lasts close to 6 months. The soil profiles are only 2–3 m deep. They still have a certain reserve of weatherable residual minerals because the chemical weathering is less intense and the outwash of silicic acids lower than in other tropical soils. The major part of the clay minerals consists of kaolinites, but each of the various three-layered clay minerals appears in considerable proportions. Thus, the exchange capacity is higher than in ferrisols with the same clay content.

Fersiallitic soils are less colour-intensive than ferrallitic soils; they are usually yellowish-brown to reddish-yellow. The occasional very red soils that do occur in areas commonly occupied by fersiallitic soils are probably fossilized ferrallitic soils.

With 1–2.5 per cent the humus content is still rather low. Schmidt-Lorenz (1971:57) comments: 'Clayey topsoils tend to create hard soil aggregates during the arid periods. . . . Typical is a tendency to lose fine fractions through sheet-floods.'

According to Pullan (1969:533)

> The elevated nutrient status is not always obvious, but the freeing of mineral nutrients from the parent rocks allows their relative rapid transport to the surface in the course of bush fallow. However, these soils are depleted of these nutrients by the yearly fires which also reduce the biologic activity and destroy the nitrogen reserves; consequently, no long-lasting effects can be expected from fallow periods. The application of specific chemical fertilizers has proven economical in various cultivated crops.

Equivalents are alfisols, eutric cambisols, *sols fersiallitiques non lessivés*.

3.12.4 *Eutrophic brown tropical soils*

These exceptional soils develop in the same climatic areas as weak ferrallitic and fersiallitic soils, but over basalt or other basic rocks. With a depth of 1–2 m eutrophic brown soils are relatively shallow; they are also rich in debris. The fine soil is characterized by a high proportion of silt. Weathering creates foremost three-layered clay minerals. Humus content varies between 4 and 7 per cent, and soil reaction between neutral and slightly acidic. These soils are remarkable for nutrient-rich residual mineral content and high exchange capacity. 'The eutrophic brown soils cover only 0.5 per cent of Africa. On other continents they represent a larger proportion of the soils that have developed particularly in recent tectonic areas and over volcanites' (Schmidt-Lorenz, 1971:65). In the FAO classification these soils can be referred to eutric nitosols, eutric cambisols.

120

3.12.5   *Red-brown and brown soils*

These soils occur mainly in the arid tropical margins where the rainless period lasts longer than 6 months; accordingly, the weathering is less intense. Soil profiles are superficial and contain a large reserve of weatherable minerals. The type of minerals depends on the nature of the parent rock. Outwash of bases is rare, and the content of three-layered minerals, like montmorillonite, is remarkable.

Red-brown and brown soils are well endowed with organic substances because of the root density created by weeds, shrubs and grasses; nevertheless, mineralization of the humic substances is high.

They show a neutral or weak alkaline reaction, and it is not uncommon to find accumulations of free calcium in the deeper horizons. Due to the weak chemical weathering, the dependency of these soils on the parent rock is much more pronounced than that of soils in the humid tropics, exemplified by the large variety of soils in that group.

Under normal circumstances, the red-brown soils reach down deeper than the brown soils, possess a lower content of organic matter and suffer from a modest outwash of bases.

When taken into cultivation, the red-brown and brown soils 'lose their structure, and erosion damages through wind and water are serious. Sandy varieties tend to drain excessively and become moisture deficient. Irrigation of these soils remains a possibility; but where this is not feasible, the cultivation of crops is risky: it might be wiser to develop pastures rather than agricultural crops' (Pullan, 1969:534)

Comparable soil classes are aridosols, *sols bruns eutrophés tropicaux*, eutrophic brown soils, alfisols, entisols, eutric cambisols, eutric arenosols.

3.12.6   *Vertisols*

These are juvenile soils that have developed over calcium-rich rocks (lithomorphic vertisols) or in topographic depressions with impoundments of base-rich waters (topomorphic vertisols) where drainage deficiencies produce alternate periods of water pooling and desiccation.

Vertisols consist of fine materials, clayey loams or clayey fine sands (clay proportion from 30 to 80 per cent) and they contain no stone fragments. Within the clay fraction, the swelling-prone montmorillonite is dominant. During the rainy season, the vertisols develop into a sticky, compact and poorly aerated mass, while during the dry season, desiccation cracks and polygonal tiles tend to develop. Residues from weeds and herbs that fall into the cracks are worked into the soils during the ensuing wet season, so as to form a self-mulching A horizon with depths up to 1 m. The relatively high humus content varies between 0.5 and 4 per cent, and gives the soil a peculiar dark-grey colouration. Due to their high content of montmorillonites, vertisols possess high CEC.

In Sudan, where they are known as *bardobes* or *terres noires*, vertisols have developed over basic gneisses and over lacustrine clays in Chad. Their widest distribution occurs in Australia, where they are called *black earths*, in the Deccan Plateau *regur* or *black cotton soils*, and in insular Southeast Asia. Pullan (1969:535) remarks that 'great difficulties have been encountered in the utilization of these soils owing to their stickiness during the humid season and extreme hardness in the dry season. Shifting cultivators tend to ignore these soils, but mechanical plowing might hold some promise'.

3.12.7 *Andosols*

Andosols are added to this list because they are found in certain highland locations of East Africa, also in special areas of recent volcanism in South-east Asia and in the mountainous regions of the Andes where the geological substratum favours the development of these soils.

Andosols occur over fine cinders of high basic mineral content from Recent or sub-Recent volcanic eruptions that have affected the humid to subhumid tropical highlands.

Characteristic is their high content of allophanes, an inorganic colloid of great exchange capacity which has the same components as clay minerals but no crystalline structure, and a high percentage of humus (10–25 per cent). The basic character of the ashes from which they originate and the high content of humic matter are the reason for the dark-grey to black colour of their upper horizons. In fact, these soils are called *an do* (dark soils) in Japan. The content of residual minerals, the CEC and the soil moisture storage capacity are very high.

With increasing age the allophanes are transformed into kaolinites and halloysites. The resulting andosol derivates differ from normal ferrallitic soils in their comparatively high content of residual minerals.

4

The seeming contradiction between lush natural vegetation and poor agricultural crops on nutrient-poor soils of the humid tropics

In the previous chapter it was established that the spatially dominant soils of the humid tropics, the xanthic and orthic ferralsols (oxisols, ultisols, highly weathered ferrallitic soils), are extremely poor in mineral nutrients and, being kaolisols, possess a very low cation exchange capacity (CEC). Their content of organic matter is normal as long as the soil is provided with large amounts of organic fragments. Vast areas of those soils are covered by lush tropical rain forests with an extremely large net production of biomass (see Table 2.1). This contrasts with the poor agricultural crops and the need for shifting fields on the same soils. The reason for this discrepancy resides in the specific features of the natural forest formation which prevent the loss of nutrients and warrant a closed mineral cycling. As a proof one has to compare the chemistry of the running waters to that of the precipitation.

4.1   The stratification of nutrients in tropical soils

The functioning of the ecological system depends to a great extent on the localization of the available nutrient content and exchange capacity in the soil profile. In extratropical soils they are often concentrated in the deeper horizons, but microstratigraphic analyses of volcanic soils in the humid regions of Ethiopia conducted by van Baren (1961) show the results given in Table 4.1. The top 3 cm, which are rich in organic substances, contain the highest concentration of nutrient elements. The next layer down to 15 cm has higher amounts of P_2O_5, Ca, Mg, NH_4 and Mn than the horizons further below which show, conversely, an increase in the toxic Al, in Na and in sulphates. Thus, at the 15 cm level there is a sharp discontinuity.

The research by Sombroek (1966, 1984) on Amazonian soils revealed the dependency of CEC on organic matter and on clay content respectively that was discussed in Chapter 3 (see Figs 3.7 and 3.8). Both analyses demonstrate that most of the available nutrients as well as the CEC are concentrated in the uppermost layer (10–20 cm) of the soil profile and that both are linked to organic matter.

Table 4.1   Nutrient content in the topsoil and in deeper layers of tropical soils. Examples from the rhyolitic soils in the Gossi forest (Ethiopia)

Aspects investigated	Horizon							
	Profile 70				Profile 76			
	0–3	3–15	15–40	40–90	0–3	3–15	15–40	40–90
Granulometry (%)								
2000–200	12	8	8	4	5	4	4	3.0
200–20	11	15.5	12	18	9.5	15.5	8.5	19.5
20–2	27	41.5	29.5	28.5	24.5	20.5	23.5	23
2	50	35	50.5	39.5	61.0	60	64	54.5
Acidity								
pH-H_2O	4.6	4.9	4.8	4.6	5.8	5.6	5.3	4.9
pH-KCl	3.8	3.8	3.6	3.6	5.3	4.4	4.1	3.7
Carbon (%)	10.9	5.5	1.4	0.6	10.5	2.6	1.1	0.5
P_2O_5 ppm/soil	56	12	3	6	43	10	21	16
Nutrient elements ppm								
Calcium	510	300	80	20	1520	750	450	200
Magnesium	160	110	20	5	300	170	130	60
Potassium	100	8	10	10	70	20	10	10
Ammonium	77	18	10	15	65	55	13	17
Iron	5	9.8	10	6	0.4	1.8	2.7	1.9
Aluminium	58	141	161	210	4	20	61	87
Manganese	0	8.5	8.5	0	10	7	3.5	7.5
Sulphate	2	2	14	12	6	2	3	26
Sodium	10	9	13	12	8	8	8	10

Source: Van Baren (1961).

4.2   Nutrient losses from natural forest ecosystems. The chemistry of natural waters

The concentration of available nutrients in the uppermost layers of rain forest soils implies that, aside from the normal leaching by percolation, there is the danger of additional nutrient loss by superficial erosion due to the frequent intensive tropical rains. The nutrients washed out in both ways should be traceable in creeks and rivers. We must, then, turn our attention to the chemical characteristics of tropical rain forest rivers.

The first chemical analyses of Amazon waters were made in the late nineteenth century when Katzer (1897) found these waters to be of extraordinary purity. More substantial information can be gained from the comprehensive investigation programmes on the limnology and landscape ecology conducted since the mid-1940s in the Brazilian Amazon region by the members of the Instituto Nacional de Pesquisas

Table 4.2 Chemical composition of black, clear and white waters from different geological regions of Amazonia as compared to North-American rivers

	Black waters		Clear waters			White-water Middle Amazon		North American rivers
	Upper Rio Negro	Rio Negro (Manaus)	Creeks in Archaean and Tertiary lithological region	Creeks in Archaean region	Creeks in Tertiary 'Barreiras' formation			
	(2)	(1)	(2)	(3)	(3)	(2)	(4)	(5)
pH	4.2–4.3	4.8	4.0–4.6	4.0–6.6	4.2–5.5	6.5–7.3	7.4	
HCO_3	0		0	0–0.174	0.00–0.04	13.2–30.8	53.1	
Ca^{2+}		0.336	0.02–9.2	0–18.4	0–±5	5.45–9.9	14.5	1–54.0
Mg^{2+}		0.172	0–0.42	0–5.6	0–0.38	1.15–2.9	1.72	0–45.0
Na^+		0.719	0.25–0.54	0.245–2.060	0.847–2.530		8.35	1.9–79.0
K^+		0.435	0.14–1.00	0.143–1.00	0.534–1.52		2.00	0.4–26.0
$Fe^{2+}+Fe^{3+}$	0.19	0.431	0.11–0.25	0–0.250	0–0.143		3.05	0.001–0.95
Mn^{2+}	0	0.009	0	0–0.212	0–0.082		0.08	0.00–3.23
Al^{3+}	Trace	0.016	0.26–0.31	0–0.314	0–0.488		1.70	
Cl^-	0.5	2.20	0.01–1.12	0–2.5	0–3.5	2.0–5.0	2.5	1.0–118
SO_4^{2-}	0			0–2.69	0.0–0.480		10.0	S 6.0–12.0
$P(PO_4^{3-})$		0.009	0–0.03	0–0.11	0–0.052	0–0.03	0.92	0.002–5.040
$N(NO_3^-)$	0	0.036	0–0.02	0–0.15	0–0.200	0.02–0.08	0.007	
N(Kjeldahl.)	Trace	0.354	0.15–2.62	0.14–0.72	0.138–0.724	0.33–0.77	0.41	1.0–54.0
Si dissolved	1.2	2.44	0.8–2.7	±0.5–4.5	0.5–4.5	3.6–4.5	5.8	SiO_2 3.0–18.2

All values in ppm (parts per million) per litre (ppm l^{-1}), equivalent mg l^{-1}.
(1) Anon/Klinge (1972a); (2) Sioli (1975); (3) Sioli (1954); (4) Klinge and Ohle (1964); (5) Sanders (1972).

125

da Amazonia (INPA) in Manaus, Brazil, as well as by Sioli and his co-workers from the Max Planck Institute for Hydrobiology and Tropical Ecology in Kiel, Germany, who have published the hydrochemical data of a vast number of rivers.

Table 4.2 combines the data of the chemical composition of representative surface waters as collected by Klinge and Ohle (1964), Anon/Klinge (1972a) and Sioli (1954, 1975). For comparative purposes the corresponding values from North American rivers have been added. Depending on the different geomorphological and pedological conditions in the headwater areas, three types of running waters can be differentiated in the Amazon basin (Sioli, 1965, 1967):

1. White-water rivers with turbid yellowish water originate in mountainous reliefs where erosion furnishes the sediment load of the water; most come from the Andes or their foothills;
2. Clear-water rivers have their catchment areas in the flat terrains of the old crystalline massif of central Brazil or Guyana, Surinam and French Guiana, and in the Tertiary sediments of the *terra firme* plains with their characteristic brown loam soils;
3. Black-water rivers with transparent olive-brown to reddish-brown water (the colour of light tea) come from the flat regions covered with bleached sands of tropical lowland podzols; the colouring stems from dissolved or colloidal humic materials.

The first three columns of Table 4.2 represent the extreme chemical values of a vast array of water analyses from black-water creeks and rivers in the virgin rain forest region. They are characterized by a high acidity and by the lowest nutrient content anywhere in the world. Of decisive ecological consequences are their extreme poorness in calcium, phosphorous and sulphur compounds as well as their low content of fixed nitrogen.

The rivers are chemically as poor as the smaller creeks. The chemical deficiency and the dark colour of the water combine to reduce primary production of biomass to extremely low values. Therefore they have the general reputation for being 'rivers of hunger', also for the human population living on the banks, caused by the scarcity of fish and waterfowl (Sioli, 1965, 1967).

With clear and black waters occupying the greatest extension of the Amazon basin, Sioli characterized most of Amazonia as an area where the natural waters may best be compared with slightly contaminated distilled water (Sioli, 1965, 1967).

The Amazon itself is a white-water river. Although its ionic content is noticeably higher than that of the clear- and black-water rivers, it must still be classified as relatively nutrient-poor in comparison with extratropical rivers.

The data concerning the macronutrient content in the different water types are complemented by analyses of the distribution of micronutrients, specially the trace elements, as reported by Furch (1976, 1984). The results have been presented by the author in a very instructive manner (Fig. 4.1): to the diagram she added the means of metallic content of a series of rivers world-wide collected from the available literature as comparative values. The upper inset(b) includes the range of values of

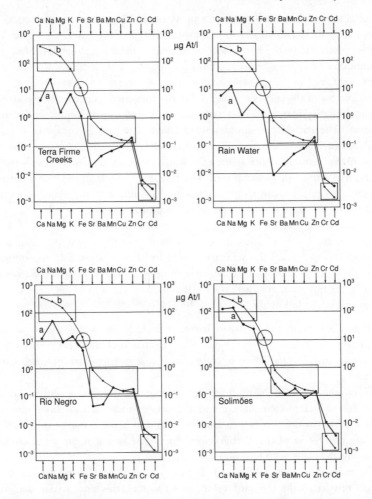

Fig. 4.1   Content of macro- and micronutrients of running waters from Rio Negro, Solimões, tierra firme creeks as compared to rain-water of the Amazon basin (after Furch, 1976).

the main cations Ca, Na, Mg, and K which are the most plentiful in running waters. By almost a three-tenths power lower is the group of the still relatively abundant trace elements such as strontium (Sr), barium (Ba), manganese (Mn), copper (Cu) and zinc (Zn). Another two-tenths power lower are the concentrations of the very rare trace elements group, such as chromium (Cr) and cadmium (Cd), molybdenum (Mo) and cobalt (Co). Iron (Fe) occupies an intermediate position between the main elements and the trace elements.

The data from three different types of Amazon runoff waters and from rain-water collected near Manaus are compared with the world's averages. The curve representing the *terra firme* creeks is actually the mean value of 10 clear- and black-

127

water creeks from the rain forest north of Manaus. The extreme sparseness of major nutrient elements shows clearly in a comparison with rain-water: the creeks contain even less calcium than the latter, the magnesium values are about identical, and the concentrations of sodium and potassium are higher. This indicates that the calcium element is retained in the rain-forest ecosystem.

The segments of the curves relating to the micronutrient concentration confirms the chemical similarity between rain-water and creek water in the undisturbed *terra firme* rain forest. The Rio Negro samples were taken at the lower course of the stream, shortly before it joins the Solimões River where the water contains almost four times as many main elements as the rain-water. Still, in comparison with the world averages, it is one-tenth power lower. The Rio Negro itself shows also a particularly low content of calcium and magnesium ions. In contrast to the major elements, the trace elements are quite similar to those of the global means.

As to the white water of the Rio Solimões, the distribution curves of the main elements and the most common trace elements mirror the world averages, but their concentration lags behind the standard values by almost one-third. As compared to the *terra firme* creeks, the white waters are roughly 10 times as rich in nutrients, but calcium and magnesium are also underrepresented.

In summary, the following points can be made:

1. All the waters in the Amazon basin, including the white waters of the Rio Solimões, lie below the world means in the macro-element contents such as Ca, Na, Mg and K. The deficit in trace elements is restricted to the small *terra firme* creeks.

2. Within the Amazon basin there are ecologically important differences in the content of macro-elements among the white water of the Rio Solimões, the black water of the Rio Negro and the clear and black water of the small creeks in the rain forest of the Tertiary *terra firme*. The water of the small creeks is so nutrient deficient as to resemble rain-water.

3. The earth-alkali ions, magnesium and especially calcium, show the largest deficit with respect to the standard values in all waters; this goes so far that in small creeks the calcium content is even less than that of rain-water.

The similarity in macronutrient content between creek waters and clear and black-water creeks and rivers, on one side, and rain-water on the other, is confirmed by the investigations of Ungemach, who analysed a total of 43 occurrences of rainfall during the rainy seasons of 1965/66 and 1967/68 near Manaus. His results were published by Klinge (Anon/Klinge, 1972a, 1972b) (see Table 4.3). Clear- and black-water creeks are evidently poorer in nutrient content than rain-water, especially in the case of primary elements such as Ca and Mg. Among the rivers themselves, the relationship between running water and rain-water changes depending on the different elements.

Reichholf (1990: 95–103) describes the notable forms of adaptation which have been displayed by Amazonian frog species in order to avoid fatal losses of ions through osmosis during long periods in this nutrient-poor water.

Thus, according to Klinge, the amazing fact emerges that rainfall can serve as a

Table 4.3 Nutrient content of rain-water and running waters near Manaus in the central Amazon basin (from Anon/Klinge, 1972a)

Nutrients (ppm l^{-1})	Precipitation water		Clear water			Black water			White water	
			Creek[1]	Stream[2]		Creek[3]	Stream[4]		Stream[5]	
	Dry*	Rainy*	Dry	Dry	Rainy	Dry	Dry	Rainy	Dry	Rainy
Total P	16	11	3	21.4	8.2	4	10	7.6	31	78.6
Total N	492	413	180	450	258	340	421	366	432	595
NH_4-N	145	169	—	—	—	—	46	35	—	—
NO_3-N	52	110	4	3	8	7	40	31	40	43
Organic N	242	118	—	—	—	—	342	304	—	—
Kjeldahl-N	438	302	170	430	248	330	372	336	390	550
Ca^{2+}	286	140	$\ll$20	1250	800	$\ll$20	335	316	5790	9880
Mg^{2+}	190	122	$\ll$10	320	270	$\ll$10	186	157	1130	1730
Na^+	—	—	190	330	—	1050	—	—	—	—
K^+	—	—	170	410	—	790	—	—	—	—

* Dry and rainy refer to season.
(1) Ig. da Enchente (Schmidt, 1972a); (2) R. Tapajós after Schmidt (unpublished); (3) Ig. Tarumãzinho (Schmidt, 1972a); (4) Rio Negro (Anon/Klinge, 1972a, 1972b); (5) Rio Solimões after Schmidt (1972b).

fertilizer of rivers, provided they have a sufficiently large open collecting area. Small creeks are usually hidden under the rain-forest canopy allowing the trees, rather than the water, to benefit from the nutrients contained in the rain.

4.3 The closed nutrient cycle in tropical rain forests

The circumstances that (1) all autochthonous streams of the Amazon tropical lowland rain forest show a marked deficiency in macronutrients, and that (2) the waters of small creeks in virgin forest areas contain even less macronutrients than rain-water clearly demonstrate that there is no net loss of macronutrients in the undisturbed ecosystem. In the case of elements such as calcium and magnesium, the input from the rain-water must be partially stored in the ecosystem.

Therefore, the ecosystem of tropical lowland rain forests over the nutrient-poor tropical ferrallitic soils of the Amazon basin is indeed a tightly closed ecological system, just as postulated by Richards (1952) in his famous work *The Tropical Rain Forest* and confirmed by the recent studies of Brinkmann (1985).

The question that arises now is: how is the prevention of nutrient losses in the rain forest ecosystem organized?

Fittkau and Klinge (1973) explain that the part of the forest above ground works like a filter system in that the nutrients supplied by the rain are used several times before they arrive at the soil surface. Most of the biomass is concentrated in the two upper storeys of the rain forest, which is also the domain of the climbers and flyers. Those live directly or indirectly on the organic matter of the trees (leaves, sprouts, fruits) and their faeces and urine are washed downwards with the rain-water. On its way down, the enriched water is used by epiphyte plants (algae, mosses, fungi) and collected by the funnel-shaped macro-epiphytes, such as Bromeliaceae, orchids and Araliaceae. Certain animals, in turn, feed on these plants. In other instances the water trickling its way along branches and trunks comes into contact with lichen covering the bark (Reichholf 1990: 152–64).

In this manner, small subcycles of nutrients are incorporated into the general nutrient circulation. Only with the decomposition of the different members of the subcycles does a certain portion of the nutritional elements finally reach the soil surface. Here, a substantial nutrient loss is essentially prevented by another agent known as the mycorrhizas.

4.4 Ensuring the nutrient cycle of the tropical rain forests – mycorrhizas

The term 'mycorrhiza' is used to describe a mutualistic association between the mycelium of certain fungi and plant roots. The mycorrhizal association is beneficial to the host plant and, under certain circumstances, is even found to be essential for its growth.

From the morphology and nutrition mode of the hosts two types of fungus–root associations have been defined: the 'endomycorrhiza' and the 'ectomycorrhiza'.

In the case of endomycorrhizas the fungal partner grows mainly inside the root cortex and its hyphae enter the cells of the host root. For non-saprophytic plants the most important of the three groups of endomycorrhizas (beside the ericoid and orchidaceous) is the group of 'vesicular–arbuscular (VA) mycorrhiza (VAM)'. 'VAM infections are seen as vesicles and arbuscules borne on the mainly aseptate hyphae occupying the root cortex. . . . The association is so well balanced that young feeder roots show no sign of damage even when densely infected' (Hayman, 1980: 487).

Ectomycorrhizas (EM), in contrast, are associations 'in which the fungus forms a sheath around the fine absorbing rootlets. Hyphae penetrate between the root cells and occasionally enter the cells, but they never penetrate beyond the cortex and any intracellular hyphae and do not cause destruction of the host cell' (Redhead, 1982: 253). Ectomycorrhizas have also been termed 'sheathing mycorrhiza'.

In contrast to VAM, ectomycorrhizas are not common in the tropics. Where EM do occur, however, they are abundant. Ectomycorrhizas are regular features of the Caesalpiniaceae and Dipterocarpaceae families and dominate the *miombo* woodland of Africa and the tropical rain forests of South-east Asia. Elsewhere in the tropical lowland rain forests of Africa and South America they are restricted to certain loci, such as those with extreme scarcity of mineral nutrients in sandy soils (white sand soils in the *campinas*). They also occur in the periodically black-water inundated 'Igapó' of the Brazilian Amazon basin, where during the inundation period, nutrient loss by leaching and export of the leaf-fall is extremely high. Under these circumstances forest development is only feasible through the presence of EM (Singer and Araujo Aguiar, 1986). The reason is that EM fungi efficiently assist in exploiting the litter layer and recycle nutrients from it. Although the fungi involved cannot break down complex organic molecules such as cellulose in the litter, once this has been done by other organisms, the mycorrhizal hyphae can absorb the available solutes more quickly than most other microflora even to the extent of starving such decomposing organisms. This rapid absorption is especially important where nutrient cycling and leaching take place very fast (Redhead, 1982).

Vesicular-arbuscular mycorrhizas are almost universal in the host range. Most tree species in tropical rain forests form VAM; the absence of mycorrhizas in tropical forests is exceptional (Janos, 1983).

4.5 The mutualistic relationship between mycorrhiza and rain-forest trees

The mutualistic relationship between host plant and mycorrhiza was first demonstrated by tracer techniques. In their pioneering work Melin and Nilsson (1954, 1955) demonstrated the transfer of radioactive phosphorus and calcium to the rootlets of pine seedlings from a distant source by means of mycorrhizal hyphae. From the roots the isotopes were distributed throughout the seedling. The reciprocal exchange, the transport of assimilating compounds from the mycotrophic plant to its fungal associate, was proved by the same Swedish authors in 1957.

The results of the Swedish botanists have been confirmed in similar studies subsequently conducted by several investigators, and other comparative experiments have demonstrated that VAM does indeed greatly increase the utilization of nutrients present in the soil. Experiments in laboratories showed, primarily, a very efficient phosphate response, but increased efficiency of potassium, sulphate and zinc uptake by mycorrhizal plants has also been confirmed experimentally. Differences have sometimes been revealed between mycorrhizal and non-mycorrhizal plants in the concentration of nitrogen, sodium, magnesium, iron, manganese, copper, boron and aluminium (Bowen, 1980).

Experiments under natural conditions within the communities of the tropical rain forest itself are much harder to conduct and their qualitative and/or quantitative results are more equivocal. However, there are studies which confirm the laboratory results. In a Puerto Rican tropical rain forest Odum and his co-workers (1970) found that radioactive phosphorus released from decaying leaves was captured in the top 5 cm of soil and subsequently absorbed by plants; potassium was recycled very quickly, while calcium, magnesium, manganese, iron and copper were cycled more slowly. In the Rio Negro region of the Amazon basin, Herrera *et al.* (1978) showed that radioactive phosphorus was present in the living root and in the mycorrhizal fungus hyphae that connected it to a dead leaf labelled with the isotope. Stark and Jordan (1978) demonstrated that a root mat of Amazonian rain forest retained virtually all of the radioactive calcium and phosphorus solution they had applied.

Thus, experiments conducted in laboratories and nature verify the hypothesis that mycorrhizas directly facilitate nutrient cycling in tropical rain-forest ecosystems. It is generally accepted that mycorrhizas act as living 'nutrient traps' in this process.

The mechanism is explained by the intense root–soil contact and the ability to translocate nutrients to the root cortex. In case of VAM infection the hyphae penetrate the epidermic cells of the roots. Outside the mycorrhizal roots a loose network of external hyphae, continuous with the internal mycelium, invades the interstices between the soil particles. Thus, the network of mycorrhizal hyphae explores the volume of the soil more intensely than the roots especially in case of coarse, hairless roots.

In this way, the nutritional elements that have trickled down can be trapped more efficiently and the nutrient losses from the system minimized. Moreover, the external hyphae establish a closer physical contact with the surface of particles in the soil from which nutritional ions dissociate. This action is particularly important for fairly immobile ions such as phosphate.

Another aspect of the mycorrhizas' action on the nutrient cycle was formulated by Went and Stark (1968a) as 'direct mineral cycling theory'. They had observed that in the Amazon rain forest most feeder roots of higher plants do not even reach the actual soil. Instead, these roots are layered above the soil within the litter and inside the decomposing organic matter not yet mineralized. The mycorrhiza-infected root tips are attached to the decomposing matter.

Direct nutrient cycling is presented by several general texts as a property of tropical rain forests. However, the validity of this thesis has been questioned by Janos (1983) who states that direct mineral cycling in a strict sense requires decomposition of organic material by mycorrhizal fungi. This can be demonstrated only by proving the enzymatic capability of VAM or EM fungi to mineralize litter. Available data suggest that the majority of EM fungi seem to have little or no ability to break down complex carbon compounds, such as cellulose and lignin; however, some species do. Evidence is scant, though, about the decomposition of these compounds by EM fungi when associated with a host (Janos, 1983). VAM fungi are less likely to decompose litter than EM fungi although some authors imply that VAM fungi have this ability. Thus, Janos (1983) concluded that direct nutrient cycling, in its strict sense, is probably restricted to some EM and to specific ecological conditions, but that it is not a general tropical phenomenon.

However, 'direct nutrient cycling' must not be confused with 'closed mineral nutrient cycling', since the latter does not require a decomposing activity of mycorrhizal fungi. It is not decisive whether it is the mycorrhizal fungi themselves that have the ability to decompose organic matter. What is significant is that the mycorrhizas achieve an almost complete closure of tropical mineral nutrient cycles, especially for some extremely scarce elements, such as phosphorus (Redhead, 1980). Access to poorly soluble forms of phosphate is of major relevance in tropical soils, especially where phosphorous fixation occurs.

From field-work observations and experiments conducted, the following key points can be underscored (Janos, 1983):

1. Most tropical rain-forest trees live from the mutualistic relationship between fungi and their feeding rootlets. The decisive benefit which the host plant derives from the mutualism is the increased rate of nutrient uptake from the soil. In return, the fungi are provided by the host with the required photosynthates. The balanced mycorrhizal symbiosis works for as long as there is a continuous supply of essential metabolites in the reciprocal exchange.
2. Mycorrhizas are important for the mineral nutrition, growth and survival of lowland rain forests. Most mature forest trees are mycotrophic.
3. Mycorrhizas especially improve the uptake of immobile nutrients such as phosphorus.
4. Mycorrhizas almost completely close tropical mineral cycles by efficient uptake of mineral nutrients from the soil and by their ability to translocate them to the feeding roots of their host.

4.6 Structural adaptations to the closed mineral cycle

The visible effects of the closed nutrient cycle in the biomass of tropical rain forests are manifested by two characteristic properties of the root systems of the trees: (1) the concentration of most of the roots in the superficial soil zone, and (2) the phenomenon of the buttressed trunks.

Klinge (1973a) summarized the existing data on fine root masses and their vertical distribution in the soils of tropical forests (Table 4.4). Despite some differences in total mass of fine roots there is a strong similarity between the different types of forests regarding the percentage of root mass at different soil depths. All the samples contain more than 45 per cent of the total mass measured in the upper 20 cm of the soil, 6 out of 11 samples even show between 74 and 97 per cent. In 8 of 11 samples 97–100 per cent of the root mass is found in the upper 50 cm of the soil.

A concentration of the bulk of feeding roots in the uppermost 20 or 30 cm of the soil represents the best adaptation to extremely low mineral content since it optimizes the uptake of nutrients from organic material decomposing on the soil surface. At the same time, this type of root system is advantageous for the intake of interception water and the stemflow (Fittkau and Klinge, 1973).

The buttressed trunks which are frequently found in tall rain-forest trees can be regarded as stabilizers against wind stress. In these trees without deep root anchoring, the buttresses at the base of the trunk strengthen the transition between trunk and shallow root mat (Reichholf 1990: 156).

Table 4.4 Root mass of forests in tropical Africa and in Amazonia (Klinge, 1973a)

Locality	Vegetation and soils	Depth of sampling (cm)	Root mass						
			Total (t ha⁻¹)	In upper 20 cm (t ha⁻¹)	(%)	In upper 50 cm (t ha⁻¹)	(%)	< 2 mm diam. (t ha⁻¹)	(%)
Kade, Ghana (Lawson et al., 1970)	Humid semi-deciduous forest								
	Bekwai series, upper slope	50	80	70.5	88	80	100	40	50
	Nzima series, middle slope	50	204	197	97	204	100	103.6	51
	Temang series,	50	79	50	63	79	100	40.3	51
Kade, Ghana (Greenland and Kowal, 1960)	Celtis-Friplochiton association of Bekwai Nzima-Kokofu-Odacatena								
	about 50 years old	122	25	18.5	74	23.8	95	5	20
	about 15 years old	122	18.6	14	75	16.5	89	—	—
Yangambi, Congo (Bartholomew after Greenland and Kowal, 1960)									
	2-year-old fallow	61	6.9	6.6	96	6.9	100	—	—
	5-year-old fallow	152	25.8	—	—	24.1	93	—	—
	8-year-old fallow	152	22.6	15.2	67	19.2	85	—	—
	18-year-old fallow	?	31.4	—	—	—	—	—	—
Banco, Ivory Coast (Huttel, 1969)	Tropical rain forest								
	plateau	250	24	10.9	45	18.4	77	6.8	28
	valley	130	21.6	14	65	20.5	95	5.4	25
Amazonia (Klinge, 1973a)	*Terra firme* forest on latosol	107	39.7	22.4	57	28.6	72	8.4	21
	Campina on humus podzol	89	28.1	26.1	93	27.2	97	5.6	20

4.7 The human impact on the mineral cycle

It is obviously quite risky to clear a forest that is growing on soils of which only a few centimetres of the top layer contain the nutrient elements in adequate quantities, and in which grave losses of essential nutrients can only be prevented by mycorrhizas functioning as nutrient traps.

The apparently lush tropical rain forest is an ecosystem as vulnerable as a haemophiliac. Clearing the forest by burning the forest biomass rips wide open the normally tightly closed nutrient cycle at a decisive point. The humic material on the soil surface is destroyed directly and the mycorrhizas die off gradually because, with the removal of the trees, they have lost their mutualistic partners.

Through the destruction of the superficial biomass, the input of organic matter into the soil ceases and, as a consequence of the high mineralization rate, the humus content of the lower horizons decreases rapidly. Thus the burning of the biomass destroys the most important part of the soil's CEC. With the mycorrhizas dying off, the nutrient traps disappear; erosion and leaching of nutrients can now proceed unchecked.

The short life of cultivated plants will not allow regeneration of the mycorrhizas and thus halt the nutrient loss. Only in the case of the long-living cassava is there time for the mutualistic association with mycorrhizas to be re-established. However, due not so much to nutrient uptake by the crops but mainly to erosion and leaching, impoverishment of the nutrient system cannot be averted. The situation can be corrected only through a return to quasi-natural conditions over a long period of time, i.e. by allowing the secondary vegetation to build up a new nutrient potential.

The above argument should shed light on the seeming contradiction between lush natural vegetation and poor agricultural crops and it should have made clear that rotational crop–fallow land use on tropical kaolisols is an ecological adaptation to sustained natural conditions rather than a persistent agro-economic anachronism.

The deduction of the difficult ecological conditions for agricultural development in most parts of the humid tropics can be further substantiated by a comparison with the different natural conditions in the deciduous forests and steppes of the temperate extratropics and the agro-historical development in these regions.

4.8 The contrasting conditions in temperate regions

A thousand years ago, in the temperate regions of Europe, the transition took place from an unregulated field-fallow economy to a regulated *Dreifelderwirtschaft* (three-field tending system), which in turn became the basis for subsequent agricultural innovations.

A three-field tending system works as a continuous rotation of two cereal crops on three permanently open fields within stable boundaries in the following manner: after clearance and preparation of the soil through ploughing and dragging, the first field is

135

sown with winter seeds (wheat or rye) in the autumn. Following the summer harvest this field is left fallow until the next spring when summer grains (barley or oats) are cultivated. (Meanwhile the second field has been sown with winter seeds.) After two crops the first field remains fallow all through the third year (when wheat or rye is grown on the third field), but it serves as livestock pasture. In the fourth year the cycle is renewed on the first field.

This procedure ensures that two winters and an additional full year of fallow suffice to maintain continuous crop rotation on permanently open fields even without manuring (except for the incidental droppings of the grazing animals during the fallow periods).

The reason for the success of the three-field system rests, during the initial stages of occupancy, on the way the nutrient reserves are distributed in the ecosystem of extratropical deciduous forests and steppes. In contrast with the tropical forests, in which most of the nutrients are stored in the biomass, in the temperate regions more than half of the organic carbon, some 90 per cent of the nitrogen and the most important minerals are contained in the soil. There they remain intact, even when the forest is cleared and crops can draw from these reserves for quite a while. In addition, the extratropical soils recover very readily, enabling them to replace the nutrients used up by the crops. This ability relates to the following facts: (1) soils in the temperate regions are usually much younger than those in the tropics, (2) hydrolitic weathering is weaker and (3) the soils contain, therefore, a considerable amount of rock debris with unaltered primary minerals which continue to release elements into the soil solution through chemical weathering. Moreover, nutrient elements in the temperate soil solution are more efficiently protected against leaching than in tropical soils because of the higher CEC which is guaranteed by the three-layer clay minerals and the slowly mineralizing humic substances in the extratropical soils. While the CEC of the tropical ferrallitic soils is normally less than 4 meq per 100 g fine soil, in European brown forest soils it reaches 20–30 meq per 100 g in the uppermost 30 cm, and in the black earths (chernozems) of the steppes more than 40 meq per 100 g.

All these facts considered, the soil conditions in the extratropical woodlands facilitate the innovation of a simple, though stable and continuous cropping system on former forest soils. For the further intensification of the agricultural systems, in its early stage by means of manuring and later on of fertilizing, the naturally high CEC of the soils was a decisive advantage.

5

The ecologically exceptional regions of the humid tropics

So far we have considered the dominant regions of the humid tropics where potential agrarian utilization is hampered by ecological conditions. These basically are certain soil conditions, such as scarcity of weatherable primary minerals, high mineralization rates of organic matter, and dominance of kaolinites and gibbsite in the clay fraction of the soils, which cause extreme nutrient deficiency and a very low cation exchange capacity (CEC) in the mineral soil, coupled with high acidity and a great aluminium toxicity.

As primary causes were recognized, first, the intensive hydrolytic weathering, second, the old age of the soils, and third the dominance of acidic rocks in the parent material.

Areas that can be considered as ecologically exceptional in the humid tropics are those where these disadvantages do not exist and, where the preconditions for agricultural use are, accordingly, more readily given. Following our trend of reasoning, these areas should stand out through their exceptionally high agrarian population density.

Considering the previously summarized soil properties and their causes it should not be difficult to determine the conditions for the exceptions from the rule. In the ever-humid inner tropics, ecologically exceptional regions are to be expected:

1. Where the parent rocks are particularly rich in bases;
2. Where rock debris with nutrient-rich weatherable primary materials is periodically or episodically added to the soil;
3. Where the efficiency of hydrolysis and the mineralization of organic matter is much less than in normal lowland locations;
4. Where the soils are relatively young.

These conditions are not mutually exclusive, some are even coupled with each other so that especially favourable soil properties may result.

5.1   Base-rich and base-poor rock groups

The beneficial effects of exceptionally base-rich parent rocks are locally or regionally restricted. Table 3.2 quantifies the differences between base-rich and base-poor groups of rocks. Basic rocks such as gabbros, basalts and diorites contain between 15 and 19 weight percentage of manganese (Mn), magnesium (Mg) and calcium (Ca) oxides, which is 7–10 times more than in the particularly base-poor granites which have only 2 per cent alkali compounds. The richness in bases comes chiefly from calcium, which later on plays an important ecological role in the weathering process, in soil formation and plant growth.

The base-rich rocks also contain much iron oxide. The high content of bases and iron is compensated by a relatively low content of silica (SiO_2), an element which is so abundant that differences have no impact on plant growth.

Not only the chemical composition but also the abundance of the different rock types is of relevance for the ecological assessment of the different regions of the tropics. In a general view, the rocks of the earth's crust may be classified into three major groups: igneous, sedimentary and metamorphic rocks. Igneous rocks were formed by consolidation of magma which was injected into the upper earth's crust (intrusive igneous rocks = plutonites), or ejected upon it (effusive or extrusive igneous rocks = volcanites).

Sedimentary rocks consist of fragments of rock debris (clastic origin such as breccia, conglomerate, sandstone or shale) commonly cemented by a matrix, or of precipitated and lithified chemical substances (chemical origin such as limestone, dolomite, rock salt, gypsum or bituminous coal).

Metamorphic rocks are formed by the physical and chemical alteration and transformation of any pre-existing rock under conditions of extreme heat and pressure. The process of metamorphism transforms granite into gneiss, shale or basalt into schist, shale into slate, limestone or dolomite into marble, and sandstone into quartzite. While the chemical composition remains about the same, it is, above all, the crystalline structure that is affected by the changes.

Plutonites, their metamorphic transformations (gneiss), and schists form the ancient crystalline shields and the basement of the continental crust. They are partially covered by a veneer of younger sedimentary or volcanic rocks.

Figure 5.1 offers a generalized regional survey of the extension of crystalline socles (massifs) and the sedimentary or volcanic covers in the tropics. Almost everywhere in tropical Africa and frequently in South America, the parent rocks on the surface consist of crystallines from the basement. A tentative standard section of the intrusive rocks of the upper continental crust (Table 5.1) suggests that 85 per cent of these parent rocks consist of granites, granodiorites and quartz diorites, all of them base-poor acidic rocks. Only 15 per cent are base-rich plutonites. Comparing areas from different parts of the world consisting of specific intrusive rock types in crystalline shields and basements, Wedepohl (1969: 244) states: 'It is surprising that all the values agree within a small range in the abundance of granitic rocks: 85–88 per cent

138

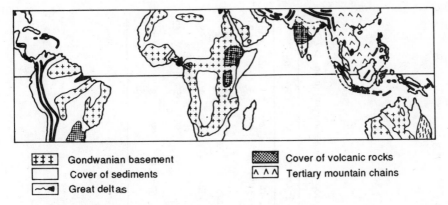

Fig. 5.1 Overview of the distribution of crystalline rocks and the sedimentary or volcanic covers in the tropics (after Benchetrit *et al.*, 1971).

Table 5.1 A tentative standard section of the upper continental earth's crust (intrusive rocks) (vol. %) (after Wedepohl, 1969)

Granites quartz monzonites	44
Granodiorites	34
Quartz diorites	8
Diorites	1
Gabbros	13
Peridotites	<0.5
Syenites, alkalic rocks	<0.5
Anorchosites	<0.5

(granites, granodiorites, quartz monzonites, and quartz diorites) and of gabbroic rocks: 7–13 per cent.'

It can therefore be assumed that the soils over crystalline socles in tropical Africa and South America have developed predominantly from acidic parent rocks. A well-known example of a locally restricted exception to the rule is the area around Altamira in the eastern segment of the Amazonia Colonization Project along the Amazon Highway, where favourable soil conditions prevail due to base-rich parent rocks (see section 9.2). Characteristic chemical properties of these soils on base-rich plutonic parent rock are shown in No. 16 of the soil list in Appendix 2C.

In the sedimentary zone of tropical lowlands, the proportions of base-poor and base-rich material are about the same as in the basement rocks. The zone contains terrestrial deposits in the form of sandstones which, by nature, are poor in bases. Where deposits are of lacustrine or marine origin, there is a noticeably higher base content in the weathering cover. For the Amazon basin this difference can be quantified with the help of data from Irion (1976) (Table 5.2). The formations of Alter do Chão (IV) and of the Estado do Acre (V) are both of Tertiary age, though

139

Table 5.2 Average values of the chemical composition of fractions smaller than 0.002 mm, in Amazonic top soils (data in ppm) (Irion, 1976)

	Na	K	Ca	Mg	Fe	Mn	Cu	Zn	Cr	Li	Sr	Co	Pb	V
I. Soils of the Andes	2 680	19 080	7 360	30 120	50 700	630	102	147	76	93	125	15	23	131
II. Soils on the Guyana shield	600	700	700	280	66 000	84	44	50	24	8	90	3	36	24
III. Soils on palaeozoic rocks	1 125	750	375	290	44 800	45	27	38	81	8	135	4	35	112
IV. Soils on the Alter do Chão formation	160	225	350	100	27 600	33	37	27	122	4	75	1.5	18	97
V. Soils in the territory of the state of Acre	1 630	15 100	1 300	5 000	55 800	330	79	126	77	53	95	5	42	162
VI. Soils on the Pleistocene *varzea*	1 650	15 200	940	5 600	51 000	98	54	115	67	51	80	8	39	139

the former is of terrestrial and the latter of lacustrine–marine origin. Accordingly, in the state of Acre, in the upper courses of the Rio Madeira, Rio Purus and Rio Jurua, the amounts of important basic nutrients are 4–10 times larger than in the adjacent regions of Alter do Chão in the central part of the Amazon basin. The Territorio do Acre, then, as the present focus of the Brazilian colonization thrust, is a relatively favourable region by reason of the chemical nature of its parent rocks. Despite the relatively good values of the Estado do Acre, Table 5.2 clearly shows a great gap in chemical conditions between these soils and those in areas beyond the lowlands within the domain of the Andes.

Among the extrusive igneous rocks (volcanites) there are base-poor rock types, such as quartz-porphyry and porphyry or trachyte and syenite, as well as base-rich rock types such as diabase and basalt. In contrast to plutonic rocks, however, the base-rich rocks are the most abundant among the volcanites and occupy by far the largest area. As basalt in all its varieties is the absolutely dominant volcanic rock, the volcanic covers, indicated on the sketch map (Fig. 5.1), can be assumed to be predominantly basaltic. They belong to the most important ecologically exceptional regions of the tropics.

5.2 Exceptional realms on basaltic parent rocks

Aside from the abundance of bases, there are two additional reasons for the ecologically favourable conditions in areas with basaltic rocks:

1. Basalt effusions occur, for tectogenetic reasons, mostly in elevated parts of the earth's crust where hydrolytic weathering is less vigorous;
2. The soils are generally younger than in the crystalline shields and basements.

It is a concurrence of all these reasons that makes the pedological conditions in areas of basaltic rocks in the humid tropics so outstanding (see Nos 15 and 16, Appendix 2C). An example of such an ecologically exceptional realm in the permanently humid inner tropical zone of Africa is the area around the Cameroon Mountains. In the semihumid part of East Africa (where the dry season lasts less than 4 months) the most striking examples are the states of Rwanda and Burundi along the central African trench, which form an island of exceptionally high population density, as mentioned in Chapter 1.

It is challenging to make a comparison between regions of different natural resources and occupation intensities on the Indonesian islands. The main island of Java (including Madura) is the most densely populated region of this size in the tropics: an area over 132 000 km^2 has an average population density of more than 500 people per square kilometre. Numerous districts support 800–1000 people per square kilometre on an agrarian basis. The situation in Bali is similar.

Even under conditions of elevated cultural standards, extreme population densities and pressing land demands, there are apparent regional congruencies between population density patterns and volcanic (or non-volcanic) environments, as seen on Pelzer's map of the Jakarta region (Fig. 5.2). 'The soils of the Gunung Sewu limestone region have a much lower carrying capacity than the rich soils derived from volcanic ejecta of Mt Merapi' (Pelzer, 1964: 18).

The map shows that the crest of Mt Merapi lies at an elevation of almost 3000 m and that the district of high population density extends up the flanks of the volcano. The corresponding land utilization map in the *World Atlas of Agriculture* (Committee for the World Atlas, 1973) shows the Merapi region, as well as central and east Java in general, as having a continuous agrarian land occupation from coast to

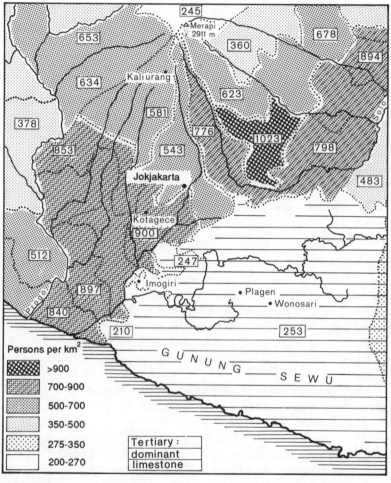

Fig. 5.2 Population density patterns in volcanic and non-volcanic environments in Java (after Pelzer, 1964).

coast, interrupted by only a few forested areas. This is due to the fact that the numerous volcanoes, instead of being imposed on top of a mountain chain, arise from the lowlands, under 500 m, as single cones. (More than half of the forested area, which comprises only about 23 per cent of the island, is found in south-west Java, in the lowlands, in the foothills and at the southern flank of a volcano-free mountain chain.)

As a matter of fact, extended parts of Java rest on a volcanic underground. The island is topped by over 100 young, active volcanoes with basic extrusions and effusions (see analysis of samples in Tables 5.3 and 5.4) whose soils are extremely young, base-rich, and of a high CEC. Since run-off eroded materials from the weathered rocks and the base-rich mountain soils, the alluvial deposits contain a high percentage of volcanic rock debris. The same is true of Bali and, to a lesser degree, of Lombok.

Things look quite different in the province of Kalimantan on the island of Borneo, only some 300–400 km away from Java, right across the Java Sea. This part of Indonesia (539 000 km^2) has at present an average population density of no more than eight people per square kilometre. A sharper contrast and greater inequality can hardly be imagined between two regions of the same political and economic unit. It could be argued that the southern rim of Borneo is taken up by vast tropical swamps. However, this southern rim is precisely the natural region with the highest population density (40 people per square kilometre) of Kalimantan. The well-drained interior tableland and the highlands of central and eastern Kalimantan are as sparsely populated as the Amazon jungle in South America. In fact, it is a jungle (in large areas virgin rain forest) inhabited only by widely scattered tribes of different ethnic groups who practise shifting cultivation.

The demographic contrast between Java and Kalimantan is equivalent to the geological–petrological contrast between both. The island of Borneo has almost no volcanism. The prevailing rocks consist mainly of sandstones or siltstones, which 'themselves are inherently impoverished having been formed from sediments that have undergone previous cycles of weathering and deposition' (Transmigration Area Development Project (TAD), 1982: 36). The soil survey (Tables 5.5 and 5.6) revealed that most of the rolling foothills, as well as of the sloping mountain terrain of the interior, are covered by base-poor, very acid orthic acrisols over non–calcareous clastic sediments, and metamorphic rocks. In the USDA Soil Taxonomy these soils are classified as ultisols. The Transmigration Feasibility Study for east Kalimantan (TAD, 1982) presents a range of mean values for some important soil chemical properties in the Muara Wahau region, shown in Tables 5.5 and 5.6.

In comparison with the corresponding values of Java the data show the strong acidity, the high levels of exchangeable aluminium, the low values of exchangeable bases and the low CEC values of Kalimantan soils.

Sumatra (473 000 km^2) has an average population density of around 40 people per square kilometre. But there are considerable differences between the various regional units. The most densely populated districts lie in the hinterland of Padang (over 125 people per square kilometre), in the Batakland around the Toba Sea in the hinterland

143

Table 5.3 Analyses of ash from Kloet and Merapi volcanoes and of rock and weathering products (after Mohr, 1938/44)

	Kloet	Kloet	Kloet	Kloet	Merapi	Rock kernel	Weathering crust
Time of the eruption	23 May 1901	23 May 1901	20 May 1919	20 May 1919	(M. Java) 1896(?)		
Place of collection	Madioen	Pekal-longan	Gambar uppermost layer	Malang			
Distance from the point of eruption (km)	95	310	10	36			
Constituents in % of the preparations							
SiO_2	56.4	66.70	54.62	54.98	56.68	52.89	15.43
Al_2O_3	18.5	—	21.20	21.58	—	18.93	38.57
Fe_2O_3	—	12.82	—	—	26.32	4.06	14.82
FeO	7.5	—	9.20	8.64	—	4.80	1.89
MnO	0.4	Not detected	0.37	0.32	0.23	0.43	0.22
MgO	3.7	2.90	4.49	3.36	1.82	3.72	0.41
CaO	10.3	10.80	9.20	9.30	7.62	8.56	0.19
Na_2O	1.8	4.24	Not detected	Not detected	6.14	3.98	0.78
K_2O	0.7	0.71	Not detected	Not detected	2.09	0.94	0.08
H_2O+	0.4	—	0.40	0.50	Not detected	—	—
H_2O-	0.2	0.42	0.24	0.25	Not detected	—	—
TiO_2	Not detected	Not detected	Trace	Trace	Not detected	0.96	1.82
P_2O_5	0.5	0.07	0.06	0.06	0.27	0.22	0.35
SO_3	0.4	0.66	0.22	0.26	Trace	—	—
Cl	Not detected	0.05	Trace	Trace	Trace	Trace	Trace
S	Not detected	Not detected	Not detected	Not detected	Not detected	0.04	—
	100.8	99.57	—	—	101.17	101.05	99.36

Table 5.4 Chemical composition (%) of different rock types in Java and Sumatra (after Mohr 1938/1944)

Rock type Origin	Basalts Java	Andesites Batakland Sumatra	Biotite dacite Batakland Sumatra	Liparit Southern Sumatra
SiO_2	49.45–59.13	53.94–62.23	66.71	72.52–75.48
Al_2O_3	14.76–19.59	16.36–18.93	15.82	12.94–15.22
Fe_2O_3	1.56– 6.10	1.27– 4.20	0.71	0.56– 0.99
FeO	4.72– 7.48	2.18– 5.08	0.32	0.18– 0.34
MgO	1.54– 6.83	2.02– 3.81	2.05	0.05– 0.34
CaO	4.40–10.00	5.75– 8.27	3.92	0.49– 1.09
Na_2O	2.41– 4.31	2.60– 3.28	7.12	3.09– 4.10
K_2O	0.44– 3.62	1.07– 2.56	2.42	3.49– 5.14
H_2O+	0.11– 1.62	0.92– 2.47	1.01	1.03– 3.25
H_2O-	0.06– 0.91	0.16– 0.81		0.37– 1.14
TiO_2	0.86– 1.47	0.62– 1.06		0.09– 0.19
P_2O_4	0.14– 1.59	Trace– 0.14	(?)	0.01– 0.12

Table 5.5 Chemical soil properties in the Muara Wahau region of East Kalimantan

	No. of samples	Acidity (pH H_2O)	Exchangeable cations (meq per 100 g)		Phosphorus (ppm)
			A1	K	
Surface	26	4.5–4.7	4.5–6.3	0.09–0.19	104–160
Subsoil	26	4.6–4.8	6.5–8.7	0.05–0.20	115–125

Source: Transmigration Area Development Project, 1982.

Table 5.6 The cation exchange capacity distribution in the three project areas of east Kalimantan (%)

Area and no. of samples	0–10 meq	10–16 meq	16–24 meq	Above 24 meq
Wahau (54)	18	52	26	4
Pantun (52)	38	37	21	4
Pesab (52)	67	21	10	2

Source: Transmigration Area Development Project, 1982

of Medan (50–125 people per square kilometre), and in the north-east corner around Sigh (50–125 people per square kilometre). Population densities elsewhere are as low as 5–25 people per square kilometre. A look at the geological map in the *Atlas von Tropisch Nederland* (1938) shows that the densely populated area between the Toba Sea and Medan corresponds to an extended patch of base-rich basalts, andesites and dacites (for chemical composition see Table 5.4). In the hinterland of Padang there are several smaller outcrops of these rocks. Such formations do not appear in the north-west corner. Instead, there is another regionally limited outcrop of basalts in 'the great longitudinal valley of Sumatra, Upper Valley Ketahoen–Selikat, bounded in the NE by the highest Barisan Plateau and in the SW by a ridge of young volcanoes.' (*Atlas von Tropisch Nederland*, Plate 6b). The longitudinal valley (at about 400 m) is a limited, densely populated cultivated plain around Moeraaman, which is uncommon in southern Sumatra where acid volcanic extrusions, such as liparites (see Table 5.4), prevail. The valley, however, is too small to stand out amid the overall district values of population densities.

A detailed description of the regional distribution of rocks, soils and population in Sumatra can be found in the study work of E. C. Jules Mohr, former director of the General Experimental Station at Buitenzorg (Java), *De Grond van Java en Sumatra*, translated from the Dutch by R. L. Pendleton (1930) under the title *Tropical Soil Forming Processes and the Development of Tropical Soils, with Special Reference to Java and Sumatra*.

For Sulawesi the same procedure of comparing population distribution with the appearance of basic volcanic rock can be conducted with the help of the *Atlas von Tropisch Nederland*: again, relatively high population densities coincide with the occurrence of basic volcanites as parent rocks.

In conclusion it may be said that if a region shows a certain relationship between population density and volcanic or non-volcanic underground, the former need not necessarily be caused by the latter. But if such coincidence appears to be the rule in various regional examples, then one cannot but accept it as a causative relationship. The way in which this relationship operates will be discussed in the following pages.

5.3 Exceptional realms on alluvial accumulations

Concerning the above-mentioned second favourable condition, namely the periodic, or episodic addition of rock debris with weatherable nutrient-rich primary minerals to the soil, this occurs in the reaches of volcanic ash deposits or in the floodplains of white-water rivers formed during the Holocene.

Mineral provision by means of volcanic ashes play an important role only where volcanoes are abundant as is the case in Java. The population is well aware of the positive effects of ashfall, and after such events the ashes are collected from roofs and other flat surfaces to be scattered over the rice fields, as reported by Uhlig (1984).

Alluvial accumulations like those of the central Amazon basin show in river-bank cuts along the levees a sequence of millimetre-thick flood deposits of silty materials that look like the pages of a book. These deposits consist of base-rich debris. Genuine soil profiles have not yet developed in these well-drained recent sandy entisols. Nevertheless, the sites are well suited for continuous cropping. Beyond the levees, in the topographically lower and more level parts of the floodplains (*varzea*), hydromorphic and humic gley soils prevail. When compared to the soils of mountainous or basaltic regions none of these alluvial soils can be classified as chemically rich (Table 5.2), but they are still several times richer than the normal ferralitic upland soils above the floodplains (II, III and IV in Table 5.2).

The main problem of most fluvisols is their poor drainage. While this problem was solved by technical means in vast regions of the Asian tropics so that the soils are very well suited for intensive agricultural land use under permanent wet-rice cultivation (for more details see section 5.5), in South America well-managed irrigation and drainage conditions are restricted to relatively small areas, as in the interior delta of the Upper Orinoco River, for example. This discrepancy has to do with the much less favourable river flow circumstances. On the Amazon River, for instance, the difference in elevation between Iquitos and the Atlantic coast is only 105 m over a distance of more than 2000 km. The average slope of the water-level is as little as 5 cm km^{-1}. This is coupled with a difference of up to 10 m between high and low waters in the middle course near Manaus. These circumstances make a rational water management in these floodplains far more difficult than on the much steeper gradient of South-east Asian rivers. Consequently, traditional irrigation in the form of ridged fields in Amazonian wetlands have been limited to those places where the flood range was relatively low. Therefore the outlook for future agricultural valorization must focus, first of all, on the floodplains of the white-water rivers in the Andean borderlands, where the gradient of river-beds is steeper than in the central Amazon basin.

Still, the sedimentation domains along the white-water rivers in the central part of the Amazon basin have always been ecologically favoured sites as compared to the uplands, even without the application of technological know-how. Today it is evident that the area influenced by white-water rivers represents an agricultural corridor with corresponding concentration of people in an otherwise practically uninhabited tropical forest. Even as far back as in pre-Columbian times anthropologists differentiate between the more densely populated realm of the *gente del rio* (peoples of the river) and the sparsely populated region of the *moradores de terra firme* (inhabitants of the interior upland) (Frank, 1987).

The deltas that appear on the sketch map (Fig. 5.1) are the outlets of white-water rivers into the ocean. Their topographic conditions are almost the same world-wide. Since there is less suspended material in that part of the river than upstream, the sedimentation and consequently rejuvenation effect on the soils are relatively small. According to Kawaguchi and Kyuma (1977), *The Nederlands Delta Development Team*, which estimated silting in the Mekong delta of Vietnam, noted that even if the entire silt load of the river were deposited in the delta, the annual addition would be no more than 0.2–0.7 mm. Further upstream, in the river floodplain of Cambodia, accretion was estimated at about 1 mm per year. The example confirms the long-term effect of silting. One millimetre of silt per year builds up 10 cm of surface soil in 100 years, which is the plow-layer depth most commonly observed in the surveys of tropical Asian paddy soils. Even where the accumulation of 10 cm takes ten times as long, the effect of silting on soil rejuvenation should still be regarded as positive (reported

147

from Kawaguchi and Kyuma, 1977, p. 188). The implications of these findings for agricultural use will be dealt with on p. 149, Section 5.5.

5.4 Ecological advantages of mountainous areas

Disregarding the local variations due to different drainage conditions in toposequences (see soil catenas in Ch. 3) and focusing on the changes on a greater scale, the effectiveness of hydrolysis and mineralization in the humid tropics is essentially related to the altitudinal decrease in temperature. Due to an average lapse rate of 0.5 °C per 100 m, the heat-induced extreme conditions for hydrolytic weathering of minerals and for humification and mineralization of organic matter which are characteristic of the hot lowlands, are considerably lessened at elevations around 1000–1200 m. The soil layer, in general, is less thick, the residual mineral content is higher, and desilicification is less pronounced than in the lowlands. This leaves the soils with more nutrients and with clay minerals consisting not only of kaolinites, but also of a certain amount of 2 : 1 clay minerals with higher CEC. Strongly affected by the temperature decrease is the mineralization rate of organic matter (see Ch. 3), which lowers the turnover of organic acids and lessens the input of fresh organic substances required for maintaining an acceptable level of humic acids as exchange material in the soil.

The temperature-dependent changes in soil properties create more favourable ecological conditions in mountainous areas of the humid tropics. An additional influence, in many places, results from the relief conditions. A mountain relief consists of slopes of different steepness which stimulate soil erosion. This erosion continuously removes the upper parts of the soil profiles. The remainder is a lithosol in which the physically fragmented and chemically slightly weathered regolith of the parent rock is covered by no more than a few centimetres of solum. The coarse particles of the eroded material reappear as colluvium downslope or at the bottom of depressions, and silty and clayey material is carried away into white-water rivers.

The common land-use system in those parts of oxisol- and ultisol-covered lowland hills, which are obviously heavily eroded, is not rotational bush fallow, but a more intensive form of cultivation. It seems that where the upper parts of the oxisols and ultisols have been removed by erosion, the rest of the soil profile is richer in nutrients and more productive.

While lithosols in the extratropics are generally poorer than the regular brown earths, most *lithosols in tropical mountain regions lend themselves better for agricultural use than the heavily leached ferrallitic lowland soils (oxisols or ultisols)*.

Testimony to the ecologically favourable situation of elevated parts of the humid tropics is offered by Ward (1973; 659) in the *World Atlas of Agriculture* using New Guinea as example. Based on the 1961–62 survey of indigenous agriculture, he states with respect to the geographical distribution of different land-use forms:

> Three major types of indigenous cultivation may be recognized, two of which
> occur in the lowlands: 1. In lowland forest and savanna areas a simple form of

148

shifting cultivation is practiced with a long fallow under regenerated forest. 2. In the swampy areas below about 500 m sago becomes the main staple with some supplementary cropping. 3. In highland areas between 1200 and 2200 m cultivation periods are longer and fallow periods shorter than in the lowlands. In some highland areas the system approaches permanent field cultivation.

The details of land use in the highlands of New Guinea are well documented in Eric Waddell's *The Mound Builders of New Guinea* (1972).

Altitudinal and relief-induced change in soil conditions are often coupled with the influence of a third factor, namely the occurrence of base-rich parent rocks. In the areas of Archaean crystalline shields, isolated massifs such as the Jos Plateau in central Nigeria or the Cameroon Mountains result from Tertiary tectonic uplifts of certain parts of the crystalline crust along fault lines. Coupled with the faults is the intrusion or extrusion of magma which occurs either in the form of dykes and intrusive plugs within a granitic or gneissic crystalline massif (as in the Jos Plateau) or in the form of extrusive volcanites (as in the Cameroon Mountains).

Applying the reasonings about the ecological advantage of mountain realms to the altitudinal levels of the Jos Plateau between 800 and 1200 m leads to a more plausible hypothesis concerning the exceptional intensive agricultural use and remarkable population density at these heights than those based on socio-economic, historical or behavioural reasoning as mentioned in Chapter 2. Both the Jos Plateau and the Cameroon Mountains are ecologically determined exceptional areas within the humid tropics of West Africa.

5.5 The special conditions of paddy soils and wet-rice cultivation

For more than half of the world's population rice is the main staple food. Nearly 70 per cent of the world area planted with rice lie in the tropics, with Asia dominating to such an extent that African and South American contributions are almost negligible (Table 5.7).

In the following pages reference will be made primarily to the subregion of South-east Asia where rice is dominant over other cereals, unlike India and Pakistan where the production of rice is surpassed by that of other cereals.

The principal systems of rice cultivation and their systematic classification are explained in the schematized cross-section taken from Uhlig (1984) (Fig. 5.3).

Comparing upland-rice and wet-rice cultivation, the latter is by far the more important not only in terms of area, but also because its yields are practically twice as large as those of upland rice. Sanchez (Table 5.7) estimates that roughly three-quarters of the planted area are naturally or artificially flooded.

Rice is the only staple food that can be produced on submerged fields. Other cereals, such as wheat or maize, will not survive several days under water. Rice, however, can create, within the anaerobic layer of the soil, its own aerobic rhizosphere by

149

Table 5.7 Extent and distribution of rice cultivation and principal cropping systems in the tropics (after Sanchez, 1976)

Extent and area distribution	Tropical Asia	Tropical America	Tropical Africa	Total
Production (million t)	153	11	5	169
Grain yield (t ha^{-1})	1.81	1.84	1.43	1.80
Area planted (million ha)	84	6	3	94
Areal distribution (gross estimates) (%)				
Rainfed lowland (paddy)	50	0	20	46
Irrigated lowland (paddy)	20	4	18	19
Upland (dryland, *secano*)	20	75	72	25
Deep-water (floating)	10	0		9
Direct-seeded, irrigated		21		1

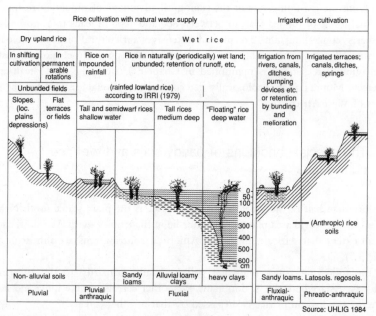

Source: UHLIG 1984

Fig. 5.3 Cross-section of rice habitats and predominant rice types (after Uhlig, 1984).

internal oxygen diffusion from the leaves to the roots, thus guaranteeing the normal functioning of the feeder roots.

The first advantage achieved from flooding is the input of nutrients: silt and clay, along with adsorbed bases, as well as the soluble elements are supplied to the paddy fields.

Table 5.8 Water quality of rivers in tropical Asia (mg l⁻¹)

River	Sampling sites	Ca	Mg	Na	K	HCO$_3$	SO$_4$	Cl	SiO$_2$	Residue on evaporation	Floating matter	pH	Hardness
*Thailand**													
Mekong	Nong Khai	31.1	5.7	7.7	1.6	115.6	14.7	6.2	15.0	139.2	174.1	6.9	101.0
	Mukdahan	26.8	4.9	7.5	1.4	100.3	12.2	6.6	13.8	124.3	99.9	6.9	87.3
Chi	Khon Kaen	21.1	4.0	57.1	3.9	69.7	5.2	94.5	10.1	244.7	60.5	6.7	69.0
Mun	Ubon Ratchathani	10.9	2.3	40.0	2.8	42.4	2.0	61.6	10.8	165.2	46.6	6.4	36.8
Wang	Lampang	28.3	4.0	5.2	3.6	119.0	1.7	0.8	22.0	125.8	228.9	6.9	87.5
Chao Phraya	Uthai Thani	13.3	2.7	8.8	4.4	71.6	0.1	6.1	18.2	94.7	50.0	6.6	44.4
West Malaysia†													
Kelantan	Pasir Mas, Kelantan	4.7	0.8	2.8	1.3	26.4	1.1	0.1	16.2	42.7	26.3	6.7	
Kesang	Muar, Johore	9.9	7.1	11.4	4.5	6.1	102.8	16.5	14.8	221.8	46.9	4.4	
Tengi	Kuala Selangor, Selangor	1.2	0.2	2.3	1.1	0.7	2.7	2.3	5.3	71.4	40.7	4.3	
Muda	Pinang Tunggal, Penang	3.6	0.6	1.9	2.1	19.6	1.8	0.1	13.7	44.2	38.4	6.0	
Cambodia‡													
Mekong	Phnom Penh	19.4	4.3	6.2	2.1	77.5	8.3	12.7	—	—	—	7.9	

151

Table 5.8 Continued.

River	Sampling sites	Ca	Mg	Na	K	HCO$_3$	SO$_4$	Cl	SiO$_2$	Residue on evaporation	Floating matter	pH	Hardness
Indonesia§													
	Geology of river catchment												
	Volcanic	12.2	4.0	10.5	4.1	—	11.4	4.0	42.3				
	Marl	101.3	16.3	57.6	5.7	—	131.5	26.8	30.2				
	Lime	73.7	8.3	6.6	1.7	—	10.1	3.8	18.2				
	Acid tuff loam	6.7	1.9	6.8	3.6	—	6.6	1.4	28.6				
Japan¶													
	Japan average of 225 rivers	8.8	1.9	6.7	1.2	31.0	10.6	5.8	19.0	74.8	29.2		
World‖													
	World average	15.0	4.1	6.3	2.3	58.4	11.2	7.9	13.1	—	—		

* Annual average data (Kobayashi, 1958).
† Annual average data; cited from Kobayashi, J., unpublished.
‡ Simple averages of the data obtained by the Institut de Pasteur for January–May 1961–63, and June–December 1961–63; cited from Netherlands Delta Development Team (1974b).
§ Simple average of unsystematic measurements (Kojima et al., 1962); (originally from van der Giessen, 1943).
¶ Kobayashi (1961).
‖ Kitano (1969) (originally from Livingstone, 1963).

Source: Kawaguchi and Kyuma (1977).

Kawaguchi and Kyuma (1977) published data on the quality of river water in different regions of tropical Asia (Table 5.8). The quality is strongly influenced by the orographic, petrographic and weathering conditions in the catchment areas. In regions that have undergone intensive tropical weathering and leaching, such as west Malaysia, river water is relatively dilute, but, compared with rivers of the Amazon basin, is nutrient-rich.

Rivers draining areas of volcanic activity are rich in silica and moderate to rich in bases, depending on the nature of the parent rock. Interesting in this regard is the comparison between Malaysia and Indonesia.

Concerning the original mineral content of water from Javanese volcanic areas, analyses of spring waters from different districts of the island offer relevant information. (Table 5.9).

The second advantage of wet-rice cultivation results from working the soil under a layer of water. It creates, in the long run, an artificial hydromorphic soil, the so-called *paddy soil* (sometimes spelled *padi* soil from the Indonesian word for wet rice, *padi*). Although they are treated as a unit, certain differentiations concerning the original natural soil and the cultivation techniques apply (Table 5.10).

Some 80 per cent of the soils in paddy cultivation are alluvial soils by origin (entisols, gleysols, organic soils of flood or coastal plains) or young soils (inceptisols),

Table 5.9 Analyses of spring waters originating from young volcanic materials (mg l⁻¹) (after Mohr 1938/1944)

Name of the spring For the water works of Analyses	Kota Batoe Buitenzorg 1928, p. 8	1928, p. 8	1929, p. 12	Tjiomas Batavia 1929, p. 11	1930, p. 17	Oemboelan Pasoeroean 1929, p. 15
1. Dry residue	132	140	140	154	151	180
2. Residue after ignition	112	112	120	134	125	158
3. SiO_2	46	55	52	46	36	48
4. SO_4	0	0	0	Trace	0	0
5. Cl	Trace	Trace	Trace	Trace	Trace	Trace
6. CO_2	12.4	14.3	16.5	20.7	28.8	17.8
7. HCO_3	72.5	70.0	68.0	75.7	81.0	122.0
8. Ca	9.7	10.3	9.2	9.2	13.2	19.6
9. Mg	Trace	Trace	2.6	5.7	5.0	8.7
10. Fe	0.05	0	0.04	0.06	0	0.02
11. Mn	0	0	0	0	0	0
12. Organic matter ($KMnO_4$)	0.22	0.44	0.16	1.26	1.00	0.30
13. NH_3	0.02	0.03	0.02	0.02	0.04	0.04
14. Protein-NH_3	0	0	0.01	0	0	0
15. NO_2	0	0	0	0	0	Trace
16. NO_3	0	0	0	0.1	0.4	1.5
17. pH	6.55	6.5	6.4	6.5	6.7	6.7

Table 5.10 Chemical characteristics of the plough layer of paddy soils

Major soil groups areal extent	%	depth (cm)	pH H₂O	Organic carbon	Exchange cations (meq per 100 g)				CEC (meq per 100 g)	Base saturation (%)
					K	Ca	Mg	Na		
*Indonesia**										
Alluvial soils (entisols, inceptisols)	55		6.8	1.3	0.8	23.2	9.4	1.1	41.4	83.4
Gley soils (aquepts)			6.6	1.0	0.4	32.2	11.3	0.8	49.7	87.9
Regosols (entisols, inceptisols)	8		6.4	0.9	0.2	13.6	5.5	0.3	25.6	76.6
Latosols (ultisols, nitosols)	17		5.5	1.8	1.0	8.3	3.8	0.2	25.4	52.3
Malaysia†										
Kangkong (tropaquept on marine sediment)		0–10		3.43	0.27	7.85	8.7		42.7	42
		10–20		2.87	0.27	7.90	8.6		41.7	44
Tualang (tropaquept on riverine alluvium)		0–7		1.63	0.24	3.17	3.0		20.9	32
		7–17		1.54	0.38	3.48	2.4		19.9	31
Lundang (tropadult on well-drained upper river terraces)		0–15		0.90	0.44	0.80	0.67		8.7	24
		15–22		0.63	0.20	0.85	0.98		8.2	26

* Adapted from Table 2 in Soepratohardjo and Suhardjo (1978). † Adapted from Tables 2 and 3 in Paramananthan (1978).

which, as is the case in Java, are of volcanic origin. Some 20 per cent of the paddy soils originate from vertisols, ultisols or alfisols and, occasionally, also from oxisols.

The transformation of these natural soils into paddy soils occurs through the following procedures used in wet-rice cultivation:

1. Artificial flooding;
2. A series of tillage operations with the purpose of destroying the original topsoil structure (soil pulverization: puddling), such as ploughing and repeated stamping, first at soil moisture contents above saturation and, later on, closer to field capacity;
3. Through enrichment of the puddled soil with organic materials (plant debris, manure) or inorganic materials (household ashes or volcanic ash or fertilizer).

These procedures create a 10–20 cm thick layer of homogeneous mud, often called the plough layer.

After many years of cultivation the characteristic paddy soil profile has developed (Fig. 5.4). While dry, it consists of a 10–20 cm horizon of greyish-brown reduced soil material, overlain by a thin (a few millimetres) brownish oxidized horizon. The greyish-brown matrix is interspersed with brown, yellowish or dark-bluish mottles of iron- and manganese-rich concentrations. After puddling the mottles disappear within the uniform mud of the plough layer. In originally poorly drained soils, such as ground-water gleys, the reduced horizon spreads into the equally reduced

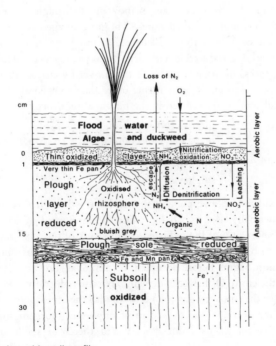

Fig. 5.4 Schematic paddy soil profile.

deeper layers of the profile. Inversely, in originally well-drained soils, the reduced horizon is underlain by an illuvial layer of precipitated manganic or ferric oxides which sometimes has developed into an illuvial pan. Underneath the illuvial horizon the normal oxidized soil profile follows. Thus, the originally freely drained soil profile has been changed, by flooding and waterlogging through puddling, into an artificial gley which, just as a natural gley, has reduced permeability and downward leaching. (Experience in Japan has shown that a moderate rate of downward water movement, say 10 or 20 mm day^{-1}, is essential for a high rice yield, because it removes the toxic substances from the rhizosphere.) In addition, ploughing of paddy soils under waterlogged conditions often forms the so-called plough sole or plough pan below the plough layer due to the intensive mechanical compaction to which this part of the soil profile is subjected.

Thus, all paddy soils are effectively protected against nutrient loss through leaching by an almost impermeable floor, while their cultivated layer consists of a reduced matrix in which the soil aggregated and non-capillary pores have been destroyed.

Of great importance are the chemical consequences of the reduction processes. In their explanation we follow the presentations of Patrick and Reddy (1978) and Sanchez (1976).

Most chemical transformations are set in motion by biological redox (oxidation–reduction) processes that result from oxygen depletion. In the process of microbial respiration some element must act as an electron acceptor. In well-aerated soils oxygen performs this function. When a soil is flooded and puddled, its oxygen supply decreases rapidly since the oxygen movement through water is about 10 000 times slower than through air-filled pores. The little oxygen that reaches the soil is quickly utilized at, or below, the soil surface, forming the thin aerobic brown film on top of the underlying reduced layer where no oxygen is available. In this layer the decomposition process of organic matter is taken over by anaerobic micro-organisms which use mineral compounds or elements other than oxygen as electron acceptors. The following substances are utilized in the indicated sequence:

nitrates (NO_3^-)	to N_2
manganic compounds (Mn^{4+})	to Mn^{2+}
ferric compounds (Fe^{3+})	to Fe^{2+}
sulphates (SO_4^{2-})	to SO_3^{2-}
sulphates (SO_3^{2-})	to S^{2-}

Some of the oxidized soil components that undergo reduction after oxygen depletion are reduced sequentially, that is, all the oxidized component of one system will be totally reduced before any of the oxidized components of the next one, in the general sequence, begin to be reduced. Some others overlap during reduction.

Nitrate reduction begins before complete O_2 removal, but not all of the NO_3^- is removed unless the entire O_2 has been depleted. Nitrate reduction causes denitrification of the soil because the nitrogen gas (N_2) escapes into the atmosphere. Unlike nitrates, the ammonium ion (NH_4^+), the second most important inorganic form of nitrogen in the soil, is present in its reduced state and therefore stable under anaerobic conditions.

156

However, in case of alternate flooding and drying as is common in extended areas of the rice-growing regions, part of the ammonium (NH_4^+) is transformed during the dry phase into nitrate (NO_3^-), the stable form in aerated soils. During the next submergence these nitrate ions are lost through denitrification: 'Alternate flooding and drying cycles result in tremendous nitrogen losses. These losses are greatest during the first cycle and decrease progressively afterwards' (Sanchez, 1976: 433). There is a certain compensation due to the fact that organic materials not suitable for mineralization under aerobic conditions can be mineralized under flooded conditions. This is particularly true of rice straw. The rate of organic nitrogen mineralization is often higher in flooded than in aerobic soils (Sanchez, 1976: 433).

The reduction of manganic compounds to the manganous form takes place during the reduction of O_2 and NO_3 although it appears that the reduction of Mn^{4+} to Mn^{2+} lags somewhat behind the NO_3^- reduction.

The next inorganic redox system to be reduced is ferric iron. For various reasons the reduction from Fe^{3+} to Fe^{2+} is probably the most important reaction taking place in flooded soils.

First, the process provides adequate Fe^{2+} for the nutritional requirements of the cultivated plants.

Second, iron compounds are usually more abundant than nitrates and manganic compounds; they can act as buffers with respect to hydrogen sulphide (H_2S) which is toxic for plants. SO_4^{2-} reduction, next in the sequence of redox processes, is carried out only by true anaerobic bacteria and occurs only after the reduction of all ferric iron. Normally the abundant hydrated ferric oxides prevent the reduction of SO_4^{2-} and thus the production of hydrogen sulphate.

Third, ferric iron reduction is involved in the formation of the iron and manganese pan on formerly well-drained soils. The ferrous forms, as well as the manganous forms, are moderately soluble. These compounds are taken into the solution and carried downward. On contact with the oxygen still present in non-puddled horizons of the subsoil they are oxidized and precipitated. It has been estimated that it takes 50 – 100 years for an iron and manganese pan to develop (Young, 1976). In soils subjected to seasonal flooding and draining, reduction and oxidation of iron compounds as well as of Mn compounds is cyclical. The reoxidation of reduced iron and manganese compounds enables these compounds to act as buffers during the next flooding.

Fourth, iron reduction is also of great importance because the reduction of ferric compounds to a more soluble ferrous form releases into the soil solution phosphorus from the iron–phosphate compounds which, in aerobic soils, are insoluble (see phosphorous fixation in Ch. 3).

Fifth, there is an increase in the pH of acid soils through the reduction of iron compounds because of the OH^- ions release in the course of the reduction of $Fe(OH)_3$ or similar compounds to $Fe(OH)_2$. Other reduction processes, such as the nitrate and the manganic–manganese reduction, have the same effect, though slightly lessened, because of bounding H^+ ions. As a whole, the acidity changes to an almost neutral reaction when acidic soils, furbished with sufficient reducible iron compounds, are submerged for a few weeks. This *self-liming process* optimizes the CEC and the

157

availability of most nutrients, while at the same time eliminating aluminium toxicity because the exchangeable Al is precipitated at pH 5.5. Consequently, in contrast to cultivation under upland conditions on normally drained tropical soils, there is no need for additional liming with limestone or dolomite in the case of submerged rice cultivation.

Potassium, calcium and magnesium compounds are not directly submitted to reduction processes; they are only indirectly affected in that 'the large quantities of NH_4^+, Fe^{2+}, and Mn_2^{2+} ions released upon flooding may displace substantial quantities of Ca^{2+}, Mg^{2+}, and K^+ from the exchangeable sites into the soil solution. Consequently, these ions have become more and more susceptible to leaching' (Sanchez, 1976). (On the other hand, in most paddy fields leaching is physically impeded due to the impermeability of the underlying plough pan or by the naturally impermeable gleys.)

Summarizing the consequences of submergence, one can clearly affirm that most of the ensuing changes in the chemical soil conditions are advantageous for the nutrition of rice plants. The relatively easy maintenance of soil fertility is one of the great advantages of paddy fields.

Kawaguchi and Kyuma (1977) published very informative histograms of different chemical soil properties from 410 paddy-soil samples from countries of tropical Asia, from Bangladesh to the Philippines. Figure 5.5 reproduces the most important histograms. The mean values of all samples are indicated and the corresponding values for Japanese and Mediterranean paddy soils have been added for comparative purposes. The values underline the relatively favourable soil properties in comparison to normally drained dry soils.

A critical issue in wet-rice cultivation is the provision of nitrogen to the plants. Although some of the nitrogen taken up by the rice plants is returned to the soil as organic matter (stubble, roots, straw and animal excreta), the nitrogen status of paddy rice soils is inevitably lowered by cropping. Since other natural nitrogen sources, such as rainfall and irrigation water, are negligible, the continuous production of rice over hundreds of years without artificial fertilization would not have been possible without microbial nitrogen fixation. Waterlogged paddy soils teem with all sorts of nitrogen fixers, blue-green algae and photosynthetic bacteria being the most effective. Blue-green algae flourish in the surface water and the uppermost part of the plough layer. The amount of nitrogen fixation by them is estimated between 3.2 kg ha^{-1} in the presence of rice plants and 10.9 kg ha^{-1} in their absence. Photosynthetic bacteria are strict anaerobes that thrive under the harshly reduced conditions of paddy soils during the cropping season. In the absence of light they use organic acids as energy sources.

Estimates of total nitrogen fixation in different Thailand soils ranged from 0.5 to 54 kg ha^{-1} N, but in more than half of the cases it was less than 5 kg ha^{-1} N. Nevertheless, a comparison with upland soils clearly shows a much higher efficiency of nitrogen fixation in paddy soils (Patrick and Reddy, 1978).

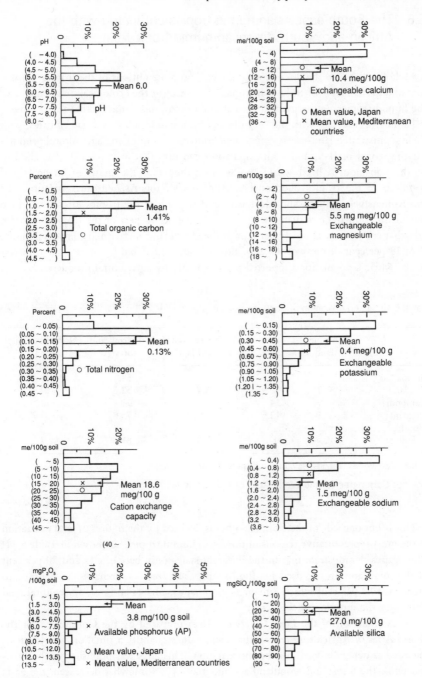

Fig. 5.5 Histograms of chemical properties of Asian paddy soils (after Kawaguchi and Kyuma, 1977).

5.6 The South-east Asian humid tropics compared with the African and South American humid tropics

To scholars who, in the discussion of the ecological conditions of the tropics, think – first and preferably – of the Asian tropics, the thesis of the ecological disadvantage of the humid tropics with regard to agricultural development must sound provocative.

Familiar as they are with the very old and high standard South Asian cultures, with their sophisticated land-use systems, their contribution of exotic agricultural products for the world market, and with population densities that are unheard of in the rest of the non-industrialized world, no wonder they find the thesis of an ecological handicap of the tropics questionable. And indeed, the Asian tropics comprise the most intensively utilized and most valorized regions of the tropics in general. Two out of three inhabitants of the tropics live in the Asian part of the tropics, and this area that covers only about 17 per cent of the global land surface and contains no more than 18 per cent of the world's agricultural land is inhabited by more than a quarter of the world's population supported pre-eminently by agricultural products.

Table 5.11   Land utilization in South-east Asian states (%)

	Arable land	Rice land	Fruit trees and orchard land	Woodland and forests	Non-agricultural land*
Burma		17		54.1	28.4
Thailand	3.2	12.9	3.4	52.8	27.7
Cambodia		14		71	12
Vietnam (South)	2.1	12.5	1.9	75.8	7.7
Indonesia		11	0.3	81	7

* Includes woods, forests, savanna and unused cultivable land.

Source: Committee for the World Atlas of Agriculture (1973).

This is but one side of the coin; the other is denoted by the figures of land utilization in the most representative regions of the Asian humid tropics, as shown in Table 5.11. The figures demonstrate that in the different states only between 17 and 20 per cent of the total land area is used for food crops, fruit trees or orchards, in contrast to more than 80 per cent which remain as woodlands, forests and non-agricultural land (whatever the meaning of this term).

An examination of the *Land Utilization Maps* prepared by the Committee for the World Atlas of Agriculture reveals that these values quantify a situation which can be regarded as generally valid for the entire region. This atlas was published in the 1970s and while the forest and woodland acreage may have somewhat decreased, the basic principle remains: *the cultivated parts of the South-east Asian tropics are still only restricted areas within a natural region dominated by forests.*

160

Table 5.12 Numbers of operational farms (over 0.1 ha) by size and area, Java, 1963 (after Palmer, 1976)

Farm size (ha)	Jakarta	West Java	Central Java	East Java	Jokjakarta	Total
0.10–0.49	14 640	1 205 857	1 378 675	1 358 650	194 612	4 152 434
0.50–0.99	4 130	525 412	731 946	813 603	82 617	2 157 708
1.00–1.49	1 970	216 462	276 163	335 062	28 830	858 487
1.50–1.99	960	83 567	113 844	140 627	11 891	350 889
2.00–2.99	860	74 409	79 289	112 531	6 824	273 913
3.00–3.99	300	23 034	25 314	38 230	1 758	88 636
4.00–4.99	140	10 714	10 092	14 583	454	35 983
5.00 and over	120	11 929	8 020	13 147	651	33 867
Total	23 120	2 151 384	2 623 343	2 826 433	327 637	7 951 917

Much of the cultivated land is intensively used by continuous cropping of mainly wet-land rice, often with two crops per year or even five in 2 years. These 'rice bowls' are very densely populated; in fact, most of them should be termed 'critically overpopulated'. The following situation may serve as an example: 10 per cent of Java's population are distributed over 345 administrative districts which can be classified as eminently 'rural' in spite of population densities above 1200 inhabitants per square kilometre. The agricultural holdings are correspondingly small: 47 per cent lie under 0.5 ha and the rest normally do not exceed 0.7 ha (Uhlig, 1984). Table 5.12 describing operational farms (over 0.1 ha) in size and area (Palmer, 1976) shows that more than half of Javanese agricultural holdings are no bigger than 0.1–0.5 ha and that six out of eight have less than 1 ha.

A direct indication of the overpopulation in some traditionally intensively used areas must be seen in the land opening and resettlement policies pursued by South-east Asian governments. The effectiveness of these measures will be discussed later.

At this point, however, the following findings should be highlighted:

1. By far the largest part of the population of the Asian humid tropics is concentrated in isolated areas which comprise less than 20 per cent of the total land area.

2. The pedological conditions in these areas correspond to those that were previously mentioned as exceptional and advantageous for agricultural purposes.

3. Under these conditions the special climatic advantages of the tropics can be brought to bear and high-yielding agricultural land use becomes possible.

4. When there is overlapping of several of the different exceptional conditions, such as extremely base-rich rocks, young soils, supply of mineral nutrients through irrigation water or volcanic ashes, as well as regular flooding, the results are extraordinarily favourable for crop production.

5. The favoured regions are usually intensively populated. Many of them have been overcrowded for generations.

6. In contrast to tropical South America, and especially tropical Africa, the ecologically exceptional regions in Asia are numerous and relatively extensive, including

161

20 per cent of the total land area. The reason for this lies in the special geologic–tectonic design of South-east Asia. While Africa and South America east of the Andes consist of Pre-Cambrian crystalline shields with a carpet of acidic sediments, South-east Asia is part of the Mediterranean–South-east Asian folded mountain chain of Tertiary age. Where this orogenic system meets the circum-Pacific mountain chain, recent volcanic eruptions are widespread and frequent. In addition to the orogenic and volcanic processes, large streams have built alluvial plains and deltas by depositing enormous quantities of eroded rock material in intramountainous basins and in plains outside the mountains. Today's rivers continue to provide new mineral substances for the man-made landscapes of the floodplains.

7.   The extraordinary population density in the 20 per cent of the Asian land surface mentioned above makes the overall average population density in most of the South-east Asian states comparable to those of extratropical countries. These average values, though, obscure the fact that the favoured regions are interspersed amid larger areas in which the usual conditions of the humid tropics prevail, and where the development of a continuously productive agriculture is as difficult as in Africa or South America.

8.   Since the high population densities are restricted to ecologically exceptional regions, any extrapolation of their actual carrying capacities as potential capacities for the almost unpopulated neighbouring areas is scientifically unsound and might be misleading for decision-makers. The documentation of the Science Advisory Committee of the American President on the World Food Problem may be quoted as testimony: 'In Asia, if we subtract the potentially arable land area in which water is so short that one 4-month growing season is impossible, there is essentially no excess of potentially arable land over the actually cultivated' (Revelle *et al.*, 1967).

5.7   High population densities on tropical low-base status soils. Counter-arguments to the general low carrying capacity?

There exist, however, certain areas in the humid tropics which show remarkably high population densities in spite of apparently not being favoured by ecologically exceptional conditions and which are quoted by reputed scientists as a demonstration that the establishment of productive agriculture on base-poor tropical soils is possible after all. For example, referring to the humid tropical soils of the older land surfaces of Africa, Greenland (1981, p. 1) affirms: 'That they are able to support relatively high densities of population is shown by the fact that on some of the poorest of these soils in south-eastern Nigeria population densities of up to 500 persons per square kilometre exist, and have existed for many years, with virtually no dependence on external sources of foodstuffs'.

The schematic sketches of Fig. 5.6 and Table 5.13 were taken from the original work of Lagemann (1977) *Traditional African Farming Systems in Eastern Nigeria. An*

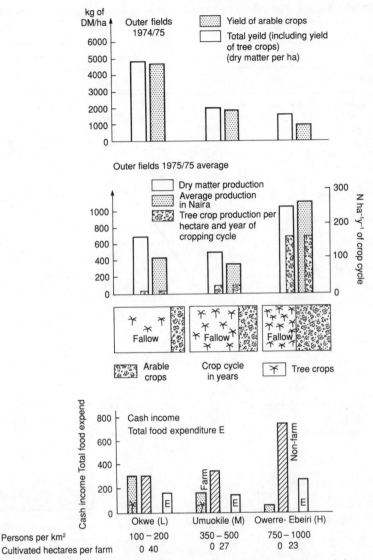

Fig. 5.6   Schematic overview of crop cycles, dry-matter production, yields, cash income and food expenditure in three villages with low, intermediate and high population densities in eastern Nigeria (values taken from Lagemann, 1977).

Analysis of Reaction to Increasing Population Pressure. For the three villages under consideration the different population densities, the average of cultivated surfaces per farm, the crop/fallow cycle, the different densities of crop trees, the average dry-matter production in the outer fields and the cash income and expenditures of the average family in the three villages are presented. The data prove clearly and leave no doubt that even the families of the village with the relatively low population

163

Table 5.13 Average cash income (US $), expenditure and saving patterns (in Naira) in three villages in eastern Nigeria, 1974/75

Village	Okwe (L)	Umuokile (M)	Owerre-Ebeiri (H)
Population density	Low	Medium	High
Cash income			
Cash farm income	307	173	76
Non-farm income	300	347	721
Total cash income	607	520	797
Farm expenses			
Labour	37	20	14
Planting material	12	2	29
Others	9	1	8
Total farm expenditure	58	23	51
Food expenses			
Fish and meat	74	40	52
Cassava products	6	42	83
Yam	6	6	24
Rice	11	5	27
Others	75	46	75
Total food expenditure	172	139	261
Other family expenses			
Trading	60	97	119
School fees	39	38	117
Medical	24	27	50
Clothes	25	18	32
Housing	32	13	27
Others	41	41	153
Total other expenditure	221	234	498
Total cash expenses	451	396	810
Surplus of receipts over outlays	156	124	−13

Source: Lagemann (1977).

density of 100–200 persons per square kilometre have a non-farm income that nearly equals the cash farm income and that they depend on food imports for subsistence. In contrast to the assumption of Greenland, the inhabitants of the densely populated area of south-eastern Nigeria do depend on external sources of foodstuffs. The region must be regarded as a net import area of biomass and can therefore not serve as an example for high population-supporting capacity on poor humid tropical lowland soils.

The poor agrarian production capacity of the soils in question becomes obvious when a comparison is made of the dry-matter production on the outer fields of the three villages. The yields per hectare decrease drastically with shortening of the fallow period, and the average dry-matter production per hectare and year of the cropping cycle in Owerri-Ebeiri (fallow period only one-third of the fallow period in Okwe), is no more than 50 per cent higher.

In fact, there is a whole group of case studies which turn out to be similarly misleading in that they induce an improper generalization and areal extrapolation of the findings gained in a limited area, especially when neglecting the crucial point of mass- and nutrient balances. Close inspection of the requirements of the described highly productive agrarian systems reveals that the importation of organic matter from outside, at times even combined with artificial irrigation, amounts to the chief factor.

In a representative example, Grenzebach (1984, p. 372) lists 'lime fertilization (exploitation of shell banks), fertilization with fish leftovers, and mixed fertilizers made with the help of imported cattle manure from the hinterland'. These, among others, are the necessary measures which warrant intensive agriculture with onion and pepper as cash crops which in turn allow purchasing the staple food for a rather dense population.

Ecologically, such intensive cultures in limited areas are similar to the home gardens of the shifting cultivators, where organic matter, mainly derived from the outer field, is concentrated on a few square metres. Obviously, they cannot be used as an argument against a geo-ecological evaluation of larger natural regions.

Particularly in relation to the moist tropical areas of South-east Asia, the often practised tree- or shrub cultures are recommended as a means of relieving shifting cultivation or rotational bush fallow on red or yellow ferrallitic soils. Scholz (1984), for example, describes this land-use system as 'from an ecological viewpoint, the most advantageous form of land use', (translated) under moist tropical soil and climate conditions. The most important cultures are rubber trees, coffee and pepper bushes and spice herbs, namely cloves and cinnamon. All products are cash crops. 'This form of land use combined with paddy rice has proved to be a sustainable small scale farming system and has replaced the traditional shifting cultivation in many areas of South-east Asia . . . In Sumatra, more than half of the agricultural area is cultivated in this way' (Scholz, 1984: 366 translated).

From an ecological viewpoint, this form of land use meets two decisive prerequisites for sustainable agro-ecosystems on tropical low-base status soils. Firstly, the biological outfit of the system is based on a large volume of biomass and biomass turnover, while the portion of biomass extracted through harvest remains very small. (Detailed information about the importance of this fact may be found in Section 9.1.) The high market price of the harvested products ensures the purchase of staple food, which in turn means an import of organic matter. Secondly, the system copies to a certain degree the structural characteristics of the natural forest with the implied consequences which are described in Chapters 4 and Section 9.4.4

However, it is obvious that tree- and shrub cultivation cannot be seen as the general method of overcoming the constraints of agricultural production on tropical ferrallitic soils. The farmers depend on external sources of foodstuffs, and as there is only a limited market for rubber and spices, only a restricted number of holdings with limited areal extent can make use of this advantageous system.

Having reviewed the reasons why limited areas are often favoured in terms of agricultural yields and population density, we must direct our attention to other regions of the tropics where soils contain intrinsically the minerals and physical attributes that make them potentially useful for cultivation, but where the limiting factor is water availibility. In these areas, management of the scant water supply assumes the dominant role in securing agrarian success. In other instances success is difficult to achieve, not because of water shortage but because some peculiarities in the landforms of the subhumid tropics hinder the development of waterworks and the unpredictable variations of river regimes do not allow optimal management.

6

The ecological conditions in the seasonally wet-and-dry outer tropics

6.1 The zonal change of soil types

In both hemispheres there is a transition from the permanently humid equatorial climates without a major dry period, to the alternating wet-and-dry climates of the outer tropics (Section 1.2, Figs 1.3 and 1.4). In its general lines, the greater the distance from the Equator, the shorter the rainy season and the lower the annual precipitation amounts (Fig. 1.2). The changes in climatic conditions are reflected in the succession of natural vegetation zones from the rain forest to the desert as described in Section 1.3 (Fig. 1.9).

The following pages deal with the basic zonal sequence in soil types that are associated with the changing climate patterns. Similar to natural vegetation, soil formation and soil type distribution show a corresponding dependency on the changing climate patterns. The principal reason is that with declining precipitation and prolonged dry periods the effectiveness of chemical weathering, especially hydrolysis, is reduced. The overall results are: (1) lesser soil depth than in the ever-humid tropics and increasing soil skeleton with weatherable primary minerals; and (2) declining desilicification leading to a rise in 2 : 1 clay minerals and cation exchange capacity (CEC) of the clay substances (see Ch. 3).

The reduced rain-water input and the more effective CEC diminish the leaching of nutrient cations in general. Thus during the dry season there is a reversal in the soil-water movement. Forced by strong evaporation, the soil solution is drawn towards the surface and leaves the originally dissolved compounds of the different elements in the upper soil horizons.

The zonal soil type which ultimately emerges depends upon the intensity of the various processes and on their mutual interactions. Departing from the viewpoint of climate dependency, two extremes of soil types can be distinguished. The luvisols of the FAO/Unesco classification (*sols lessivés* in the French nomenclature, alfisols in the US soil taxonomy) represent the transition from the ferralsols of the equatorial region and comprise soils in which leaching has been so strong as to transfer nutrients and clays into the deeper soil horizons. The other extremes appear in arid areas of high evaporation and no horizontal water input where chemical weathering is so weak that there are not enough soluble products in the soil solution to be concentrated in the upper soil horizons. The final products are immature arenosols (L. *arena* = sand), regosols (G. *rhegos* = mantle) or yermosols (G. *yermos* = barren).

In locations with horizontal water input, strong evaporation leads to the concentration of soluble salts near the surface or in the upper soil. The resulting solonchak and solonetz soils (L. *sole* = salt) are unusable for agricultural purposes.

Between these two extremes extends a wide zone with moderate weathering intensity and leaching, which is characterized by higher dependency of soil formation on parent rock chemistry and soil-water regimes along toposequences (for details see Sections 5.1 and 3.11). Rock-induced as well as catena-induced changes are implicit in the term cambisol (L. *cambiare* = to change). Within the wide range of cambisols is found the differentiation between ferrallitics and fersiallitics as a consequence of different superposition of climatic, petrographic and topographic influences. Of special interest in terms of their agricultural potential are the tropical black earths, which develop in this zone over base-rich young volcanites (black cotton soils of the wet-and-dry Deccan Plateau in India). In Appendices 2A–C are listed several soil types of the wet-and-dry tropics with their characteristics as found in the explanatory volumes of the *Soil Map of the World* (FAO/Unesco, 1971, 1977, 1978).

As a whole the fersiallitic soils (*sols fersialitiques non lessivés*, eutric cambisols, alfisols) of the dry savanna and the brown soils (*sols bruns eutrophés tropicaux*, eutric cambisols, aridosols) of the thorn savanna possess a considerably higher potential fertility than the ferrallitic soils of the humid savanna (dystric cambisols and acrisols, luvisols, *sols ferralitiques moyennement désaturés*, ultisols) and, certainly, more than the leached ferrallitic soils (ultisols and oxisols, orthic or xanthic ferralsols, *sols ferrallitiques fortement désaturés*) of the rain-forest region.

The causative antagonism between pedological and climatic growing conditions leads to the situation that the same climatic circumstances which in the outer tropics caused the formation of favourable soils become now a limiting factor as far as agricultural exploitation is concerned.

6.2 Climatic restrictions and the need for water management

There are several aspects of the climatic growing conditions in the outer tropics. First, the annual dry period which restricts the length of cultivation and leaves the land fallow during the remainder of the year; second, the great interannual rainfall variability; and third, the recurrent catastrophic droughts (see Section 1.2, Figs 1.5 and 1.6).

Particularly disastrous are the long-term climatic variabilities (Giessner, 1985; Mensching, 1986). Taking into account the sequence of annual precipitation amounts from 1917 to 1975 in El Fasher (Sudan) as an example, one can realize the hazards of agricultural life in the marginal zone of rainfed agriculture (Fig. 6.1). Between 1919 and 1940 rainfall conditions were so favourable that the cropped areas expanded

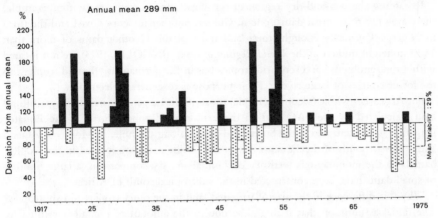

Fig. 6.1 Variability of annual precipitation in El Fasher, Sudan (after Mensching and Ibrahim, 1977).

towards the semi-desert and the number of people and cattle increased due to an ample food supply. To those born at the beginning of that 'fat' period it would never occur that conditions might take a turn for the worse, as happened after 1940. In 15 years of extreme interannual rainfall variability their crops failed repeatedly while the cattle as ultimate reserve increased. Foreign aid was funnelled to maintain this reserve and to compensate for the deficits in staple foods. Then, in 1955, the long-lasting drought began: crops failed, the natural vegetation vanished and so did overabundance of cattle. The ensuing sequels of famine and desertification are inferred in the well-known concept of the 'Sahel problem'. (For additional information on the desertification of this region, see Mensching and Ibrahim, 1977; Mensching, 1986).

However, the phenomenon of periodic long droughts is not restricted to the sub-Saharan zone. Indeed, it affects the *noreste* of Brazil, the Sudan, East Africa, Ethiopia, Zimbabwe, Mozambique, Pakistan and India: precisely the tropical areas that are much more densely populated than the humid inner tropics, as mentioned in Section 1.4

6.3   Actual lack of water management

The superposition of potentially fertile soils, relatively intensive agricultural use and great population density on one side and general water shortage and recurrent catastrophic droughts on the other, demand careful water management and extension of irrigation systems. But the disillusioning fact is that in the semiarid and semihumid tropics, only a very small percentage of the total cultivated land consists of irrigated cropland while rivers drain large quantities of unused water out of this region into the ocean.

169

Regarding the wet-and-dry region of the West African states as the first example, there are only three great dams: one in Guinea, another in Ivory Coast and the third in Nigeria. Curiously enough, these and most of the 17 other dams of more than 15 m in height quoted in the *World Register of Dams* (INCOLD, 1973) are not located within the semihumid and semiarid tropics but in the permanently humid tropics, on the lower courses of large rivers. Their purpose is to generate electricity.

For Africa as a whole (second example) most of the 4.2 per cent irrigated agricultural land, calculated by the FAO, lies in the extratropical zone. North of the Equator irrigation is concentrated in the Sudan and Egypt along the Nile, in Morocco, Algeria and Tunisia. South of the Equator only three major dams are located in the tropics, namely in the mountainous territory of Zimbabwe. By comparison, a large number of small dams have been constructed in the subtropical south of Africa.

Why this difference in the size of irrigated land in the outer tropics and subtropics? The initial assumption that it might be due to the amount of rainfall is refuted by the fact that annual rainfall in the North African mountains, with a yearly mean of 250–500 mm, is clearly less than in the dry savanna of the Sudan with annual averages between 600 and 800 mm. The same holds for the southern part of Africa.

Explanations of a socio-economic or political nature that link the absence of dams to underdevelopment must be rejected, as can be demonstrated in the case of Brazil (third example). In this country, the densely populated area of the north-east with its potential fertility is threatened by recurrent dry spells and catastrophic droughts (*secas*). Several decades ago, the government of the formerly wealthy republic came up with a development plan for the *noreste*, deciding that each year a certain percentage of the state budget was to be used to finance irrigation projects. But in the long term the effect has been minimal. The reason is that the cost–benefit relationship was very unfavourable: extended dams contained water bodies of large area and small depth, which under the conditions of tropical rains and denudation processes silted up and lost their effectiveness very fast.

6.4 The technical difficulties of dam construction in the ectropical zone of excessive planation: the Deccan Plateau example

The whole problematic which has been elucidated in different examples can best be analysed as a complex of interrelated factors in the case of monsoon India. The Deccan Plateau comprises, especially in the states of Madhya Pradesh, Maharashstra and Andhra Pradesh, extended areas of mineral-rich deep black soils. Long ago people advanced from shifting cultivation to semi-intensive (semipermanent) rainfed agriculture on permanent farming units. The agrarian population density in many areas surpassed 100–120 people per square kilometre by the end of the nineteenth century and is higher than elsewhere in the wet-and-dry outer tropics (Clarke, 1971). However, because of the effects of the monsoon, in large areas there is only one harvest per year, the rainy season being restricted to 4 or 5 months.

Recurrent failures of the monsoon caused real famines in the abundant rural population. According to Carlson (1982), between 1750 and 1950 India suffered at least 11 catastrophic famines (1769–70, 1790–92, 1803–4, 1837–38, 1866–70, 1873, 1876–78, 1896–97, 1899–1900, 1943–44), with estimated deaths ranging from 1.25 to 5 million each time. These facts alone make clear that there was an urgent need to introduce irrigation in order to secure crops, to lengthen the cropping period and to expand the farming area dedicated to more productive land use.

It is a fact, however, that even today the irrigated areas in the interior states of the

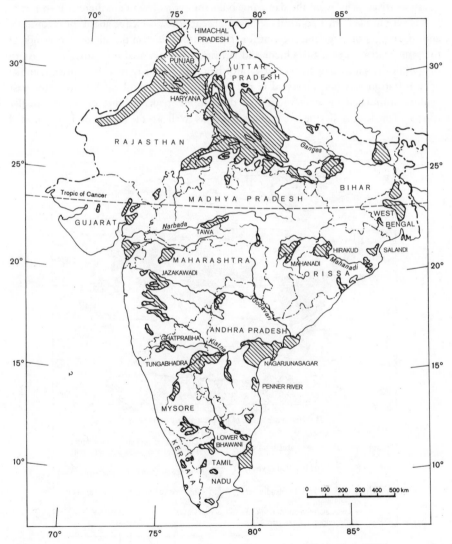

Fig. 6.2   Major irrigation areas in the Indian subcontinent (after Chatterjee, 1973).

171

Deccan Plateau comprise less than 5 per cent of the cultivated land. Chatterjee's map in the *World Atlas of Agriculture* offers an informative overview of dams and irrigation areas that existed at the beginning of the 1970s (see Fig. 6.2).

A remarkable detail in this respect is that out of the 40 dams that existed in 1961, 29 were built only after 1951 (Rouvé, 1965). Yet another fact is that in the Deccan Plateau there is a good number of rivers whose flows are in large measure unused. Chatterjee (1973) estimates that more than 42 per cent of the rainfall over the whole of India is lost in this way.

In view of the population pressure and food shortages of the past 200 years the question arises: why were the dams not built sooner and why are there still so few?

It is very unlikely that political factors or absence of knowledge about the techniques and advantages of irrigation account for the situation, since in the adjacent lowlands of the same states water management and irrigation have existed for generations. As to the politically motivated answer, it cannot be taken seriously that the British colonial administration was not interested in the problem. It was the British who in the past century stimulated the construction of the famous 'tanks' in the reaches of smaller creeks. The length of the containing dams as well as the shape of the dammed

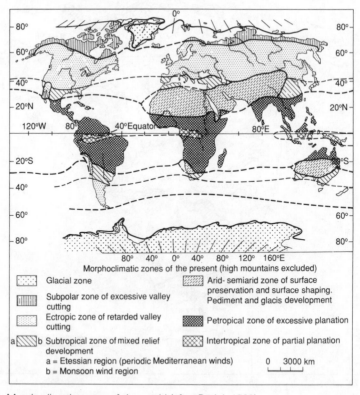

Fig. 6.3   Morphoclimatic zones of the world (after Büdel, 1982).

reservoirs resemble in minor scale very much their larger successors of later decades (see Fig. 6.5).

The length and height of the dams in connection with the shape of the water surface reflect the special topographic character of the climagenetic relief that dominates the Deccan Plateau. The consideration of the particular fluvial erosion relief coupled with peculiar hydrological regimes must be proposed as the fundamentals to answer the question posed above. These natural conditions presented great technical problems first to dam construction itself and led to very unfavourable cost–benefit relations after their completion.

According to Büdel's classification of the 'morphoclimatic zones of the present' (Büdel, 1982) the Deccan Plateau, as all tropical dry savannas, belongs to the 'zone of excessive planation' (Fig. 6.3). In English textbooks this zone is also referred to as 'savanna planation' (Cotton, 1961; Thomas, 1974; Butzer, 1976).

For inhabitants of the extratropics ('ectropical zone of retarded valley-formation', according to Büdel), it is only natural that big rivers cut deep water gaps into the bedrock. Consequently, maps published in extratropical countries show signatures along the rivers of tropical savannas in accordance with such a perception of a valley. Reality in the semihumid and semiarid tropics differs fundamentally from perception. In the Deccan Plateau even the big rivers flow in extremely wide and shallow concavities between gently sloping convex interfluves on the upper surface of the highlands, regardless of their absolute elevation. In tectonically stable blocks, 'etchplains' (Fairbridge, 1968) or 'planation surfaces' are dominant, even at heights above 1000 m above sea-level. These are conspicuously sloping plains with gently convex interfluves and complexes of higher residual topography (Butzer, 1976). Under the special environmental conditions of the tropical dry savannas, the weathering–erosion ratio leads to a three-dimensional denudation rather than to an

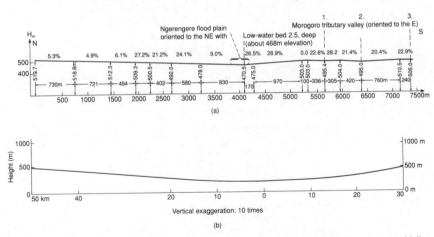

Fig. 6.4 Cross-sections of tropical trough valleys: (a) profile of the Ngerengere Valley, Tanzania; (b) cross-section through the flat trough valley of the Ruvama River, Tanzania, 200 km upstream from the mouth (after Louis, 1964).

erosional incision of river-beds, clear-cut valley slopes and interfluves. The scientific term for such a feature is *Flachmuldentäler*, illustrated in Fig. 6.4 (after Louis, 1964) or 'trough valleys' (Young, 1972).

In the field such slight depressions are hardly recognizable as valleys, and it is not difficult to imagine the length of a dam that would span across such a trough valley. Exceptions occur only where parts of the earth's crust have been subjected to tectonic uplifting. In those areas etchplain escarpments are formed and the segment between the upper and the lower etchplain is crossed by the rivers in short V-shaped and stepped valleys. However, due to the steep longitudinal gradient and numerous cataracts, these sections of the river course are ill-suited as sites for water reservoir construction. For dams, the most advantageous sites on the etchplains must be chosen.

In the case of the Hirakud Irrigation Project in India (Fig. 6.5), the dam across the Mahanadi trough valley stretches over 26 km while reaching a maximum height of only 64 m above the former river-bed. More than 1 million m^3 of stone and concrete and about 17 million m^3 – i.e. about 2 million truck loads – of compacted loam were needed in the construction of that dam (Ministry of Irrigation and Power, 1957).

Fig. 6.5 The Hirakud Dam Project in Orissa, India. (Source: India, Ministry of Irrigation and Power, 1957).

174

Given enough time, a community would be able to master the construction of the outer parts of the dam away from the actual river-bed. Completing the inner part, however, is extremely difficult in view of the hydrological regime of the monsoon rivers with their seasonal run-off and periodic extreme discharges. The Mahanadi, in the area of the dam near Baramul (Fig. 6.6), has a base flow of no more than some tens of cubic metres per second during the dry season when only a small fraction of the channel is occupied by water. With the onset of the monsoon rains the discharge swells to 20 000 and even 30 000 m³ s⁻¹. The flood spills over the river banks, and with the rise of the water-level and volume, the flow velocity increases dramatically.

The differences in the annual range and discharge volumes of the rivers in the wet-and-dry tropics and the monsoon regions as compared to those of rivers in the winter-rain subtropics or permanently humid extratropics are illustrated in Fig. 6.7 (values from Unesco, 1971). The diagrams show maximum and minimum monthly averages of daily discharges (1931–60) and the absolute daily maxima for the Krishna and Godavari rivers in the Deccan Plateau of India, the Rhine and Danube rivers in the extratropics of Central Europe, and the Columbia River in semihumid western North

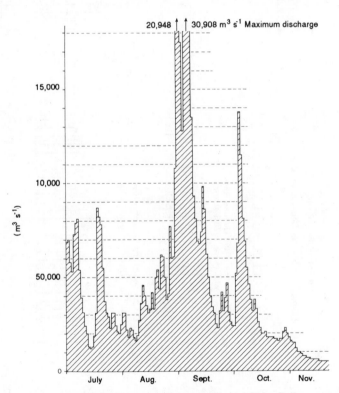

Fig. 6.6 Runoff of the Mahanadi River near Baramul, Hirakud Dam. Daily mean values, July to November 1955. (Source: *Water Year Book*, New Dehli, 1969).

175

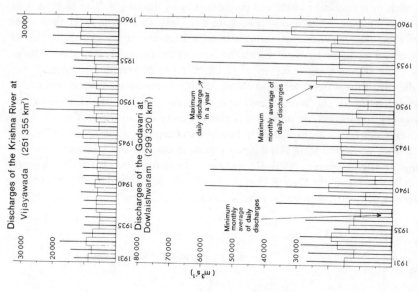

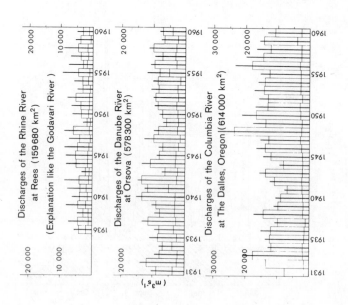

Fig. 6.7 Discharges of extratropical rivers (Rhine, Danube) and the Columbia River in the semihumid North American West as compared to the discharges of the Krishna and Godavari in the Deccan Plateau of India (values from Unesco, 1971).

America. Often the low-water values for the Krishna and Godavari are so small that they do not show up in the diagram.

When comparing the discharges it is necessary to keep in mind the dimension of the drainage basins. Those of the Krishna and the Godavari are only half the size of the Danube and the Columbia, yet, their 'normal' maximum daily discharge is much higher and their extremes reach volumes unheard of outside the tropics.

Under the environmental restrictions mentioned above and given the limited technical equipment, the people of the Deccan Plateau might have started construction of big river dams from the rims. But the last million cubic metres or so of material would have had to be put in place during the few months between two rainy seasons. This task can only be accomplished with an enormous transport technology, as was developed in the industrialized world during the Second World War.

Compared with the enormous technical and financial effort applied, the gains in irrigation potential are relatively limited. They are unprofitable in comparison with corresponding undertakings in subtropical regions. In the case of the Hirakud dam as the longest dam in the world, a lake with 750 km^2 surface contains less than 6 billion m^3 of usable water. The irrigated area is only three times the size of the lake itself, not an advantageous cost–benefit relation at all (Fig. 6.5). As a consequence of the tropical intensive deep weathering and downwash of weathered material, the transport of huge amounts of loam and clay implies the danger of diminishing the storage capacity of the rather shallow reservoir.

The values of the Hirakud dam may be somewhat extreme, but they do not principally convey a wrong impression when compared with corresponding works in other climatic regions. Dams in the Deccan states (see Appendix 3B) average almost 6 km in length and 61 m in height, those in Mysore 4.7 km and 65 m, and those in Maharashtra nearly 3 km and 40 m. In comparison with these, the Sotonero dam with 3859 m is an exception among the hundreds built in subtropical Spain. Three dams are between 2100 and 1500 m long, five about 1000 m. All the others run only a few hundred metres. Hoover dam and the Grand Coulee in the North American subtropics with a length of 379 and 1242 m respectively are considered minor dams, but the retained water reservoir is several times larger than that of the Hirakud.

The construction of dams has been much easier in the summer-dry subtropics surrounding the Mediterranean Sea, first, because in this area valleys, and not plains, have been formed since the recent geological past, and second, because of more moderate annual river discharge.

In general, it appears that for the Deccan Plateau of India we hold the following view: the favourable soil conditions allowed an early development of semi-intensive rainfed agriculture on permanent farming units. The progress to more intensive irrigation practices, needed many generations ago, was hampered until recently by the unfavourable conditions of the fluvial erosion relief and the hydrologic characteristics of the monsoon rivers. Only a highly effective transportation technology, developed abroad, allowed the construction of the exceptionally extended dams which are characteristic of the Deccan Plateau.

What applies to the Deccan Plateau can be extrapolated in principle to all upland regions of the semiarid outer tropics. According to the findings of climagenetic geomorphology, three-dimensional denudation, etchplains and trough valleys are causatively related to climate-induced forcing factors like sheet flows on sloping surfaces and extreme peaks in river discharge. The following examples confirm this thesis. For West Africa the morphographic sketch in Louis (1988) illustrates the surface form complex mentioned, especially the wide trough valleys in uplands above 250 m. The medium-size valleys have widths of 20, 30 or even 40 km between two interfluves. In the semiarid part of East Africa the conditions are just the same, as revealed by the detailed field studies of Louis (1964). The Ruvuma Valley is 70–80 km wide with a depth of 300 m between the interfluve divides and the river-bed.

The same principles apply to the *planaltos* of the Brazilian Shield where trough valleys similar to those of eastern Africa and the Deccan Plateau have been sculptured on crystalline massifs. The numerous reservoirs that were built in north-east Brazil to cope with the water shortages caused by recurrent droughts are extended but shallow and tend to be filled up by silt just a few decades after their construction, thus, rendering these palliative measures against drought stress very inefficient. Other parts which correspond to the same climagenetic zone are poorly documented. The vast semihumid and semiarid interior highlands of Brazil are also included by geomorphologists in this zone. However, human occupation and land use are still so scarce that there are only a few major installations which could demonstrate, as in India, their dependence on relief conditions. A general idea is, nevertheless, conveyed by the meaning of the local name *planalto* (high plain), which is given with particular regional attributes to different parts of central Brazil (Planalto de Borborema, Planalto dos Parecis, Planalto Central being good examples).

In general it must be acknowledged that the seasonally wet-and-dry outer tropics suffer a region-specific disadvantage: the building of dams and irrigation facilities in their uplands is hampered by particular natural geographic characteristics that are reflected in the landforms and river discharges.

Part II

Modern attempts at solving the problems of tropical agriculture; the persistence of the ecological constraints

Introduction

The preceding chapters addressed the thesis that agricultural activities in the greater part of the tropics are subject to certain ecological constraints which are unknown in extratropical countries. Consequently the often quoted North–South gradient in the socio-economic development of the present world is, in fact, inherited and has its roots in differing natural geographic conditions.

The key arguments of the pertinent discussions may be briefly summarized as follows:

1. The persisting shifting cultivation or field/fallow rotation practices on tropical ferrallitic soils cannot be rated as underdeveloped, backward and basically modifiable types of land use: they are the necessary adaptation to the prevailing ecological conditions.

2. The ecologically decisive properties of the ferrallitic soils in the ever-humid tropics are, first of all, the extremely low cation exchange capacity (CEC), the very high acidity, and the high decomposition rate of organic material. The intrinsic nutrient poorness, in comparison, must be regarded as restricting rather than decisive.

3. There is no contradiction between the huge biomass production of the virgin tropical rain forests and the rigorous limitation imposed on the production of food crops on the very same ferrallitic soils. The former is made possible by specific structural properties of the natural forest formations which prevent the loss of nutrients and warrant a closed mineral cycling.

4. Well-delimited habitats with extraordinarily high agricultural productivity and high population density in the midst of field/fallow rotation landscapes exist at circumscribed locations with exceptionally favourable soil conditions, such as high CEC, low soil acidity, restricted decomposition of organic material, and high residues of weatherable primary minerals. In total these habitats amount to about 15 per cent of the area of the humid tropics.

5. There exists an antithesis between the thinly populated areas in the inner

181

tropics with optimal climatic growing conditions and the densely or even over-populated and famine-stricken zones in the wet-and-dry outer tropics which are climatically unreliable. However, the latter conditions are more favourable to the formation of productive soils (fersiallitics). They result from less intensive chemical weathering, reduced nutrients outwash and slower decomposition of organic matter. These in turn result in a higher content of weatherable primary minerals, a high CEC, favourable soil acidity and better provision of nutrients.

6.   The climatologically unreliable conditions in the potentially fertile wet-and-dry outer tropical zones call for a switch from rainfed agriculture to artifical irrigation. However, zone-specific features such as the peneplanation relief and the runoff conditions render it extremely difficult to construct dams and irrigation systems. Therefore, the lack of appropriate water management systems persists until today.

The second part of the book analyses the approaches, perspectives and results that emerged in the field of tropical agriculture as a consequence of the implementation of international development strategies.

Two phases must be distinguished here. The first is connected with the spread of the Green Revolution technological package, a revolutionary development, indeed, which started after the Second World War and within a few decades dominated those domains where it proved applicable. In most parts of the world the spreading process was already finished by the mid-1970s, and what remains to be analysed are the reasons why it succeeded in some areas and not in others.

The second phase is less revolutionary. It comprises the various approaches of numerous institutions of different shapes and administrative stature. While some approaches are of a more scientific–theoretical nature aiming at long-term results, others search for short-term practical solutions through farming and colonization projects.

Prominent in the field of long-term research are the 'International Institutes of Tropical Agriculture', such as IITA, ICRISAT, CIAT and YAES, but other minor ones will also be mentioned. Decisive for a somewhat detailed discussion of the long-term approaches and results is their importance in an ecological context (Ch. 8). Of the numerous attempts to overcome the constraints in agricultural production on low-base-status soils by means of farming practices, those of supraregional and local significance must be considered.

Moreover, the persistence of strong geoecological constraints in the cultivation of tropical ferrallitic soils has to be acknowledged. Regardless of the kind of approach taken, there is one basic message: the key role for overcoming the limitations of the soils mentioned is a careful treatment of all organic matter by means of a waste-free recycling.

Finally, agroforestry will be discussed in order to see if it really is as revolutionary an approach as is often claimed (Ch. 9).

7

The Green Revolution in the tropical realm

As indicated, the progress which is associated with the term 'Green Revolution technological package' represented a real revolution in agrotechnical and production development. Beginning in the late 1940s, it expanded within two or three decades to all environments where its applicability was proved. It is not the purpose of the following exposition to discuss in detail the positive and negative effects of this new technology. Rather in the ecological context, which is our concern, an attempt is made to evaluate the regional differences in progress that has actually been achieved in the tropics. At the same time, we want to discuss the limits that are imposed on it by the tropical environment.

7.1 The main characteristics of the Green Revolution technological package

The new agrotechnology consists basically of the adoption of high-yielding varieties (HYVs) of wheat and rice and an associated set of improved production techniques. Without the latter, a successful cultivation of HYVs would be impossible.

The new wheat and rice varieties were developed by the International Maize and Wheat Improvement Centre in Mexico and the International Rice Research Institute in the Philippines, respectively, both with grants from the Rockefeller and the Ford foundations. They were ready for world-wide use at the beginning and end of the 1960s, respectively, and environmental conditions permitting, they found rapid and widespread adoption.

The HYVs are (1) endowed with a 20 per cent higher photosynthetic efficiency than traditional varieties, (2) generally semi-dwarf with short, stiff stalks that do not fall down (lodge) with high fertilizer applications as traditional varieties often do, and (3) usually early maturing. Some new rice and wheat varieties mature within 100 days.

The associated improved production techniques comprise basically four measures: first, there is intensive soil preparation aimed at establishing optimal physical soil structure for aeration, water reception, and rooting. This demands a lot of energy and fuel-consuming machinery. Second, applications of herbicides are required to eradicate competitive weed growth, of pesticides to wipe out disease-causing agents, and of insecticides to eliminate the transmitters of disease and nutrient competitors. Third, care is taken to space the crop plants as closely as light conditions and root competition will allow in order to maximize the effectiveness of the aforementioned measures. Fourth, in response to the resulting extreme nutrient demand per soil volume, the nutrient supply is increased with the help of intensive fertilization. The yield response is greatly affected by the quantity, time and method of fertilizer application.

All these measures require an input of energy which is almost exclusively derived from fossil energy sources. However, deployment of this energy and correct management of the technological package result in amazingly high yield increments for the HYVs. The greatest successes were achieved in West and Central Europe and in certain regions of the United States. Table 7.1 reflects the revolutionary development which took place in Germany and France after 1950 in the context of the Green Revolution style of production techniques.

In the United States the average yields and the yield increase of wheat lag behind Europe since wheat is grown partially in drought-prone areas. The absolute yield values are clearly lower than in Europe, but the increase from 16 to 32–34 bushels per acre during the 1950s and 1960s is just as striking as the yield increases experienced in European countries. If grain production takes place under relatively reliable climatic conditions and on good soils, as in the case of corn in the Middle West and the eastern Central Plains, the production increase is even more dramatic (Fig. 7.1). Between 1950 and 1970 there was a two-and-a-half fold average yield increase for corn: from 36–38 to 80–83 bushels per acre (1 bushel of corn equals 31.8 kg). This level has been maintained until the 1990s.

Table 7.1   Wheat yields in Europe, 1850 to 1977–79 (100 kg ha^{-1})

	*c.*1850	1909–13	1934–35	1948–52	1977–79
Denmark	c.12	33.1	28.1	36.5	52.3
Belgium	10.5	25.3	27.7	32.2	48.2
Netherlands	10.5	23.5	30.9	36.5	59.0
Germany	9.9	24.2	20.0	26.2	45.6
United Kingdom	9.9	21.2	21.5	27.2	51.2
Austria	7.7	13.7	18.0	17.1	36.9
France	7.0	13.1	14.1	18.3	46.7
Italy	6.7	10.5	10.1	15.2	25.4
Norway	5.7	16.6	19.5	20.6	40.7
Romania	—	12.9	8.0	10.2	25.8
Hungary	—	13.2	11.3	13.8	38.6
Bulgaria	—	6.2	9.2	12.4	38.3
Spain	4.6	9.2	14.7	8.7	16.3
Greece	4.6	9.8	9.2	10.2	23.9
Russia	4.5	6.9	8.1	8.4	16.5

Source: Grigg (1984).

184

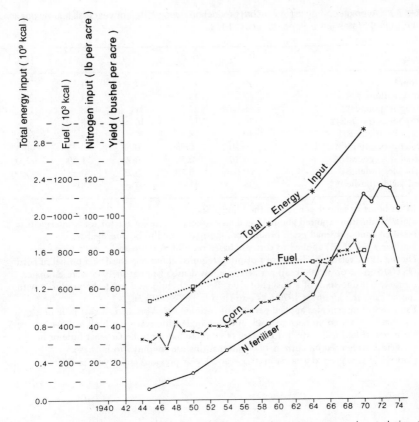

Fig. 7.1 Increase in corn yields, fertilizer application and average energy input during the 1960s and 1970s in the United States. (Source: Weischet, 1978).

Aside from the new hybrid seeds, all this could only be achieved with an impressive deployment of energy and fertilizers. In 1973, Pimentel and his co-workers from the New York State College of Agriculture and Life Science related the increase in corn production with the corresponding rise in average energy input during the decisive phase of revolutionary agricultural development (Table 7.2).

Considering the possibility of transferring the Green Revolution technological package to the countries of the tropics one should realize that, between 1950 and 1970, consumption rose 4.5 times in phosphates, 10 times in nitrates and 12 times in potassium.

185

Table 7.2   Average energy inputs in corn production during different years (all figures per acre) in the United States (after Pimentel *et al.*, 1973)

Inputs	1950	1954	1959	1964	1970
Labour*	18	17	14	11	9
Petrol (gallons)†	17	19	20	21	22
Nitrogen (pounds)‡	15	27	41	58	112
Phosphorus (pounds)‡	10	12	16	18	31
Potassium (pounds)‡	10	18	30	29	60
Seeds for planting (bushels)§	0.20	0.25	0.30	0.33	0.33
Insecticides (pounds)¶	0.10	0.30	0.70	1.00	1.00
Herbicides (pounds)‖	0.05	0.10	0.25	0.38	1.00
Corn yields (bushel)**	38	41	54	68	81

*   Mean hours of labour per acre in the United States.

†   Corn production required about 15 gallons of fuel per acre for tractor use – intermediate between fruit and small grain production. Corn appeared to be intermediate.

‡   Fertilizers (N, P, K) applied to corn are based on USDA estimates.

§   During 1970, relatively dense corn planting required about one-third of a bushel of corn (25 000 kernels or 34 000 kcal) per acre; the less dense plantings in 1945 were estimated to use about one-sixth of a bushel of seed. Because hybrid seed has to be produced with special care, the input for 1970 was estimated to be 68 000 kcal.

¶   Estimates of insecticides applied per acre of corn are based on the fact that little or no insecticide was used on corn in 1945, and that application reached a high in 1964.

‖   Estimates of herbicides applied per acre of corn are based on the fact that little or no herbicides were used on corn in 1945 and that this use continues to increase.

**   Corn yield is expressed as a mean of 3 years, 1 year previous and 1 year past.

7.2   Regional differentiation in the adoption of the new technology

Dana G. Dalrymple of the US Department of Agriculture, probably the best informed agricultural economist with respect to the spread of the new technology in developing countries of the tropics, quotes from the latest update of the 1976/77 crop year the following numbers regarding the proposition of crop area planted to HYVs in different parts of the tropics (Dalrymple 1979, p. 705; Table 7.3): somewhat more than 24 million ha had been planted with the new varieties of wheat in Asia (south, east and west) and in North Africa, but only 225 000 ha (0.6 per cent) in the tropical part of Africa. The data may be more regionally indicative when expressed as percentages of the total area planted with HYVs. The greatest acceptance (72.4 per cent) was in South and East Asia, where the HYV wheat areas are concentrated in four countries: India (14.7 million ha), Pakistan (4.6 million ha), Nepal (0.25 million ha) and Bangladesh (0.17 million ha). 'India has by far the largest HYV area representing over half the total world area' (Dalrymple, 1979: 708). All four countries lie in the outer tropics, in semihumid or semiarid climates. This agrees with the statement of Ruttan (1977)

Table 7.3 Estimated area of high-yielding varieties (HYVs) of wheat and rice, and proportion of crop area planted with HYVs, in less developed nations, 1976/77*

Region	Wheat	Rice	Total
HYV area (ha)			
Asia (South and East)	19 672 300	24 199 900	43 872 200
Near East			
(West Asia and North Africa)†	4 400 000	40 000	4 440 000
Africa (excluding N. Africa)†	225 000	115 000	340 000
Latin America	5 100 000	920 000	6 020 000
Total	29 397 300	25 274 900	54 672 200
HYV area/total area (%)			
Asia	72.4	30.4	41.1
Near East	17.0	3.6	16.5
Africa	22.5	2.7	6.5
Latin America	41	13.0	30.8
Total	44.2	27.5	34.5

* Excluding Communist nations, Taiwan, Israel and South Africa.
† Particularly rough estimate of area.
Note: 88 per cent of the tropical HYV wheat area was in Asia.

Source: Dalrymple (1978).

that the HYV wheats were adopted at exceptionally fast rates in relatively arid areas where farmers have had access to effective tube-well or gravity irrigation systems.

Based on this, it can be assumed that in Africa the HYV wheat areas lie in the semihumid or semiarid climatic regions of the East African Highlands. In South America most of the HYV wheat is grown in the extratropical countries of Argentina and Chile, in the temperate highlands of the Andes, and in the drier regions of Brazil and Venezuela.

The estimated HYV rice area of 24.2 million ha in Asia also includes (in contrast to the HYV wheat area) parts of the humid inner tropics. In the crop year of 1976–77, the area of HYV rice amounted to 2.4 million ha (68 per cent) in the Philippines, 3.4 million ha (41 per cent) in Indonesia, 0.22 million ha (37 per cent) in west Malaysia, 0.7 million ha (11 per cent) in Thailand, and 0.35 million ha (7 per cent) in Burma (Dalrymple, 1979). The percentages in parentheses represent the proportions of the total rice crop area, which are well correlated with the proportion of wet-rice cultivation in the different countries.

Figure 7.2 clearly shows that the ratio of irrigated land to total cropland (irrigation ratio) in Cambodia, Burma and Malaysia is noticeably smaller than in Indonesia, that the land productivity rises in proportion with the irrigation ratio, and that the most productive regions lie in the outer tropical and subtropical parts of East Asia. 'Japanese economists distinguish between two landmarks in land productivity in Asia. The first is an average rice yield of about 2.3 tons/ha, the second, of about 3.8 tons/ha' (Bray,

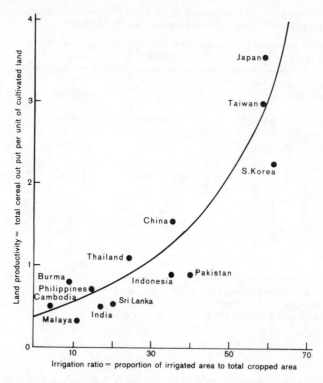

Fig. 7.2 Ratio of irrigated land to total cropland in relation to land productivity in Asian countries (after Ishikawa, 1967).

1986: 62). In all likelihood, the first productivity level is typical of the inner tropical regions.

For an assessment of the success the yield increment is at least as important as the areal extension of HYVs and the new technology. Dalrymple (1979, p. 716) summarizes the experiences in this respect as follows:

> It is commonly thought that the effect of the HYV package on yields has been more significant for wheat than for rice. In an earlier study I assumed that, as of 1972/73, the HYV package resulted in a 50 per cent increase in wheat yields and a 25 per cent increase in rice yields in Asia. A subsequent report on rice suggested that 'the real South Asian margin of superiority is in the order of 25 per cent, while the Southeast Asian margin is in the 15 to 25 per cent range'. A recent survey primarily based on data from India and Pakistan suggests average yield increases of 100 per cent for wheat and 40 per cent for rice.

Thus the largest increments in yields have been achieved in the outer tropical countries of South Asia, such as India, Pakistan and Bangladesh, while the increments in the humid tropics of South-east Asia are much lower.

188

As to the actual rice-yield figures, studies sponsored by the International Rice Research Institute (IRRI) found a considerable difference between (1) biological potential yields of about 14 t ha^{-1}, (2) practical potential yields of 6–8 t ha^{-1}, and (3) actual yields of about 2 t ha^{-1}. 'Much of the gap between (1) and (2) is due to largely uncontrollable environmental constraints (such as weather and soil characteristics). The gap between (2) and (3) is composed of physical and biological constraints subject to management and control by farmers; the degree to which these limitations are overcome is influenced by socio-economic constraints' (Darymple, 1979, p. 712).

7.3 Soil-related constraints on the application of the new technology in the humid tropics

The first three of the above-mentioned conditioning production techniques (intensive soil preparation, application of herbicides, pesticides and insecticides, and dense seed spacing) might, *in principle*, be implemented with international assistance to such an extent that all climatically suitable regions of the world, including the tropics, could benefit from enhanced production of grains. The actual difficulties consist in solving the problems of the propagation of the know-how, of the distribution of chemicals, and of the supply of capital.

However, the three mentioned conditions remain useless if the fourth prerequisite is not met, namely the supply of the necessary nutrients with the help of chemical fertilizers.

Efficient fertilization depends on the existence of sufficient cation exchange capacity (CEC) in the soil. As explained in Chapter 3, most of the well-drained soils of the humid tropical lowlands have CEC values of less than 4–5 meq per 100 g fine soil as compared to normally 20–30 meq in 100 g fine soil in the extratropics. The details of the difficulties involved in applying the new technology to tropical soils of low CEC (oxisols and ultisols) will be discussed in section 8.1 in connection with the long-term continuous cropping experiments in Yurimaguas. These experiments revealed that, at present, this technology is applicable in only a few specific regions of the humid tropics.

Quite different are the prospects for the ecologically exceptional regions of the humid tropics (Ch. 5), especially alluvial plains used for the cultivation of wetland rice on submerged soils. The ecological advantage and superiority of the artificial paddy soils as compared to the well-drained soils of the tropics pertaining to fertility maintenance have been analysed in section 5.5. The problem is the low nitrogen fertilizer effectiveness: 'Numerous nitrogen-use experiments in different Asian countries have shown that the recovery of fertilizer nitrogen applied to the rice crop is normally less than 30 to 40 per cent. Even with the best agronomic practices and strictly controlled conditions, the recovery seldom exceeds 60 to 65 per cent' (de Datta, 1978). The most important pathways of nitrogen losses are nitrification and denitrification as well as ammonia volatization (see section 5.5).

189

Table 7.4 Production and index of principal foodstuffs in Indonesia, 1952–73 (1,000 tons: 1960=100) (after Palmer, 1976)

		1952	1960	1966	1967	1968	1969	1970	1971	1972	1973
Rice (paddy):	Production	6 600	8 767	9 339	9 046	11 606*	12 249	13 140	13 724	13 291	14 702
	Index	75	100	107	103	133*	140	150	157	152	168
Maize:	Production	1 637	2 460	3 717	2 369	3 166	2 293	2 825	2 632	2 254	2 912
	Index	67	100	151	96	129	93	115	107	92	118
Cassava:	Production	7 535	11 377	11 232	10 747	11 356	10 917	10 478	10 042	10 385	9 399
	Index	66	100	98	92	100	96	92	88	91	83
Sweet potatoes:	Production	2 290	2 670	2 476	2 144	2 364	2 260	2 175	2 254	2 066	2 180
	Index	86	100	93	80	86	85	81	84	77	82
Peanuts:	Production	167	256	263	241	287	267	281	280	n.a.	n.a.
	Index	65	100	103	94	112	104	110	109	n.a.	n.a.
Soyabeans:	Production	286	443	417	416	420	389	498	475	518	446
	Index	65	100	94	94	95	88	112	107	117	101

* A new statistical series beginning in 1968 raised substantially the estimate of rice production, which would have been 10 166 000 t (index 116) according to the old series.

Sources: 1952 and 1960, Central Bureau of Statistics, Statistical Pocketbook 1963, Jakarta, p. 53; 1966–71, Central Bureau of Statistics, Monthly Statistical Bulletin, Jakarta, November 1972, p. 107 (milled rice = 52 per cent of paddy). New Statistical Series: 1968–73, Bulletin of Indonesian Economic Studies, Vol. X, No. 3, November 1974, p. 18. Also new rice production series: 1968–73.

This is the reason why nitrate-containing fertilizers such as ammonium nitrate are unsuitable, while ammonium sulphate and ammonium chloride are more effective. Ammonia volatization becomes operative only if the NH_3 cation does not find enough changing sites in the soil.

Because of the price advantage, ammonium-producing urea is the most widely used – though not ideal – fertilizer in tropical Asia. Nitrogen losses from urea are reported to range from 60 to 80 per cent for wet rice and from 40 to 60 per cent for upland crops (de Datta, 1978).

The restriction in the nitrogen supply probably causes the gap between the actual average rice yields ($c.$ 2 t ha^{-1}) and the biological optimal values, as well as the difference between the yields of rice or wheat that are reached in the temperate zone.

The progress of agriculture in Java during the years after the adoption of the new technology is traced in Table 7.4. The total production and the production index of main food crops reveal that progress has been achieved only with rice. Smoothing the statistically founded incongruity of the data in 1968, for the 13 years during which HYVs were used, a production increase from 8.7 to 13.7 million t was achieved. Since on the island itself no considerable expansion of the cropping area was possible, the increase (roughly 57 per cent for wet rice) can only be ascribed to the new agricultural technology.

7.4 The Green Revolution in the semihumid and semiarid outer tropics

The suitability of soils for HYVs increases from the humid inner tropics to the semihumid or even semiarid outer tropics where soils have a higher content of primary minerals, are less leached, and most importantly, possess a higher clay-mineral-bounded CEC (see Ch. 6). If to these zonal changes other favourable circumstances are added, such as altitudinal changes in environmental conditions (see Section 5.4) and possibly base-rich parent rocks of volcanic origin (Section 5.2), the conditions for the application of the new technology become optimal, at least as far as soil properties are concerned.

The problem with these regions is that the climatic conditions that create relatively favourable soils are limiting for agricultural activities. It is, therefore, necessary either to achieve a working compromise between soil quality and climatic risk, or to minimize the climatic risk with the help of water management and irrigation. The particular problems faced by water management projects in the peneplanation zone of the outer tropics were examined in Chapter 6.

The most successful countries in the Asian tropics to apply the Green Revolution technology were, without doubt, India and Pakistan. Up to the 1960s these densely populated countries had suffered from episodic famines in the wake of recurring monsoon failures. The last time, in 1966 and 1967, famine was averted only through grain deliveries from the United States, which sustained over 60 million Indians for a period of 2 years. Fifteen years later India no longer depended on imports, and today the government has enough buffering reserves to compensate for crop failures. This is truly a miracle, and a miracle that should be acknowledged in the numerous critical articles on the second-generation problems published since then (see overview in Glaeser, 1987).

Table 7.5   Wheat production in representative countries of the semihumid tropics

	1954–58	1960–64/65	1967/68–71/72	1970–74	1974–78
Kenya					
Wheat area (1000 ha)		104	146	105	120
Yield (100 kg ha⁻¹)	No values	11.4	13.3	13.7	13.4
Production (1000 t)		119	207	144	164
India					
Wheat area	11 979	13 406	15 732	24 820	25 485
Yield	7.1	8.1	11.4	12.7	10.6
Production	8 587	10 816	18 102	23 370	27 124
Pakistan					
Wheat area	4 240	4 957	6 016	6 042	6 157
Yield	7.8	8.2	10.4	11.8	13.5
Production	3 515	4 052	6 291	7 105	8 301

Source: Values from International Wheat Council (1980–81).

Table 7.5 shows the production increase in figures. During the second half of the 1950s there was in India a production of 8.5 million t of wheat on less than 12 million ha of land. Twenty years later the crop area had been increased by more than 25 million ha and total production had reached more than 27 million t. This was a yield increase from 710 to more than 1200 kg ha⁻¹. The figure is all the more astonishing if one considers that the overall mean includes large areas of non-irrigated upland fields, and that wheat is frequently planted during the relatively dry winter season. Therefore, the yields from the irrigated fields must be considerably higher still.

The ecological background for the exceptional situation of Indian agriculture before, and especially after, the introduction of the HYV type of agricultural techniques can be assessed by an inspection of Chatterjee's (1973) soil map in his excellent presentation of Indian agriculture in the *World Atlas of Agriculture* (Fig. 7.3; for analyses data of representative soils, see Appendix 4.)

According to Chatterjee (1973), almost 80 per cent of the land area falls into the three broad soil groups – alluvial, red and black. Alluvial soils, by far the most extensive (101.2 million ha), are also the most productive, contributing the largest share of the agricultural wealth of the country. These immature soils generally ensure

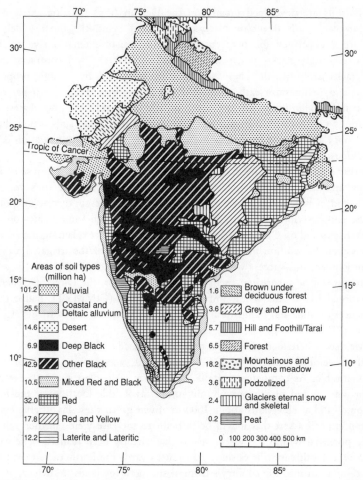

Fig. 7.3 Soil map of India (after Chatterjee, 1973).

good drainage and other conditions favouring crop production. The exchangeable base content is generally satisfactory and the mineral and organic plant food reserves are adequate and easily replenishable. (The alluvial soils are classified as a cambisol–luvisol association or fluvisols in the FAO/Unesco soil map of South Asia.)

Red soils cover over 50 million ha mainly on the Archaean crystalline rocks of the southern Deccan; depth and fertility vary considerably and they are fairly rich in potash. Although generally deficient in calcium, magnesium, phosphorus, nitrogen and humus, they are, if properly irrigated, suitable for a variety of crops. (In the FAO/Unesco soil classification these are chromic luvisols.)

Red-and-yellow soils occur in the hilly landscapes over the weathered metamorphic rocks of the Deccan and east of the trap basalt. They vary greatly in their fertility (FAO/Unesco: ferric luvisols).

193

Black soils are developed mainly on the Deccan traps (basaltic extrusions) with medium black soils (*regur* or black cotton soils) being the most widely spread. They are very clayey (62 per cent) and hold moisture tenaciously which makes them exceedingly sticky when wet. On drying out the soil contracts and large, deep cracks are formed. These soils are characterized by a high proportion of calcium and magnesium carbonates (6–8 per cent), iron oxide (9–10 per cent) and fairly constant aluminium (10 per cent); potash is variable and phosphates, nitrogen and humus content are low as a rule. Areas of these black soils are credited with high fertility and do not require manuring for long periods, but irrigation is essential (FAO/Unesco: chromic or pellic vertisols).

Lateritic soils occupy only 3.7 per cent of the total land area. Because of intensive leaching and their low base-exchange capacity, typical lateritic soils are generally infertile and of little value for crop production (FAO/Unesco: dystric nitosols).

These soil descriptions from Chatterjee, combined with some values on chemical characteristics and on CEC (see Appendix 4) are useful for explaining the areal success of the Green Revolution in India. With the expansion of the irrigated areas on the Deccan Plateau, this success will increase even further (see Ch. 6).

It should not be surprising, then, that R. D. Havener, the Director General of the International Maize and Wheat Improvement Centre (CIMMYT), hailed the wheat experiences of India and Bangladesh as real success stories achieved by his organization (Havener, 1985).

In Mexico, where increased wheat production based on improved Mexican cultivars ('locally developed varieties' or LDVs) started 10 years sooner, yields of 3.8 t ha^{-1} are remarkable. As compared to India, cultivation takes place during a better season (not in winter) and in a geographically more favourable location: at the margin of the tropics and at elevations above 1000 m. Even given these favourable conditions, Mexican experts assert that it will be difficult to increase yields further and that the genetic potential within the wheat species has already been realized. The expanding area of wheat cultivation is pushing its limit towards the northern semiarid fringe of Mexico, and not into the country's tropical southern lowlands (Martinez, 1985).

A glance at the wheat-cultivating situation in tropical Brazil reveals a similar trend. The major efforts are concentrated on incorporating the *cerrados* of central Brazil (acidic soils with highly soluble aluminium content and very low volumes of calcium, magnesium and phosphorus) into the cultivation of irrigated wheat. In good locations wheat from germplasm of old Brazilian varieties has yielded between 900 and 1700 kg ha^{-1}. It is expected that the utilization of a Mexican-type cultivar can increase the yields to 2–3.5 t ha^{-1} (da Silva, 1985).

Kenya appears in the statistics of the wheat-producing countries only after 1960. There has been no increase in cropped area and yields since then. Thus, it seems that Sudan, with 240 000 ha of wheat, is the tropical lowland African country that fares best among its neighbours. In fact, Ethiopia, which has the largest number of hectares dedicated to wheat (523 000), grows only rainfed highland varieties (Mann, 1985).

Summarizing the findings concerning the achievements of the Green Revolution in the tropical realm, it is obvious that countries in the wet-and-dry outer tropics were able to profit most from the new technology. The progress was especially considerable in the semiarid part of that climatic zone. When the water supply was secured, on the nutrient-rich fersiallitic soils of high CEC, HYVs of wheat as the most successful grain could be produced.

In the humid inner tropics where rice is the main staple production remarkable advances were also made. In total, however, they are less pronounced than wheat production in the outer tropics. Especially they are limited to the regions with paddy rice cultivation. Only the paddy soils permit the application of modern fertilizer technology, whereby nitrogen fertilization is less effective than in the case of wheat cultivation on irrigated alluvial soils.

No progress at all has been achieved by the new technology on ferrallitic upland soils of the humid inner tropics. Upland rice, as upland staples in general, must still be grown with traditional methods of land rotation as before. Since in tropical Asia the area with intensive continuous cultivation comprises only 20 per cent of the total land surface and the proportion elsewhere (particularly in tropical Africa) is still less, the largest part of the climatically favourable humid tropics has not seen progress. The principal constraints in cultivating tropical low-base-status soils still persist after the spread of Green Revolution technology.

8

International attempts at the level of research institutions

The chief constraints for cultivating tropical low-base-status soils still persist after the spread of the Green Revolution technology. Investigations in the context of the ongoing second phase of international development strategies are numerous. Many of these investigations are of region-specific scope; others are of supraregional significance and of special importance in view of the ecological argumentation. Among these, the long-term continuous cropping experiments in Yurimaguas (Peru) deserve for varied reasons, prominent attention. First, the experiments were essentially designed to reveal the conditions under which 'continuous cropping is possible for soils normally subjected to shifting cultivation' (Sanchez et al., 1981: 11). Second, after 13 years and 25 consecutive crops, Sanchez and collaborators were ready to answer the question and publicised the results in numerous publications as an agricultural alternative. And third, the authors made available in the series *Technical Reports, Agronomic-Economic Research on Soils of the Tropics* a unique scientific material on agro-ecology, on the basis of which it is now possible to discuss the problem with accurate records and solid conclusions (Soil Science Department, North Carolina State University, Technical Reports 1976 to 1983; Triennial Technical Report 1981–84).

8.1 The long-term continuous cropping experiments in Yurimaguas (Peru) and their consequences

In 1972, Pedro A. Sanchez, Professor of Soil Science at North Carolina State University, initiated in collaboration with the Tropical Soils Research Programme of North Carolina State University (NCSA), the United States Agency for International Development (USAID) and the Peruvian Ministry of Agriculture through its Instituto Nacional de Investigación y Promoción Agropequaria (INIPA), the Yurimaguas Agricultural Experimental Station (YAES).

The station is located some 80 km from the Andean foothills in the Peruvian part of the Amazon basin. The deeply weathered underground is characterized by red regolithic material overlain by reddish kaolinitic clay with inserted redox horizons

196

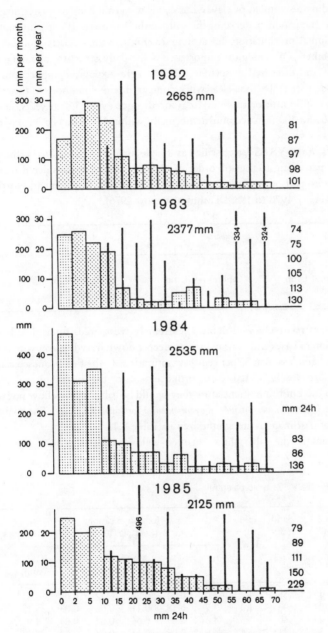

Fig. 8.1 Frequency of daily precipitation amounts, annual course of monthly rainfall and maximum daily amounts of more than 70 mm per 24 h in rainy and less rainy years in Yurimaguas (YAES) for 1982–85.

and tapped by a sediment of clayey fine sands with thin beds of quartzitic gravel. The latter form the parent material of the soils of the Yurimaguas experimental plots.

Mean annual precipitation lies around 2200 mm with a relative lull between June and September. Of ecological importance is the daily structure of rainfall as shown in Fig. 8.1 and discussed in section 1.2. The agro-ecological implications of such a precipitation pattern become clear when compared with conditions in the extratropics: downpours of 50 mm or more in 1 day occur once every 5 or 6 years, and amounts between 50 and 100 mm accumulate normally within 15–20 days of rain over a period of 4 weeks or so.

The soils of the YAES are classified as 'fine loamy, siliceous, isohyperthermic Typic Paleudult' pertaining to the soil order of ultisols, which are considered representative of the western part of the Amazon lowlands (Soil Science Department, North Carolina State University 1976 to 1983; Technical Report 1976).

Major group	Suborder	Order
Paleo	ud	ult
abbreviation of Latin	abbrev. of Latin	abbrev. of
palaios = old	*udus* = humid	ultisols

Ultisols are relatively well-drained, extremely weathered, mineral soils with very low retention of bases and poor in humus; deeper down lies a horizon with translocated silicate clays and less than 35 per cent base saturation. In the FAO/Unesco classification they are more closely related to the orthic acrisols or the dystric nitosols than to the luvisols. In the English nomenclature they would be called 'red–yellow podzolic soils', in French *sols ferralitiques lessivés, moyennement à fortement désaturés*. The analyses of the paleudult of Yurimaguas are synthesized in Table 8.1.

The values in the table show this red–yellow soil to be deep, very acid, low

Table 8.1   Profile y 17: typic paleudult

Horizon (cm)	Clay (%)	Sand (%)	pH (1:1 H_2O)	Exchange cations (meq per 100 g)						Base saturation (%)
				Al^+	Ca	Mg	K	Na	CEC (sum)	
0–4	19	58	4.2	3.0	0.24	0.20	0.12	0.04	3.6	17
4–16	24	46	4.3	3.7	0.18	0.11	0.06	0.03	4.1	9
16–25	28	40	4.3	4.6	0.09	0.06	0.04	0.04	4.9	5
25–50	30	41	4.4	4.9	0.05	0.03	0.04	0.03	5.1	3
50–100	32	39	4.4	5.4	0.09	0.03	0.04	0.03	5.6	3
100–200	39	31	4.6	7.4	0.05	0.03	0.06	0.03	7.6	2

Source: Soil Science Department, North Carolina State University, 1976 to 1983; Technical Report 1975/76.

in organic matter, low in mineral residues other than quartz, deficient in most nutrients, with a low effective cation exchange capacity (CEC) and a high aluminium toxicity. The physical properties are favourable for agriculture. In the virgin forest soils of Yurimaguas the bulk density close to the surface (down to 5 cm) lies around 1.2 g cm^{-3}, at a depth between 20 and 30 cm, near 1.3 g cm^{-3}, and near 1.5 g cm^{-3} at 60–100 cm. In slash-and-burn plots the values for the upper 20 cm change to 1.45–1.5 g cm^{-3}. Infiltration rates are variable, but generally rather high: 30–80 cm h^{-1} in forested areas and 10–15 cm h^{-1} in slash-and-burn *chacras* (Soil Science Department, North Carolina State University, 1976 to 1983; Technical Report 1978/79, Table 3.8.4). Another favourable physical property is the high stability of the soil aggregates. These favourable physical soil conditions could be preserved during the entire test period (Sanchez, 1985, p. 14).

8.1.1   Set-up of the experiments and research strategy

Originally the terrain of YAES was covered by a 17-year-old evergreen secondary forest (*purma*). In June and July 1972 the first experimental plot (*chacra* I) was opened by the traditional slash-and-burn method. *Chacras* II and III followed in 1973 and 1974 (Fig. 8.2), each *chacra* measuring 1.5 ha. Initially, four crop systems were cultivated in each *chacra* without the use of heavy machinery. Later on, only the most promising crop systems were retained:

1.   Continuous dry rice (IR 4–2);
2.   A rotation of dry rice (IR 578–8)–maize (Cuban Yellow)–soyabean (Improved Pelican);
3.   A rotation of dry rice (IR 578–8)–peanut–soyabean (Improved Pelican).

After suspension of the continuous dry-rice culture in 1976 the remaining two most promising cropping systems were continued without change until 1985.

Each of the cropping systems was subjected to three different treatments: one received no fertilization at all, one received medium fertilization, and one complete fertilization. In this way it was possible to compare the temporal effects after clearing as well as the impact of different fertilization treatments given under the same weather conditions.

In 1981, Sanchez *et al.* were ready to address the question of whether or not continuous cropping of staple food is possible on soils normally subjected to shifting cultivation: 'The answer, we have found, is unquestionably "yes" when proper management practices as dictated by a constant monitoring of the soil dynamics are used' (Soil Science Department, North Carolina State University, 1976 to 1983; Technical Report 1980–81). In the same report the authors summarise grain yields of 21 consecutive crops for the rice–corn–soyabean and the rice–peanut–soyabean rotation harvested from the same fields since they were first cultivated in October 1972, and conclude that 'these results attest to the fact that continuous production can be achieved in the Amazon with adequate fertilization'. In 1985 they published for the first time the 'Fertilizer requirements for continuous cultivation of annual rotations of rice–corn–soyabean or rice–peanut–soyabean on an acid ultisol in Yurimaguas, Peru' (Table 8.2). The authors stressed that the 'fertilizer recommendations, as are all sound ones, are site-specific. Nevertheless, they are representative of the level of fertilizer and lime inputs required for continuous crop production in ultisols' (Soil Science Department, North Carolina State University, 1976 to 1983; Technical Report 1980–81).

In the following years the main results of the continuous cropping experiments were published in numerous sources; for example Sanchez *et al.* (1982), Sanchez (1983, 1985), or Nicholaides *et al.* (1985a). The fertilizer recommendations, occasionally highlighted as 'Yurimaguas technology', are reproduced in Table 8.2.

Accepting the statement from 1981 and the findings reported in later publications, one can conclude – as many certainly do – the following: if long-term experiments in a typical location lead to an 'unquestionably yes' concerning the possibility of permanent cultivation on humid tropical lowland soils, then it was an error to assume that modern agricultural technology would not be capable of establishing continuous cropping systems for the production of food staples even on the ferrallitic soils of tropical rain forests.

It is not that simple, though. First of all, the authors of the above statement have already added to their 'unquestionably yes' the condition: 'when proper management practices as dictated by a constant monitoring of the soil dynamics are used'. What

Table 8.2 Lime and fertilizer requirements* for continuous cropping of a three-crop per-year rotation of rice–maize–soyabean, or rice–peanut–soyabean on an ultisol of Yurimaguas, Peru

Input†	Rate per ha	Frequency
Lime	3 t $CaCO_3^-$-equivalent	Once each 3 years
Nitrogen	80–100 kg N	Rice and maize only, split applications preferable
Phosphorus	25 kg P	Each crop
Potassium	100–160 kg K	Each crop, split applications preferable
Magnesium	25 kg Mg	Each crop, unless dolomitic lime is used
Copper	1 kg Cu	Once each year or two
Zinc	1 kg Zn	Once each year or two
Boron	1 kg B	Once each year or two
Molybdenum	20 g Mo	Mixed with legume seed during inoculation

* Depend on soil test analysis and recommendations.

† Calcium and sulphur requirements are satisfied by lime, simple superphosphate, and Mg, Cu and Zn carriers.

Source: Nicholaides *et al.* (1985a).

is meant by 'proper management' and by 'dictated by constant monitoring of soil dynamics'?

To probe the meaning of these formulations, it is necessary to reconstruct the course of the experiments. Most of the pertinent information appears in the annual reports, the remainder was personally gathered at the site of the experiments, in Yurimaguas.

8.1.2 *The actual performance of the continuous cropping experiments*

Figures 8.2–8.6 describe the course of the experiments by indicating the yield sequences, soil parameters, precipitation data and the amounts of fertilizers applied. The yield values refer only to the unfertilized and the completely fertilized crops. The complete treatment consisted of a base application of 80–100–80 kg ha^{-1} N–P–K

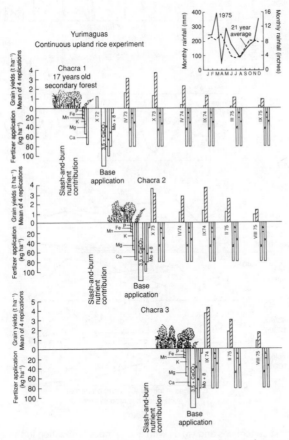

Fig. 8.2 Yields and fertilizer treatments in three *chacras* under continuous dry rice rotation. YAES, 1972–75.

(nitrogen, phosphorus and potassium) supplement to the nutrients received from the ash after burning the secondary forest. Moreover, 3.5 t ha⁻¹ of ground limestone were added in order to raise the soil pH to 5.5. Subsequent crops received 80–20–80 kg ha⁻¹ N–P–K fertilizer for each planting.

In 1973 and 1974, the harvests of *chacras* 1 and 2 yielded considerable amounts

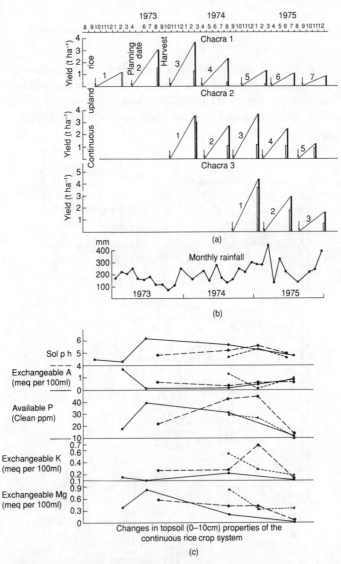

Fig. 8.3 (a) Planting sequences and yields with and without fertilization in the continuous rice rotation experiment; (b) monthly rainfall patterns; (c) variations of important soil parameters during 1973–75 in Yurimaguas.

202

of 3–4 t ha^{-1} of rice or corn. But 'a severe reversal of these trends was observed in 1975. The continuous upland rice and the rice–corn–soyabean system suffered a precipitous yield decline regardless of fertility treatments' (Soil Science Department, North Carolina State University, 1976 to 1983; Technical Report 1976). Even the second harvest of *chacra* 3, which had been cleared and planted only in July 1974, was greatly reduced despite the nutrient amounts supplied by ash and chemical fertilizers. The third harvest of this *chacra* was as poor as those of the other two (Fig. 8.2).

What were the reasons for this general decline? The corresponding soil analyses (Fig. 8.3) indicate a marked worsening of growing conditions in the upland rice culture of all *chacras* between January and July of 1975 because of an increase in soil acidity and aluminium toxicity and a decrease of available phosphorus as well as exchangeable potassium and magnesium. Exchangeable calcium showed similar deterioration after January 1975 (Sanchez, 1983).

The results presented in Figs 8.2 and 8.3 for the continuous rice cultivation are equally valid for the other cropping systems (Fig. 8.4).

Years later, evaluating the soil and crop uptake dynamics, Nicholaides *et al.*, (1983a, p. 12) makes the following statement concerning the reasons for the yield decline: 'Soil analysis identified two responsible factors: a shorter than expected residual effect of the lime applied, and the triggering of a Mg deficiency induced by the K applications.' On the same subject, Sanchez (1983, p. 53) writes that yield reduction began 'when fertilization produced a Mg/K imbalance that triggered Mg deficiency, and the unexpected short residual effect of lime terminated'.

The question arises why the trigger effect and the unexpected short residual effect of lime occurred simultaneously in all cropping systems. Since the drastic yield reduction and deterioration of soil conditions occurred regardless of the type of crops or rotation, the methods of fertilization, the time elapsed after liming (several years or just one), and independent of the length of cultivation in the different *chacras*, one must conclude that the determining factor was beyond the experimenter's control.

One cannot but suspect that this determining factor must have been the extremely high precipitation that fell between December of 1974 and May of 1975 (Fig. 8.2 and 8.3). The crops planted and fertilized in November of 1974 were exposed to 3 months of heavy rain with a total of 750 mm between December and February. The next seeds, which were planted and fertilized at the end of February and beginning of March respectively, received 400 mm of precipitation, i.e. twice the average amount, right at the beginning of their growth period. The exceptional character of the period in question shows up in a comparison of the monthly values with those of previous years and the 21-year average (Fig. 8.2).

Information can be gained from the measures that were taken in order to stop and reverse the downward trend of the yields (Fig. 8.4). In August of 1975, another 0.7 t ha^{-1} of lime were applied to the older *chacras* 1 and 2. *Chacra* 3 was excluded because it had been limed only the year before. All *chacras* received 32 kg ha^{-1} of phosphorus and 9 kg ha^{-1} of magnesium in addition to the customary 100 kg of nitrogen (20 kg for soyabeans), 26 kg of phosphorous fertilizer and 80 kg of potassium fertilizer (Soil Science Department, North Carolina State University,

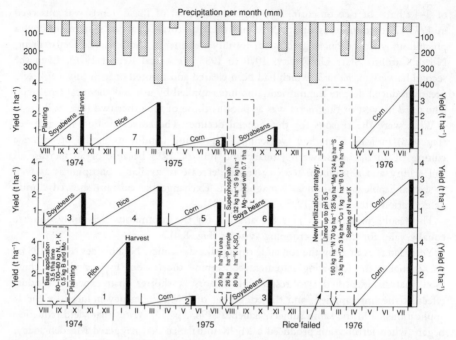

Fig. 8.4 Crop yields, fertilizer applications in three YAES *chacras* of the continuous soyabean–rice–corn rotation and monthly rainfall, 1974–76.

1976 to 1983; Technical Report 1975, p. 127). However, the November yields for soyabeans as well as upland rice were still unsatisfactory, and worse was yet to come. After more than 350 mm of rain in December 1975 and almost 400 mm of rain in January 1976, the rice stood so poorly in March of 1976 that the crop was abandoned (Soil Science Department, North Carolina State University, 1976 to 1983; Technical Report, 1976–77, p. 16). The same report states: 'It was evident that new fertility levels had to be developed for continuous cropping in all *chacras* and the decision was made to change our fertilization strategy.'

After evaluating the extensive soil and plant analyses from the experimental plots and from other special experiments, the new fertilization strategy was ready in April of 1976. It consisted of the following measures:

1. Further lime applications in order to lower soil acidity with amounts varying between 0.25 and 2 t ha^{-1}, depending on the testing site.

2. Increase of fertilizer amounts to 189 kg nitrogen (N), 70 kg phosphorus (P), 125 kg potassium (K), 30 kg magnesium (Mg), 124 kg sulphur (S), 3 kg zinc (Zn), 3 kg copper (Cu), 1 kg boron (B) and 0.1 kg molybdenum (Mo) per hectare (30 kg ha^{-1} N for soyabeans).

3. Splitting of the nitrogen and potassium amounts, the first half to be applied at planting, the other 40 days later.

204

These additional high fertilizer rates helped to stabilize yields at 2.2–2.5 t ha^{-1} for rice, soyabeans or peanuts, and at 3 t ha^{-1} for maize by the end of 1976. However, constant monitoring revealed that by 1978 the pH value of the soil had again become too low, so that another lime application became necessary, the third in 3 years!

The course of the crisis and the different steps taken to correct it clearly reveal the importance of 'constant monitoring of the soil dynamics' under the soil and climate conditions of the humid tropics. The prerequisites for this monitoring, such as know-how and technical equipment, must be taken into account when considering the transfer of a technology developed on experimental plots to on-farm conditions, especially in view of the ecologically very critical soil characteristics.

Apart from various other deficiencies such as inadequate infrastructure, competition disadvantages, low capital disposal, irregular market prices and availabilities of inputs, severity of pests and diseases, Phillip M. Fearnside (1987) stresses the lack of prerequisites for such monitoring in the actual world of tropical small farmers as the key point of his very critical paper 'Rethinking continuous cultivation in Amazonia'. However, his objections do not do justice to the scientific approach and the sophisticated experiments of many years because they do not deal with the ecological problems which form the backstage of the Yurimaguas experiments.

Next to high acidity, the aluminium toxicity and low base content, the low CEC is of utmost importance. The CEC values of 3.6–4.9 meq per 100 g given for the top 25 cm in Table 8.7 are not even representative of the actual conditions since they stand for the highest potential values under neutral soil reaction. According to McCollum (1985), the effective CEC lies below 3.6 meq per 100 g for a pH value of 4.5–5.0 and may drop as low as 2.7 with decreasing pH. Such low CEC values lead to an extremely low buffering capacity, so that one cannot simply apply high fertilizer dosages of all essential nutrients. Under low CEC conditions the cations compete with each other, creating a trigger effect, which at a certain phase in the plant growth may cause an excessive element to push another below the minimum threshold. According to the minimum law of plant growth the entire fertilization then becomes useless.

Therefore, 'monitoring' also means knowing exactly how to match, at a given stage during plant growth, the various fertilizer amounts with the plants' nutrient requirements, in order to avoid undesirable trigger effects.

After readjusting the different cropping systems in the manner described, the generally good yields gathered in the following years led to the lime and fertilizer recommendations given in Table 8.2. The nitrogen and potassium application dates were changed to 15 and 45 days after planting.

Everything went well until 1983 (Figs 8.5 and 8.6), when despite all monitoring, another severe reversal of the trend occurred. In the publications of Sanchez (1983, 1985) and Nicholaides *et al.* (1985a), the sequences of yields cease in 1981 or 1983 with the twenty-first or twenty-fifth harvest respectively, but data from the Annual Progress Report 1985 allow a follow-up (Fig. 8.6). June 1983 shows a sharp yield reduction for corn and peanuts, and January 1984 for rice. The declines are about as drastic as those of 1975. (In addition, the mid-1985 maize crop was struck by a very special catastrophe.) Corresponding to the experience with the 1973/74 yield declines,

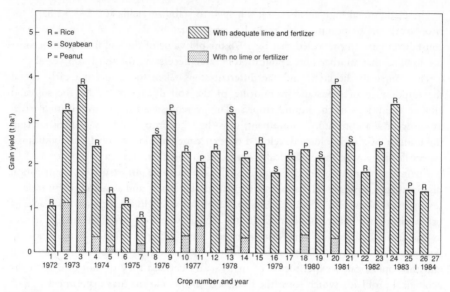

Fig. 8.5 Yield sequences from 1972 to 1984 in the Yurimaguas continuous dry rice–soyabean–peanut cropping experiments (after Nicholaides *et al.*, 1985a).

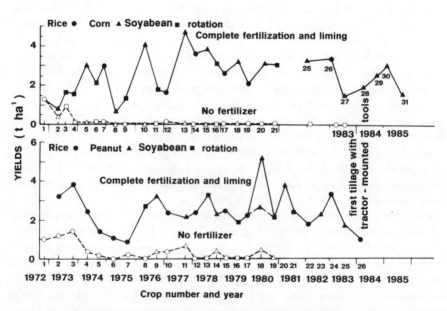

Fig. 8.6 Yield records for two continuously cultivated plots at Yurimaguas, with and without fertilization, 1972–85 (after Sanchez *et al.*, 1985a; Annual Progress Report, 1985).

206

the special rainfall conditions and the soil dynamics between 1981 and 1984 are of special interest.

Daily rainfall data are available for the years, but the exact planting dates are not. Therefore, we had to extrapolate the earliest and latest possible planting dates for each crop from the published harvest month, using the detailed records on planting and harvest dates from the years 1973–75, to calculate the normal length of growth periods of the different crops. Together with the first and second fertilizer applications, the possible planting periods were entered into the graph of rainfall dates from 1981 to 1984 (Fig. 1.7).

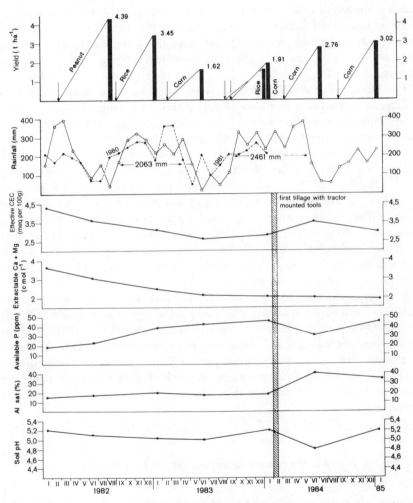

Fig. 8.7 Yields rainfall and topsoil properties on completely fertilized plots of the continuous dry rice–peanut–maize rotation, 1982–85 in YAES (data after McCollum, 1985).

207

It can be seen that at the end of 1983 the rice and maize stands received 920 mm of rain during October, November and December, and that seven times during this period they were exposed to heavy daily rains surpassing 50 mm: 100 mm fell shortly after the beginning of the growth period on 9 October, 50 mm 6 days later, 54 mm after another 10 days and 75 mm more 1 week after that.

In comparison with these severe rainfall events, the 898 mm which fell in the corresponding months of the previous year were distributed much more evenly. Only five heavy rains occurred during the whole growth period, with as little as two during the first month after planting. In the following year, when the January harvest of 1985 was twice as high as in 1984, the total precipitation amount from October to December 1984 was only 657 mm. Moreover, during the entire growth period at the end of 1984 there were only 3 days with rains of 50 mm or more.

The change of soil parameters during 1983 and 1984 (Fig. 8.7) again demonstrates the influence of the high precipitation and concentrated rains on the decrease of base elements and the increase of acidity and aluminium toxicity, in the same way as occurred in 1974/75.

The disaster with the mid-1985 maize crop was caused by an extreme rainfall event of 225 mm in 6 hours, which unfortunately occurred soon after planting and washed away many of the recently germinated seedlings (Soil Science Department, North Carolina State University, 1976 to 1983; Technical Report 1983).

It is true that such devastating floods are exceptional even in the ever-humid tropics; still, amounts between 50 and 100 mm day^{-1} are part of the climatological norm (Fig. 8.1 and Section 1.2). The impact of such high precipitation amounts is inescapable in open air cultivation in the humid tropics and must be taken into account as a natural factor. While it affects any crop, whether part of shifting cultivation or a permanent cultivation system, the latter is also subjected to the high permeability of most tropical ultisols and ferralsols. The soil's permeability combined with very low exchange capacity and considerable water input from tropical rainfall lead to a strong outwash of nutrients from the root zone of cultivated shallow-rooting plants.

A fertilization technique with water-soluble salts in tropical ultisols or oxisols resembles a hydroponic system in which the nutrient provision of the plants is carried out according to the percolator method in which the tropical rainfall constitutes an unpredictable element with often disastrous consequences. With the outwash of lime and fertilizers, a small farmer loses not only replaceable nutrients, but all the money invested in those purchases. It is unlikely that he has the capital for a repeat application, and even if he does, the entire farming venture would become unprofitable.

8.1.3   *The reality in the countryside beyond the YAES confines*

In order to test the possibility of transfer and the economic cost/benefit problem under actual on-farm conditions, a 'Yurimaguas Small Farmer Extrapolation Program'

Table 8.3 First-year yield results for the 3 rotations from the 11 small farm *chacras* of the extrapolation programme, Yurimaguas, 1979 (t ha⁻¹)

Chacra location	No.	System	Rotation 1 Corn–peanuts–corn			Rotation 2 Peanuts–rice–soyabean (IR 4-2)			Rotation 3 Soyabean–rice–peanut (Carolina)		
Low exchangeable A1											
Km. 15	1	I	1.71	0.70	1.55	1.87	—	0.46	1.16	—	1.50
		II	2.68	1.10	2.39	2.41	—	0.74	2.29	—	2.20
		III	4.90	2.20	4.47	2.91	—	1.82	2.76	—	2.60
Puerto Arturo	4	I	2.15	Left programme		0.67	Left programme		1.57	Left programme	
		II	4.31			0.95			3.76		
		III	6.22			1.21			4.19		
Km. 8	5	I	3.05	1.30	3.48	0.71	—	1.64	1.65	2.13	1.00
		II	3.53	1.50	3.48	1.19	3.89	1.66	2.01	2.44	1.90
		III	6.19	2.30	3.82	1.19	5.51	2.18	2.14	2.28	2.20
Km. 22 A	6	I	2.75	0.50	1.17	0.39	—	1.27	1.42	—	1.00
		II	2.97	0.20	0.48	0.62	3.00	2.38	1.79	0.91	2.40
		III	5.49	0.50	4.92	0.87	3.82	3.26	1.83	0.85	2.20
Apangurayacu	7	I	3.38	1.40	Left programme	0.97	—	Left programme	3.78	1.93	Left programme
		II	4.71	1.90		0.85	3.61		5.04	2.13	
		III	5.30	1.90		0.85	5.13		4.99	2.28	
Sapote	8	I	3.26	1.10	Left programme	1.16	—	Left programme	1.40	1.46	Left programme
		II	5.32	1.50		1.25	3.29		2.36	3.10	
		III	6.11	1.50		1.27	4.10		2.74	4.32	
Km. 28	9	I	4.01	0.90	1.50	1.49	—	2.23	1.10	3.19	1.60
		II	5.13	1.30	6.38	1.27	4.32	2.35	1.34	2.38	1.80
		III	5.57	1.00	6.79	1.79	5.55	4.89	3.18	0.95	1.90
Average yields 1, 4, 5, 6, 7, 8, 9		I	2.90	0.98	1.93	1.04	—	1.40	1.72	2.18	1.28
		II	4.09	1.25	3.18	1.22	3.62	1.79	2.66	2.19	2.03
		III	5.68	1.57	5.00	1.44	4.82	3.04	3.12	2.14	2.23

209

Table 8.3 continued

Chacra Location	No.	System	Rotation 1 Corn–Peanuts–Corn			Rotation 2 Peanuts–Rice–Soyabean (IR 4-2)			Rotation 3 Soyabean–Rice–Peanut (Carolina)		
High exchangeable A1											
Munichis	2	I	0.22	1.00	1.90	1.03	—	—	0.15	1.58	1.50
		II	4.67	1.50	2.57	1.37	3.66	2.36	0.49	2.07	1.80
		III	5.18	1.95	4.47	1.84	3.52	2.72	1.90	2.16	2.20
Callao	3	I	1.23	1.60	—	0.16	—	1.27	0.72	0.81	0.95
		II	2.42	1.70	—	0.51	2.42	2.81	1.20	2.42	2.00
		III	4.39	2.20	—	0.93	3.84	2.64	1.60	3.42	4.00
Shucshuyacu	10	I	1.92	1.60	1.70	1.50	4.00	1.47	2.03	—	0.50
		II	3.96	2.00	2.65	2.24	4.60	2.20	1.26	2.14	2.30
		III	4.88	1.70	6.54	1.67	5.46	2.63	2.60	2.91	1.40
Km. 22 B	11	I	3.18	0.90	1.06	0.71	—	1.01	0.73	2.32	1.10
		II	2.24	0.90	1.14	0.80	3.24	1.36	1.42	2.64	0.70
		III	2.06	0.90	1.61	1.82	3.80	1.86	2.08	3.55	1.20
Average yields 2, 3, 10, 11		I	1.64	1.28	1.55	0.85	—	1.25	0.91	1.57	1.01
		II	3.32	1.53	2.12	1.23	3.48	2.18	1.09	2.32	1.70
		III	4.13	1.69	4.21	1.57	4.16	2.46	2.05	3.01	2.20
Average yields all		I	2.44	1.10	1.77	0.97	—	1.34	1.43	1.91	1.15
		II	3.81	1.36	2.73	1.22	3.56	1.99	2.09	2.25	1.89
		III	5.12	1.62	4.66	1.49	4.53	2.75	2.73	2.53	2.22

Source: Bandy *et al.* (1980).

(Bandy *et al.*, 1980) was conducted in 1978 and 1979 with 11 local farmers in the surroundings of Yurimaguas. The main objective was 'to show the farmer how he can permanently and economically farm the same piece of land. A secondary objective was to determine which agronomic practices are most readily accepted' (Bandy *et al.*, 1980:199). To reach these objectives, three input level systems were compared (Table 8.3): (I) The traditional systems as the way each farmer had cultivated his land, using his own seed; (II) an improved agronomic practice utilizing better varieties, correct plant spacing and populating, weeding and application of pesticides when needed; and (III) the same improved agronomic practices as in system II plus lime and fertilizer applications at the following rates: lime 1 t ha^{-1} yr^{-1}, N 60 kg ha^{-1} per crop of rice and corn, P 35 kg ha^{-1} per crop, K 66 kg ha^{-1} per crop, Mg 22 kg ha^{-1} per crop. The rotations were corn–peanut–corn, peanut–rice–soyabean, soyabean–rice–peanut. Table 8.4 shows the soil chemical characteristics from the 11 *chacras*. The soil properties were basically the same as in the experimental station (Table 8.1). There was, however, one major difference in that the slash-and-burn procedure resulted in two different groups of soils: a very acidic soil with a high content of exchangeable aluminium and an extremely low content of basic cations; and another almost neutral in reaction with a remarkably high base content and little exchangeable aluminium.

While the special seed varieties, the insecticides and the pesticides were paid for by the farmers, the fertilizers were distributed to them free of charge.

Table 8.3 shows the first-year yield results for the three rotations from the seven farmers with plots on soils of low exchangeable aluminium values and the four farms on acidic soils with high exchangeable aluminium values. The data clearly reflect the

Table 8.4   Some soil chemical properties from the 11 *chacras* sampled after fire-clearing but before first planting. Soil depth 0–15 cm. Yurimaguas, Peru, 1978

Chacra Location	No.	Forest fallow (yrs)	pH (1:1 H$_2$O)	KCl extractable Al	Ca	Mg	Modified Olsen P (μg cc^{-1})	Organic carbon (%)
				(meq per 100 cc)				
Low exchangeable Al								
Km. 15	1	4	5.7	0.2	8	1	6	1.7
Puerto Arturo	4	1	6.6	0	35	5	5	3.1
Km. 8	5	10	6.2	0.2	20	2	9	3.6
Km. 22 (A)	6	4	6.3	0.1	2	1	7	0.8
Apangurayacu	7	1	6.0	0.1	34	8	10	2.2
Sapote	8	6	6.7	0	22	3	15	2.5
Km. 28	9	7	6.5	0	14	2	13	2.3
High exchangeable Al								
Munichis	2	8	4.7	1.7	8	2	10	3.2
Callao	3	4	4.3	3.5	2	1	9	1.8
Shucushyacu	10	8	4.5	1.7	6	1	7	2.3
Km. 22 (8)	11	8	4.5	3.1	2	1	13	1.9

Source: Bandy *et al.* (1980).

Table 8.5 Average cumulative grain yields (t ha^{-1}) of 11 small farmer-managed continuous cropping demonstration trials from July 1978–June 1979 near Yurimaguas, Peru

Production system	Corn–peanut–corn			Total	Peanut–rice*–soyabean			Total	Soyabean–rice†–soyabean			Total
Traditional	2.44	1.10	1.77	5.31	0.97	1.91	1.34	3.53	1.43	1.91	1.15	4.49
Improved, no lime or fertilizer	3.81	1.36	2.73	7.90	1.22	3.56	1.98	6.76	2.09	2.25	1.89	6.23
Improved, with lime and fertilizer	5.12	1.62	4.66	11.40	1.49	4.53	2.75	8.77	2.73	2.53	2.22	7.48

(Columns grouped under *Crop rotation*)

* Rice in the traditional system is the Carolina variety; in both improved systems, it is IR 4–2.
† Rice in all systems is the traditional Carolina variety.

Source: Nicholaides *et al.* (1985a).

disadvantage of the low-base-status soils in which the liming effect and the nutrient input from the ash after burning was particularly low. It is reasonable to conclude that in following years yields will decrease even on farms with initially good results since the effect of ash-liming wears off, and soil acidity and aluminium toxicity increase.

In order to assess the effectiveness and profitability of the new systems, the researchers have given mean values for all locations without considering the initial soil properties. These values published in Nicholaides *et al.* (1985a) are average cumulative grain yields (t ha^{-1}) from 11 small farmer-managed continuous cropping demonstration trials (Table 8.5).

Calculated in this way, the results are really as impressive as stated by the authors. Yields of maize and soyabeans more than doubled and peanuts and rice almost doubled after a full application of the 'improved Yurimaguas technology'. The sharpest rice increase followed replacement of the Carolina variety by the high-yielding IR 4–2 variety. But one should keep in mind that the high yields were obtained under particularly favourable soil conditions, namely in the first year after burning the secondary forest.

To arrive at an economic evaluation Hernandez and Coutu (1983) took the set of values from Table 8.5. Concerning the crucial question whether the data offer economic encouragement to farmers, the authors state: 'In general the economic analyses for the selected crops were very favorable with respect to existing slash/burn systems.' Among the more specific conclusions appears the fact that 'the most sensitive factor impacting economic viability is the ability to transfer the experimental results to on-farm-conditions. A twenty per cent yield reduction from experimental plots to farm plots is profitable to existing systems, but at a fifty per cent transfer loss the potential for a continuous cultivation system is limited'.

Experimental plots are those of the Yurimaguas station, where soil conditions were controlled by constant monitoring, while the on-farm plots are not closely monitored by the experimenters. This lack of supervision makes it difficult for farmers to stay within the 50 per cent loss range, as stated by José Benites (1981, unpublished) in his paper 'Proposal and recommendations for the review of Yurimaguas Small Farmer Extrapolation Program'. 'The limited relationship between researchers and personnel conducting the technology transfer and the lack of clear communication between the agent and the farmer resulted in a large number of farmers and schools leaving the Program after the second or third crop.'

In 1986 the senior author of this book visited 5 of the 11 farms. At best, one of them showed signs of the implementation of system III, i.e. continuous cropping with sizeable applications of lime and fertilizers. However, the large plots were cultivated by a state agency. Sandy soils sported a beautiful crop of the Africano desconocido, a rice variety that has been used exclusively on the experimental station for several years because it is very aluminium-tolerant, and also does better on acidic soils than the varieties that are normally used on the farms, such as Carolina or the high-yielding varieties. Unfortunately, this Africano rice is not accepted by the rice mills because 60 per cent of the grains crumble in the course of the threshing and peeling process.

All the small and medium-size farmers were no longer working with the demanding technology of the extension programme, though a few had joined a new programme on improved systems, which are described in section 8.2.

Even if counselling and sound supervision eventually improve the acceptance level among the farmers and overcome the difficulties in constant monitoring, a serious environmental problem remains unsolved: the huge amounts of lime as basic requirement of the Yurimaguas technology.

In the humid tropical lowlands of South America and Africa limestone or dolomite are found only in a few geologically exceptional locations. For example, in the Amazon basin not a tonne of lime can be found in areas the size of Central Europe, as can easily be verified using the petrographic map accompanying the FAO/Unesco *Soil Map of the World, vol. IV South America* (1971). And this is not accidental, but due to the geological and climatological conditions. The African and South American inner tropics belong to the world's oldest continental masses, consisting mainly of acid crystalline rocks partly covered by similarly acid sediments of terrestrial origin. In addition, for eons the land surfaces have been subject to the same extreme chemical weathering and extensive denudation that prevails today. Under these conditions, the easily soluble calcium is one of the first elements to be washed away.

In the Amazon basin, some limestone crops out in the Carboniferous sediments near Santarém and more occurs in the Andes, but even there the quantities are not comparable to the amounts found in the folded mountain systems of the Old World. Today, it is theoretically possible to transport the required quantities over long distances. But in practice it is not feasible. (The limestone for the experimental station in Yurimaguas comes from quarries in Tingo Maria, some 480 km away from Yurimaguas in the Andean foothills. In 1985, it cost approximately $US 2500 to transport a truck load of 20 t from Tingo Maria to Yurimaguas; personal information from the YAES director of administration.).

A rough calculation based on the figures makes it evident that the transportation costs for lime alone by far exceed the financial capacity of small farmers in the Amazon basin. Even if they somehow procured the money, loss of the fertilizer through heavy rains, as happened in 1975, 1983 and 1985, would be ruinous.

The problems with the high-input technology (Yurimaguas technology) were discussed by the senior author with Dale Bandy, who had directed the Yurimaguas experiments for 10 years. In the end it was agreed that today the new technology can be applied only where the indispensable infrastructure is available, where the farmers have received intensive schooling, and where the marketing conditions for the products of the continuous cropping system exist. At present, these premisses are met in only very few locations. But the know-how for future developments is at hand. In the meantime, it is imperative to establish intensive cultivation systems on the more fertile alluvial soils and to develop transitional technologies (low-input cropping systems, improved legume-based pastures, alley cropping and agroforestry systems). This agrees with the general statement of Sanchez and Nicholaides (1985, p. 96): 'The high input approach of Yurimaguas technology was recognized as being limited to areas with ready accessibility to inputs and good market infrastructure.'

8.1.4   *General conclusions concerning the ecological frailty of the humid tropical environment*

Apart from the difficulties encountered when trying to implement the Yurimaguas recommendations, the experience gained from the long-term continuous cropping experiments brings to light more basic findings:

1. In order to achieve in humid tropical countries the agricultural land-use intensity that is normal in the extratropics, it is necessary to apply large amounts of a mineral (limestone) which, for geological and climatic reasons, hardly occurs in the humid tropics. The humid tropical region is therefore at an agro-ecological disadvantage in comparison with those areas where limestone is abundant and easily accessible. In the course of their agricultural development, the natives of the humid tropics had no opportunity to experience the growth-promoting effects of limestone or marl application, while these were among the first amelioration techniques used in the extratropics. The main source of calcium accessible to them was, and still is, the biomass (see Figs 2.6, 2.7 and 4.1).

    The utilization of the calcium and nutrient reserves contained in the phytomass is possible, in principle, by harvesting the phytomass and depositing it on plots for the slow release of the nutrients through humification and mineralization. However, the transportation work can be avoided if the phytomass is left in its original location, the crops are cultivated on the spot and only the harvests are taken away. Since the latter constitutes a nutrient export, the cultivation area becomes increasingly poorer in nutrients. The native farmers are therefore forced to shift location to make use of the nutrients available in the biomass of other areas.

    Considering the distribution of essential nutrients in biomass and soil, the forest–field rotation system is an ecological necessity on tropical low-base-status soils (the burning, however, is not!).

2. The sharp discrepancy which exists in humid tropical land-use forms between the prevailing limestone-free regions and those where the soil developed from lime sediments or from calcium-rich vulcanites, especially young basalts, is due to the influence of lime on the base status of the overlying soils. The larger expanse of intensive land-use systems and the higher population density of the Asian tropics as compared to their African or South American counterparts is rooted in the different ecological conditions, highly influenced by the rock basement.

3. When a forest is burned, a considerable portion of the nutrients contained in the ash does not end up in the soil and the biomass through recycling, but is instead washed into the rivers. In the long run, this will lead to a progressive decline of calcium and other essential nutrients in the ecosystems, if the slash-and-burn rotations follow each other too fast to allow the replacement of the necessary elements from the subsoil or from the air. When the critical point of no return is reached, depends largely on the mineralogical composition of the bedrock,

the weathering products and the length of the fallow period. In some areas of the tropics this point may already have been reached; in other areas, it will be unavoidable if the current practices continue. *Therefore, the precaution should be taken not to burn primary and secondary forests; instead the nutrients of any available biomass should be utilized in a careful and waste-free recycling process.*

4. The actual significance of the long-term experiments lies in the fact that the attempt to overcome a handicap has but revealed its true seriousness. Main efforts should therefore be directed towards identifying the serious difficulties that are connected with fertilizer-based cropping on tropical ferralsols in order to counteract the spreading wave of optimism concerning the potential agricultural capabilities of these regions. This includes the insight that everything must be avoided that might lead to further impoverishment, and eventual destruction of the tropical rain forest ecosystem. Once the rain forest is gone the ferralsol areas in the humid tropical lowlands have lost their main resource of essential elements; the ecological consequences of this process are described in Reichholt (1990).

8.2 Low-input technology at Yurimaguas

In the Triennial Technical Report 1981–1984, *TROPSOILS* (McCants, 1985), Sanchez and Nicholaides wrote (p. 96), in their capacity of principal investigators of the Soil Management Collaborative Research Support Programme (acronym TROPSOILS): 'concentration on low input technology was strongly recommended by the 1981 AID Review of the previous program'. (AID, the United States Agency for International Development, was the chief financial supporter of the previous investigation programme Agronomic–Economic Research on Soils of the Tropics.)

Low-input technology is aimed at developing soil management practices for continuous crop production in which liming and fertilization quantities are minimized. The strategies devised include primarily 'selection of acid-tolerant cultivars, minimum tillage and fertilization, weed control and managed fallows' (Sanchez and Nicholaides, 1985, p. 96). Since, on one side, tropical ferrallitic soils combine low CEC and high acidity with aluminium toxicity, and since, on the other, most farmed food crops are generally Al-sensitive, it is an important task to select varieties of annual crops or develop breeding lines for new varieties that are tolerant to the Al levels of tropical ferrallitics. Field, greenhouse and laboratory experiments in various places in the world have shown limited success. An Al-tolerant dry-rice cultivar is the so-called 'Africano desconocido', which is at present the most used experimental variety in Yurimaguas. The variety responds well to low fertilizer levels, but the harvested grains have a serious shortcoming: the local millers reject this rice because it crumbles so badly during the dehusking process. So, the small farmers return for their market production to the cultivation of their traditional Al-sensitive Carolina rice variety.

Another Al-tolerant cultivar is the Vita 3 cowpea variety, which, however, has only limited human consumption. Still, cowpeas are widely utilized for research purposes. Soyabeans, on the other hand, are generally Al-sensitive.

This leaves only a few annual crops that are economically profitable and at the same time Al-tolerant. Real progress in this respect would have far-reaching consequences for the agricultural development of the humid tropics.

A key role in the so-called minimum tillage and fertilization approach is played by the rational use of crop residuals. In the case of high-input technology, the plot is rotated after the crop residues have been removed. The tillage results in a favourable soil condition for root development and a more uniform distribution of nutrients in the seeding beds. The exact opposite management practice is the zero-tillage and no-fertilization approach using Al-tolerant cultivars and making the most of the residues.

In practice, drawing the maximum benefits from the residues in the zero-tillage approach entails the following procedures, as observed by the senior author in the field in 1986. After the first crop of dry rice had been grown in a recently cleared plot of secondary forest, the rice and the weeds stood about 70 cm high. The rice was harvested by hand, panicle by panicle. Later on, four hired hands were needed to cut the remaining biomass with machetes close to the surface, truly laborious work. Before planting cowpeas as the next crop the question confronted is how to manage the recycling of the residues: burning them to ashes, leaving them as mulch, or converting them to compost?

Burning is the most effective method in terms of weed control, but by far the least effective for nutrient recycling. With the soil remaining unprotected for a certain period of time, the ashes on top are likely to be washed away by heavy tropical rains. Without burning the farmer is confronted with the expanding weed problem. Nevertheless, during the initial cropping cycles after clearing, it is important to use the organic matter as mulch or compost. Findings in this respect have been presented by Bandy and Nicholaides (1983). After several years of experimenting with mulch from various residues the results were still variable. Soil physical properties had clearly improved, and the reduction of soil temperatures was considered favourable. As far as nutrient transfer to the crops is concerned, the yields obtained revealed a definite interaction between mulching and rainfall. When soil humidity shortage occurred during the grain-filling stage, the effect of mulching on yields was favourable, but when there was excess water in the soil, the mulch effect was negative.

Composting with residues from previous crops turned out to be by far the most effective way of nutrient recycling. The materials were decomposed under natural conditions for 4 months and then incorporated into the soil prior to planting. Comparing the nutrients supplied by the compost materials and the nutrient uptake by given yields of several crops, it became evident that the compost alone, at 10 t ha^{-1}, could provide a high proportion of the required nutrients, except for potassium and nitrogen. The compost contained enough calcium and magnesium to neutralize the high levels of exchangeable aluminium in the ultisol. Compost from kudzu vines (*Pueraria phasiloides*) produced the best results in terms of crop yields and soil nutrient status improvement, but the calcium and potassium content was insufficient.

217

The experience gained from these experiments underlines the overriding importance of proper treatment of all the organic materials. Burning is indisputably the most wasteful practice. The problem of weed control must be solved in another way as long as the cultivated plot is not engulfed by weeds – and that is exactly the weak point of this management procedure.

'Control of weeds is a minor managerial problem for about one year after cutting and burning well-forested jungle. Weed infestation increases rapidly thereafter' (McCollum and Pleasant, 1985: 163). After two crops of rice, for example, the invasion of grasses that produce seeds all year round is so powerful that the rice plants are suffocated, the germinating seeds lie all over the ground. With returns of 2 t ha^{-1} of rice investments for special herbicides are not profitable.

Experimenting with the following combination is the best procedure: use of Al-tolerant rice cultivars, no tillage, initial pre-planting weed control with suitable chemical herbicides, further manual weed control during the growth period, no burning of plant residuals, planting cowpeas as a second crop after applying chemical herbicides to the existing residues of weeds and rice ratoon, with follow-up manual weed control and subsequent application of reduced levels of potassium fertilizer (Nicholaides *et al.*, 1985b).

Managed fallow with kudzu could be a valuable alternative with respect to weed control. The fallow period has, besides soil-improving properties, in addition the function of restricting the light-dependent grass weeds by shading. The regeneration time of a secondary forest fallow ranges from 10 to 12 years, as a minimum. An artificial leguminous fallow, such as that of kudzu, may be able to exert similar effects in a shorter period of time. In practice, seeded kudzu grows fast even on acid ferralsols as to cover the soil totally, impairing the growth of other plants. In Yurimaguas experiments, accumulation of kudzu fallow biomass on plots previously dedicated to continuous cropping systems was maximum after 2 years and reached 20 per cent of easily burned biomass accumulated by a 25-year forest fallow (Bandy and Sanchez, 1985). Slash and mulch of kudzu fallows are slightly more effective than slash-and-burn. Kudzu can be cut for mulch easily because the plant possesses deep-reaching roots only at large intervals the rest of the roots being only grip-roots.

The experiments have shown that letting cropland rest for 1–3 years with kudzu fallow can subsequently produce 80 per cent of the crop yields reached on plots after clearing 25-year-old secondary forests. Managed kudzu fallows seems to be a feasible alternative to a mature forest fallow (Bandy and Sanchez, 1985).

Unfortunately, the alternative is not without flaws. The researchers themselves well realize that kudzu can grow out of control very fast. As a climbing vine with a rapid growth rate, kudzu can overgrow not only plants on the ground but also shrubs and trees.

At this point, it must be remembered that the low-input approaches mentioned above are aimed at continuous cultivation of annual crops and that this continuous cultivation must function under the worst of soil conditions, namely, on nutrient-poor

218

acidic ultisols and oxisols, with extremely low exchange properties and high levels of Al toxicity.

8.2.1 *Alternatives to continuous cropping*

Since it seems that continuity of annual crop production cannot yet be reached with low-input technology on the less fertile soils mentioned (Sanchez and Nicholaides, 1985, p. 97), alternatives for other land-use systems and for varied soils in different landscape locations had to be tested. Hence, the scope of the Tropical Soils Programme was broadened from the original option of fertilizer-based continuous cropping by means of high- or low-input systems, to encompass three other options: agroforestry, legume-based pastures, and flooded rice on alluvial soils.

Agroforestry is dealt with in this book in Chapter 9. In Yurimaguas agroforestry techniques were restricted to alley cropping and nutrition management of peach palms (*Guilielma garipaes*) for which the Yurimaguas area is one of the origin centres.

Of greater relevance are the investigations into legume-based pasture systems. In semihumid savannas or in the semiarid tropics legumes are mainly used as forage during the dry season when the grasses are dry or dead. In humid tropical areas without a dry season, legumes are needed theoretically for their nitrogen contribution, and it must be tested how a grass–legume mixed pasture fares under permanent humid tropical conditions. The first step in the pertinent investigations involves the selection of grass and legume cultivars tolerant to the severe acidity and low fertility of the humid ferrallitic soils. An evaluation of 39 grass ecotypes and 9 legume ecotypes from different genera revealed that several ecotypes well adapted to other tropical environments failed under the Yurimaguas soil conditions; but there is still plenty of acid-tolerant germplasm that could eventually be utilized. The main constraints in humid environments are the insect and disease resistance of grass and the high tannin content of the legumes.

After 4 years of experimenting with an annual fertilization programme of 25 kg ha^{-1} P, 42 kg ha^{-1} and 8 kg ha^{-1} Mg, the most persistent mixtures appeared to be *Brachiaria humidicola/Desmodium ovalifolium*; *B. decumbrens/D. ovalifolium* and *Andropogon gayanus/ Stylosanthes guinensis* (Sanchez et al., 1985a,b).

Many problems arise from the palatability of *D. ovalifolium*. This legume is very adaptable to acid ferrallitic soils. It shows excellent yields; however, its high tannin content in the leaves (20 per cent in comparison with 2–4 per cent in other forages) leads to poor palatability. Consequently, it is consumed in low proportion, causing an imbalance in the mixture composition favourable to the legumes. Therefore, grasses are more fragile than legumes in the mixed pastures of the humid tropics (Sanchez *et al.*, 1985a).

Since the consumption of legumes is low, their nitrogen contribution to the soil via animal excreta is also likely to be low. In addition, the high tannin content suggests a reduced rate of litter decomposition, so that the nitrogen transfer to the grasses

219

Table 8.6 Chemical characteristics of topsoil (0–15 cm) and underlaying layer (15–20 cm) of some alluvial soils (inceptisols) on the Shanusi River, Yurimaguas, Peru

Soil No.	Depth (cm)	pH (1:1 (H_2O))	Ca	Mg	K	Effective CEC	Base saturation (%)	Organic carbon (%)	P (ppm)
				(meq per 100 g)					
1	0–15	5.0	33.50	10.71	0.31	43.09	96	1.6	6.1
	15–20	4.7	36.01	7.54	0.22	41.61	84	0.8	2.8
2	0–15	5.1	24.86	6.45	0.29	32.91	93	1.3	3.5
	15–20	4.9	25.14	7.17	0.24	38.95	81	0.4	2.5
3	0–15	5.2	35.60	10.15	0.27	46.54	99	1.4	3.4
	15–20	5.1	32.01	10.28	0.24	47.05	90	0.4	1.6
4	0–15	5.3	37.41	8.92	0.35	46.06	99	1.3	12.5
	15–20	5.2	33.91	9.26	0.27	44.47	98	0.5	5.8
5	0–15	5.4	34.13	9.75	0.34	44.67	99	1.3	3.8
	15–20	5.2	34.36	10.10	0.33	46.34	96	0.5	4.5
6	0–15	5.7	42.06	10.23	0.37	52.74	100	1.3	15.8
	15–20	5.1	40.04	10.78	0.33	52.13	94	0.5	13.0

Source: Bandy and Benites (1983).

through this path remains limited. New breeds of *D. ovalifolium*, selected for their lower tannin content, offer the chance to alleviate the palatability problem.

A problem that has not been addressed in the Yurimaguas reports pertains to the necessity of supplying the cattle with minerals and trace elements that are insufficient in forage grown on tropical ferrallitic soils. They are indispensable for the satisfactory nourishment of the cattle. In the experimental plots of Yurimaguas salt tabs containing calcium, phosphorus and a variety of trace elements are available for the animals to be licked. Daily consumption was reported to be 40–60 g per head.

It is remarkable that among the alternatives of the Yurimaguas technology research activities, aiming at wet-rice production on alluvial soils only began in 1981 (Bandy and Benites, 1983, p. 101) as one of the last initiatives. There is almost no feature more illustrative of the contrast in land-use intensity between tropical Asia and tropical South America than the situation on alluvial soils along the rivers. In South-east Asia every square kilometre of alluvial plains has been converted into cropland in a highly sophisticated manner for generations. Due to the population pressures in some regions, even the slopes of the bordering hills have been transformed into terraces for paddy rice cultivation. In contrast, the alluvial plains in the forelands of the tropical Andes are almost unutilized. Benites *et al.* (1985) estimate that in the Peruvian *selva* alone around 4 million ha of fertile soils on flat topography are available for expansion of irrigated rice. According to Cochrane and Sanchez (1982) nearly 8 per cent of the Amazon basin consist of relatively fertile, usually flood-free, well-drained soils (alfisols, mollisols, vertisols, tropepts, fluvents).

Table 8.6 depicts the high base status, moderate acidity, non-aluminium toxicity and high CEC values of these soils which stand in strong contrast to the widespread upland ferrallitic soils. The remark of Bandy and Benites (1983: 101) that 'there has been interest, recently, in developing rice production under irrigation in this area', highlights the underuse of the South American tropical alluvial plains when compared to their Asian counterparts. The technology for wet-rice cultivation known in South-east Asia could be adopted for the site-specific conditions of the Andean forelands.

Five crops were grown in 2 years after the research began in 1981, the annual average yield being 16.5 t ha^{-1} (Benites *et al.*, 1985). A group of farmers, some with college degrees, began production of paddy rice across the Shanusi River in the vicinity of Yurimaguas. Three hundred hectares are the promissory beginning of a great potential for the future.

8.3   Soil and land-use experiments in the semihumid tropics (IITA, ICRISAT, CIAT)

8.3.1.   *Institutional concepts and geographic location of IITA*

The International Institute of Tropical Agriculture (IITA), the first of its kind, was established in 1967. Generous contributions from the Ford and Rockefeller foundations and an informal group of donor countries, joined since 1971 in the Consultative Group of International Agricultural Research (CGIAR), helped launch the institute at the newly created University of Ibadan, Nigeria. Today, CGIAR supports no less than 13 agricultural research centres around the world, among them the International Crops Research Institute for the Semi-Arid Tropics (ICRISAT) in Patancheru, Andhra Pradesh, India, where environmental conditions are comparable to those at IITA.

The main research conducted at IITA was geared towards improving land use and agricultural production in the transitional belt between the humid tropical forest and the semiarid dry savanna. The wet-and-dry tropical savanna belt has its special problems resulting from mounting soil erosion due to the vulnerability of the soils and to exploitative and degradative land use in response to increasing population pressure.

The IITA main station is centrally located with access to four ecological zones: the humid tropical forest, the dry forest, the humid savanna and the Guinea savanna (Fig. 8.8). These four zones are representative of climate and soil conditions in West Africa.

The mean annual precipitation of 1258 mm is distributed over two rainy seasons: the major one from April to July, accounting for approximately 48 per cent of the yearly mean, and the minor one from mid-October to mid-November with 33 per

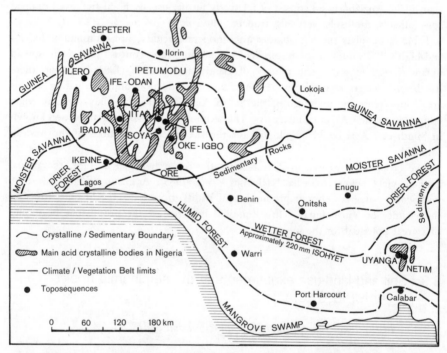

Fig. 8.8 Geological sketch map of western Nigeria. Vegetation and climate boundaries also included. (Source: IITA, 1973).

222

cent of the annual rainfall. Perhaps more relevant than the mean total amount is the frequency distribution of rainfall intensities. Records taken by Lal (1976a) revealed that 8, 12 and 14 storms in 1972, 1973 and 1974 respectively had 7.5 minute maximum intensities above 75 mmh^{-1}. Storms with rainfall intensities above 130 mmh^{-1} were also observed. A 213 mmh^{-1} intensity was recorded during the first 5 minutes of a singular storm event. Downpours of 41 mm within 15 minutes are not uncommon.

The mean daily temperature of the dry month of February is 34 °C, and those of June, July and August lie between 26 and 28 °C. The high evapotranspiration leads to moisture stress on crops, to crust formation on the soil surface and to surface sealing.

Associated with the insolation and high air temperature is the temperature of the upper soil, a variable that tends to limit the growth of certain crops in the tropics and to create further stress on vegetal species (Monteith, 1979). Seedlings grow optimally at soil temperatures between 32 and 34°C. However, during the spring growing season, and even under conditions of relatively high pluviosity, soil temperatures 5 cm below the surface soar into the mid-forties, which makes it necessary to take measures such as mulching and keeping tillage to a minimum.

8.3.2 Soil properties

The overall classification of the soils in the IITA experimental region is alfisols (i.e. ferric or chromic luvisols, leached ferruginous soils, *sois ferrugineux tropicaux lessivés*). They are juvenile soils that have developed mostly over colluvial layers of intermediate or acid crystalline rocks and over arenaceous sedimentary rocks. Deep weathered oxisols or ultisols do not occur.

Due to their location in the climatic soil zone with decreasing weathering intensity, alfisols display a wide diversity and spatial variability depending on their parent rocks and/or topographic location (soil catenas, toposequences). Still, there are general characteristics of ecological relevance. The first is the argillic horizon some 30–60 cm below the soil surface and the relative enrichment of surface layers with coarse particles. Both are assumed to be the result of clay migration via percolating water, termite activity and/or selective removal of fine particles by erosion. The second characteristic is the dominance of clay minerals of low CEC (kaolinites) with a small proportion of 2 : 1 layer lattice clays. This occurs in well-drained soils in upper slope positions, while in lower positions the clay spectrum displays higher montmorillonite proportions (see Section 3.11).

Soil chemical conditions in general can be qualified as satisfactory. Acidity or aluminium toxicity does not influence the ecological processes as it does in oxisols or ultisols of the humid tropics. The quantity of extractable bases and base saturation are moderate. The CEC values, which range from 3.5 to 15 meq per 100 g fine soil, also prove adequate for agricultural use. The soil skeleton in the gravel and coarser sand fractions offer a certain reserve of weatherable primary materials. Carbon content differs widely from site to site. For an example of an alfisol profile see Table 8.7.

223

Table 8.7 Characteristic physical and chemical properties of an alfisol profile

Horizon	Depth	Coarse fragments >2mm (%)	Size class % of < 2 mm diameter			pH H₂O	Organic carbon (%)	Extractable bases (meq per 100 g)				CEC (meq per 100 g)	Base saturation (%)	Clay fraction mineralogy				Sand fraction mineralogy				
			Sand 2–0.02	Silt 0.02–0.002	Clay 0.002			Ca	Mg	Na	K			AM	KK	Mi	SM	QZ	FDM	FE	HE	FDP
Ap	0–5	17	79.3	6.4	14.3	6.0	0.55	2.6	0.5	—	0.4	4.8	74	11	37	12	17	35	25	10	10	5
B1	5–18	17	66.7	5.5	27.8	6.9	0.52	3.8	0.9	—	0.5	8.2	64	12	37	10	19	45	20	5	5	10
B2₁	18–36	36	41.6	6.8	51.6	6.9	0.63	5.8	3.8	—	0.6	14.8	69	14	37	10	23	40	30	10	—	10
B2₂	36–71	54	45.0	4.4	50.6	6.8	0.40	7.9	3.1	0.3	0.6	14.1	82	12	38	11	20	30	30	5	5	15
B2₃	71–112	50	54.1	7.4	38.5	6.5	0.10	5.4	2.5	0.3	0.4	9.8	88	12	39	8	18	40	20	5	—	10
B3	112–140	63	70.6	4.1	25.3	6.2	0.18	5.7	1.9	0.5	0.3	9.1	92	10	39	8	21	35	25	5	—	15

Am = amphibole; KK = kaolinite; Mi = mica; SM = smectite; QZ = quartz; FDM = Feldspar-microline; FE = magnetite; HE = haematite; FDP = feldspar-plagioclase.

Source: El-Swaify *et al.* (1987).

How pronounced the properties are and what impact their varying combinations have on land use depend on site-specific differentiations. The general rule for the spatial differentiation of alfisols is mentioned in the following statement:

> Considering the upper slope members over crystalline rocks, differences between the three parent material groups (acid, intermediate and basic) are generally greater than the differences within the same group but in different bioclimatic zones. The differences in the lower members of the crystalline sequences are less between the various parent materials than between the climatic zones (IITA, 1973: 18).

Far less satisfactory than the chemical characteristics are the general physical properties of alfisols. In fact, they pose the gravest problem of the 'alfisol landscape', as Sanchez (1976) calls this environment (see Section 3.11). The prevalent soil series in the Ibadan region are of the Egdeba (oxic paleustalf, ferric luvisol), Ibada, Iwo and Apomu series. Regarding the particle size of the surface soils, the last three are commonly referred to as 'loamy sands' with sand proportions fluctuating between 65 and 79 per cent. The Egdeba series, classified as 'sandy clay loams', contains a slightly higher clay percentage (around 25 per cent) with only 59 per cent sand. The Patancheru soil series in Andhra Pradesh (a udic rhodustalf) contains 79 per cent sand and 14 per cent clay in the uppermost 5 cm, and 67 per cent sand and 28 per cent clay in the next horizon (5–18 cm). They are also classified as 'sandy clay loams'.

The first consequence of the dominance of sands and of kaolinites in the clay fraction is the low water-storage capacity of alfisols. Rainfed cropping on alfisols, therefore, faces the constant threat of lack of soil moisture even during short dry spells. Reports by Lal (1979a) indicate that after rainy episodes the water available in soils planted with maize suffices for only 5–7 days and in soils planted with dry rice, for only 3–5 days. With crop failures being common, small-scale farmers are reluctant to make substantial investments in fertilizers.

The second consequence of alfisol's physical structure is a high infiltration rate which is enhanced by the conduits and stable channels of earthworms and other animals in the soil (Lal, 1979a).

A very unfavourable physical property of alfisols is the lack of structural development in the form of soil aggregates. As a result, sandy clay loam alfisols tend to display rapid surface sealing following rainfall, and crusting during subsequent drying cycles. This crusting makes it practically impossible to work the soil before the rainy season, and so the fields appear strangely bare before the onset of the high-intensity rains. Excessive runoff and soil loss are inevitable. 'Hydrological studies conducted by ICRISAT on the traditional farming system have shown that, of the total rainfall potential available, an average of about 26 per cent is lost through runoff, 33 per cent through percolation, and only the balance of 41 per cent is utilized for evapotranspiration by crops' (El-Swaify *et al.*, 1987: 38). Crusting often extends beyond the immediate soil surface, resulting in consolidation of the soil profile to a depth that is determined by several other factors (El-Swaify *et al.*, 1987).

The combination of the soil properties mentioned heightens the alfisols' vulnerability to widespread soil erosion.

Consideration of these physical soil constraints provides some suggestions for a technological package to improve agricultural productivity on tropical alfisols. Land and soil management techniques have to be designed in such a way that they are effective in reducing runoff and erosion, providing structural stability to the soil, improving water-storage capability, and reducing sealing and crusting.

8.3.3   *The soil erosion problem and prevention strategies*

The most obvious, eye-catching, phenomenon of the wet-and-dry alfisol region is soil erosion. It is only reasonable that IITA Monograph No. 1 is entitled *Soil Erosion Problems on an Alfisol in Western Nigeria and their Control* (Lal, 1976a). Further experiences with the physical constraints have been addressed in two comprehensive volumes entitled *Soil Conservation and Management in the Tropics* (Greenland and Lal, 1977a) and *Soil Physical Properties and Crop Production in the Tropics* (Lal and Greenland, 1979). The combination of farming systems and resource management research is documented in the pamphlet *Resource and Crop Management Program at IITA* (IITA, 1986).

To collect information about erosion hazards and to develop countermeasures, a detailed study of the relationship between erosion and food crop production was started in 1970 on runoff plots over well-drained soils of the Egdeba series located along a toposequence on natural slopes of 1.5, 10 and 15 per cent inclination. The plots received different soil and crop management treatments: disc-ploughed and harrowed, no-tillage, mulched, maize and cowpeas planted in rows up and down the slopes, and bare fallow. Soil and water losses, nutrient losses, changes in soil physical characteristics, and crop yields were measured. The main findings are summarized by Lal (1976a: 38, 92) as follows:

(1) The exponential slope–runoff relation for bare-fallow treatments does not hold. The runoff losses are governed more by soil hydrological characteristics than by slope. The correlation coefficients between slope and runoff for no-tillage and mulched treatments are high, although the magnitude of the runoff losses under these treatments is small, regardless of slope.

(2) The slope–soil loss relationships for bare-fallow treatments indicate an exponential function. The exponential relationship, however, did not apply to mulching and no-tillage practices.

(3) Mulching and no-tillage treatments were most effective in preventing runoff and soil loss on the slopes investigated.

(4) Nutrient losses in runoff water from plots mulched at the rate of 4 to 6 tons/ha of straw or from the no-tillage plots were negligible.

As far as changes in soil physical characteristics and crop yield due to erosion are concerned, Lal (1976a: 124) stated that, in addition to the physical removal of surface soil, losses of organic matter and nutrients,

the moisture retention characteristics and infiltration rate are also significantly affected. The yield depression that results can be attributed to the degradation of the physical properties of the soil and losses of organic matter and soil nitrogen. Physical characteristics of the subsoil may dominate the crop response on eroded lands and addition of fertilizer may not compensate for removal of surface soil by erosion.

All these findings point to the overriding importance of covering the soil surface with crop residues or harvested organic matter in order to prevent erosion losses and worsening of the already delicate physical conditions of alfisols.

ICRISAT reports that mulching at the rate of 10 t ha^{-1} reduced seasonal runoff by 74 per cent and soil loss by 80 per cent, even in a high rainfall year. In a normal year, no significant runoff or soil loss occurred after mulch treatment, while a runoff of 205 mm and soil loss of 3.7 t ha^{-1} were recorded from the control plots (El-Swaify et al., 1987).

As time progressed, understanding about the processes connected with mulching has increased. Utilization of crop residues restores part of the organic matter. The build-up and maintenance of a certain organic-matter level are compulsory in order to elicit biotic activity and increase the exchange capacity. In addition, adequate crop residues on the soil surface keep soil temperatures down, protect against splash, especially during heavy erosive storms in the planting season, increase rainfall acceptance, stabilize the infiltration rate, lessen erosion, and have proven conducive to maintaining the aggregation of the soil (Falagi and Lal, 1979).

An alternative to mulching is fallowing with selected grasses, partially coupled with mulching. Some of these grasses can be replaced by leguminosae that contribute to raise soil nitrogen. Maximum infiltration rates were observed in plots planted with *Brachiaria* grass and in fields covered with legumes, such as *Glycine, Stylosanthes* and *Pueraria*. The benefits of such grass/leguminosae fallowing last up to 5 years. Most of the improvements in infiltration and nutrient provision disappear rapidly after the first mouldboard ploughing (Lal, 1979a).

Considering this experience with ploughing, another alternative was tried, namely 'zero tilling' or 'no-tillage'. In this approach all tilling is avoided and the soil is prepared for cultivation by mulching only. Seeding is done by plant stick or hoe. No-tillage plots with adequate mulch on the surface yielded larger maize crops than those subjected to conventional ploughing, especially after growing periods with drought stress. Since after a good rainy season yields from no-tillage fields were equal to yields from ploughed plots, no-tillage appears a valuable alternative for solving the grave erosion problems in well-watered years (Lal, 1976b; Falagi and Lal, 1979).

Another practical measure to control soil deterioration is the utilization of soil conditioners and soil penetrants (uresol and polyacrylamide). The effects were particularly strong in the loamy sands of the Ibadan series, but less in the sandy clay loams of the Egdeba soil series. If mulching accompanied the application of

227

soil conditioners, water diffusion was 30 per cent greater than in unmulched plots. This was interpreted as the additive effect of worm activity and a higher degree of aggregate stability (de Vleerschauwer *et al.*, 1979).

Beyond these locally restricted measures on cultivated plots, a number of management practices for modifying land surface configuration have been the subject of investigation. As indicated in Section 6.4, the land surfaces of the wet-and-dry outer tropics are characterized by vast peneplains with shallow depressions of various sizes. These act as receiving basins for erosional fine sediments. Waterlogging is common there and has led to a series of experiments aimed at monitoring crop behaviour at varying degrees of soil moisture.

Based on the premiss that water excess can have a significant influence on plant growth, some crops were subjected to controlled periods of submergence. Waterlogged cowpeas suffered a yield reduction of 15 per cent after a first flooding, and of 91 per cent after a second flooding period. The figures for soyabeans were 22 and 56 per cent, respectively. Maize grain yields showed no significant declines. Similar experiments conducted by ICRISAT showed crop yield declines of 10–35 per cent. Analyses of the nutrient concentration in leaves indicated that nitrogen, phosphorus and zinc concentrations declined with flooding, and so did the calcium content as well as the Ca : Mn ratio. The only increase observed was in manganese concentration (Lal, 1979b). Nutrient imbalances of this order are certain to have deleterious effects.

To reduce the influence of waterlogging, land levelling and installation of field drains are essential. The incompatibility between measures to ensure both maximum water entry into the soil and safe removal of excess water has prompted controversies about the value of tiered ridging and installing furrow dams, because this soil-and-water conservation system also increases waterlogging, development of anaerobic conditions in the root zone, excessive fertilizer leaching, and yield reduction in rainy years. Experiments conducted by Pathak showed the modified contour-band system, which includes gated outlets and involves land levelling and planting on grade instead of on contour, to be the most promising alternative (Pathak *et al.*, 1985). Harnessing runoff for supplementary irrigation contributes to increased, stabilized crop production. It is convenient to divert excess water to water-storage installations, such as tanks, since even improved cropping systems utilize less than half of the seasonal rainfall.

8.3.4 *IITA's crop-centred breeding programmes*

Since the end of the 1970s, IITA's research has progressively shifted its emphasis to breeding of crops and improvement of farming systems. In the *IITA Research Highlights* and in its *Annual Reports* from 1982 to 1986, three out of four research programmes are crop-centred: Root and Tuber Improvement Programme, Grain Legume Improvement, Cereal Research. The fourth is the original Farming System

Programme which, in 1986, was renamed Resource and Crop Management Programme.

The goal of the crop-centred research programmes is to increase the productivity of selected food crops and their integration into sustainable improved production systems. The root and tuber programme comprises cassava, yams and sweet potatoes. Among the grain legumes, IITA puts a special emphasis on cowpeas, but also works with soyabeans. Cereal research is concentrated on maize and dry rice.

Cassava is a remarkable plant: it adapts to different environments, requires limited input, stays in the ground for up to 24 months until required for consumption, and is relatively drought-tolerant. In spite of these advantages and of its prominence as a staple food, the average cassava yield in Africa of 6 t ha^{-1} has been lagging far behind its potential of 15–20 t ha^{-1}, due to certain diseases. IITA has therefore developed cassava varieties and breeding lines that are high-yielding, resistant against disease and insect pests, liked by the consumer, and low in cyanide content. Some of the improved varieties have been rapidly multiplied and distributed among farmers through different national organizations.

The problem with yam was the availability of planting material. Most of the yam plants would not flower and the seed set among those that did flower was low. IITA scientists have succeeded in breeding improved families and lines by hybridization. The programme also focused on increasing the multiplication ratio through the minisette technique involving the planting of pieces of yam tuber weighing only 45–90 g in nursery beds that are transplanted after sprouting into the field (*IITA Research Highlights*, 1986).

Since the local varieties of sweet potatoes had a low genetic potential, high disease infection, insect damage and nematode infestation, new lines were developed that are capable of producing consistently higher yields in less time (21–41 t ha^{-1} in 140 days) than the traditional varieties (6–14 t ha^{-1} in 180–200 days).

The major thrust of the Grain Legume Improvement Programme was directed at cowpeas for mixed and multiple cropping systems in various tropical environments. By 1980, significant progress had been made with a new high-yielding variety which, within a few years, has become very popular among farmers. Under the subhumid climatic conditions of northern Nigeria, this variety is ready for harvest in 75 days.

Yield reductions in cowpeas are caused to a large extent by disease, especially by different mosaic viruses. New lines immune to mosaic virus, of extra-early maturing and with different seed colour and quality to suit differing regional preferences, have been developed. Average yields of the most promising 60-day cowpea varieties are as high as 1600–1700 kg ha^{-1}.

Insect pests are another major constraint for cowpea production. In 1986 scientists announced a multi-purpose line by combining the available sources of resistance to disease and insect pests with high-yield potential, desirable seed types, and varying maturity. Meanwhile, it has been distributed for multilocational trials. In addition IITA scientists initiated a systematic breeding programme to develop bush-type vegetable cowpeas that need no stacking.

Soyabeans are a relatively new crop for farmers in the tropics, except in parts of

Asia. The Soyabean Improvement Programme, which started in 1982, places emphasis on the development of topically adapted high-yielding soyabean varieties with good seed storability, promiscuity, and resistance to pod shattering.

IITA's Cereal Research Programme comprises maize and rice, but more emphasis is placed on maize as a major cereal crop in Africa. Unfortunately, maize yields throughout Africa are low, averaging 1 t ha^{-1} in most countries, as compared with more than 3 t ha^{-1} in the rest of the world (*IITA Annual Report and Highlights 1986*: 101–5). While yields are depressed by many factors, such as poor soils and restricted input, the major factor has always been disease. Maize streak virus, which is spread by leafhoppers, is one of the most economically costly diseases and most common in the countries south of the Sahara. Downy mildew disease, which is a costly maize disease in Asia, also threatens crops in many African countries. Resistance to streak is not usually linked with resistance to downy mildew.

IITA researchers have succeeded in developing streak-resistant maize varieties and hybrids. In 1986, IITA received the King Baudouin Award for its research submitted as 'Solving the problem of maize streak virus: a research breakthrough to increase maize production in sub-Saharan Africa' (*IITA Annual Report and Highlights 1986*: 18). Maize with combined resistance to downy mildew and streak virus is the next aim of the plant breeders.

Following the selection of disease-resistant, high-yielding maize hybrids, seeds were distributed among farmers in different locations. The on-farm comparison between the yields from selected maize hybrids and open-pollinated varieties demonstrated that growing hybrid maize could be profitable for the farmers. The only problem was a scarcity of seeds from the selected maize hybrids.

The main reason small-scale farmers are growing open-pollinated varieties is that they can save seed from their own harvest for the next planting season. Therefore, development of varieties with this advantage is part of the maize research programme.

Consumption of rice in Africa is rising rapidly. Increase of production, however, comes from larger land areas being cultivated rather than from yield improvements. More than half of all the rice in Africa is grown under upland conditions. Breeders and virologists are working together to develop high-yielding lines of upland rice which are resistant to disease, insect pests, iron toxicity and drought stress.

Reviewing the details reported from the Root and Tuber Improvement Programme, the Grain Legume Improvement Programme and the Cereal Research Programme gives the impression of a highly successful work which promises corresponding progress in the field of tropical agriculture in Africa.

8.3.5 Transferring the innovations

Without playing down the importance of programmes aimed at breeding new varieties, we must not lose sight of the main problem, namely the transfer of these innovations into practical production under on-farm conditions in the harshness of the ecologically complex tropical world. Many of the breeding trials are conducted under artificially manipulated growing conditions on small experimental plots, in cement beds, or even in greenhouses outside the tropics. It is true that the greater food production of coming years will originate largely from improved varieties developed by the breeding programmes. But, 'there will be little impact on greater food production unless the new varieties are productive in the complex environment of the African farmer' (*IITA Annual Report and Highlights 1986*: 60). A matter of great concern is the promotion of the new commodities in order to increase short-term productivity at farm level.

Extensive trials on different farms demonstrated that grain yields did not increase significantly with improved maize varieties, as long as the crop was planted according to the traditional random method, without weeding and without fertilizer. In contrast, with row planting, rotation with groundnuts and cowpeas, fertilizer applications at rates of 80 kg ha^{-1}N + 40 kg ha^{-1} P$_2$O$_5$ + 40 kg ha^{-1} K$_2$O, and one weeding at 4 weeks after seeding, the grain yield of improved varieties was about twice that of local varieties. This confronts us with the problem of fertilizer application and weeding, which are discussed repeatedly in this book.

With the purpose of utilizing the tropical no-till farming experience and of reducing hand labour, necessary for good plant establishment, IITA's agricultural engineers designed a Farmobile that can penetrate the mulch, plant, fertilize and spray all at the same time. This minimizes compaction, although it cannot be totally avoided. An economic analysis showed that the investment for such a machine can be recovered by a farmer growing 10 ha of first-season maize followed by second-season cowpeas. But 'small-scale, mixed crop farms of less than one hectare make up the majority of farms in the humid and subhumid regions of West-Africa' (*IITA Research Highlights 1982*: 22). On the other hand, field surveys in 1983 indicate that future food supplies will come mainly from larger farms at a distance of 45 km or more from city markets. 'Small-holder production near the city presents only limited chances for expansion' (*IITA Research Highlights 1984*: 85).

Surveying the *IITA Highlights* and *Reports* reveals that most of the on-farm adaptive research concerned with alley cropping, intercropping, performance of leguminous shrubs, nitrogen contribution of hedgerow trees, effects of tillage methods, etc. is aimed at use on medium-sized farms.

As to the general results of the Farming System Programme related to increased and sustained production of upland crops, the *IITA Annual Report and Research Highlights 1985*: 17 summarizes:

> For the non-acid or high base status alfisols and inceptisols in the subhumid and humid zones, research results have shown that the no-till maize–cowpea –*mucuna* rotation system and alley farming with *Leucaena* are able to sustain high levels of maize and cowpea yields over a long period without reverting the land to varying periods of natural or planted fallow. On the coarse-textured, well-drained acidic ultisols in the high rainfall region of southern Nigeria,

231

results from cassava tillage experiments indicated that a no-till system with chemical weed control produced tuber yields equal of those of conventional and strip-tillage methods.

Currently there is no more research reported about progress related to increased and sustained production of upland crops.

Considering that the report mentioned above appeared 15 years after the beginning of the Farming Systems Programme and that the same report admits that a food crisis has been building up over the years in many of the African countries, we can only state that there is a wide gap between the existing results of plant breeders and pathologists and the application of new varieties and farming technologies by an impressive number of farmers.

8.3.6 *Centro Internacional de Agricultura Tropical (CIAT)*

Similar research to that conducted in Africa by IITA is being developed in South America by CIAT (International Centre of Tropical Agriculture). CIAT headquarters were installed on a 522 ha farm near Cali, Colombia. The main task of the organization is a systematic exploration of the possibilities of improving the agricultural utilization of the tropical savannas of semihumid South America.

The mandate of CIAT encompasses:

1. Research responsibilities for beans and cassava;
2. Improvement of tropical pastures (specifically on the acidic, infertile soils of the American tropics):
3. Regional responsibilities for the development of rice in the American tropics.

CIAT Reports and *CIAT Highlights* are subdivided into Bean Programme, Cassava Programme, Rice Programme and Tropical Pasture Programme (or Beef Programme). Occasionally a Swine Unit is added.

The objectives of the bean, cassava and rice programmes are similar to those of the IITA crop programmes: development of improved germplasm that combines resistance or tolerance to principal diseases, pests and soil constraints, efficiency in the use of applied inputs, and improved plant structure and yield potential.

The Tropical Pasture Programme has been adopted by CIAT because beef cattle production is one of the major agricultural activities in Latin America. Per capita beef consumption in Latin America is 3 times higher than in tropical Africa and about 16 times higher than in Asia. About two-thirds of Latin America's beef is produced in its tropical regions, where 71 per cent of the continent's cattle population is found. The annual productivity per head of cattle in tropical America is about one-half that of temperate South America and one-quarter that of the United States (CIAT, *Annual Report*, 1977). The objective of the beef programmes is to evaluate the beef production potential of the vast areas of acid, infertile soils classified as oxisols and ultisols, which occur naturally under savanna and forest vegetation.

232

'Improved management, particularly adequate mineral supplementation, would provide some productive increases. However, the poor growth rate of stock and, most likely, low fertility of breeding stock, can only be improved measurably by overall better nutrition' (CIAT, *Annual Report*, 1980: 69). Consequently, by far the largest part of the programme is oriented at the development of improved year-round forage production under different climate and soil conditions. This demonstrates that from an ecological viewpoint the management of plant production is the main concern, even within the cattle programme.

Among secondary considerations, such as the selection of cattle breeds best adapted to the tropical lowland climates and relatively resistant to tropical animal diseases, as well as prevention of disease and parasite infestation, it is of ecological significance that mineral supplementation is necessary, even with adequate nutrition. 'Inadequate mineral nutrition is severely limiting ruminant livestock production in tropical regions. Tropical forages frequently contain inadequate concentration of required minerals. Proper mineral supplementation of grazing livestocks is essential for maximizing production. Mineral needs will vary considerably, depending on the many factors' (McDowell *et al.*, 1983: 5).

Mineral feeders with mineral supplements must be distributed in the grazing areas, although the specific mineral requirements are difficult to define. The most common mineral deficiency in cattle is lack of phosphorus. In the publication of McDowell *et al.* (1983) photos show a cow chewing a bone and cattle ingesting soil in the llanos region of Venezuela, both indications of a severe lack of essential elements.

8.4 Balance of the accomplishments of the international centres of agricultural research in the tropics

In evaluating the efforts of the different research centres of tropical agriculture to increase the production levels of upland crops the general impression conveyed is that of far from satisfactory results. Regarding the future, the IITA's General Director's 1985 Report states: 'No one should advance misleading hopes that Africa will experience a "great revolution" similar to the one so widely publicized in Asia. Africa's traditional agriculture is much more complex. The dramatic breakthroughs in Asia with rice and wheat were based on a backlog of basic knowledge in plant sciences that is not so readily applied to important African crops.'

What is the meaning of 'backlog in basic knowledge in plant sciences'? The production progress that was achieved with wheat and rice by the Green Revolution can be put down to the improved plants themselves at only a secondary level; the key role is played by the high nutrient and energy input (see Section 7.2). The contribution of high-yielding varieties lies only in their ability to optimize these high inputs for the production of staple food seeds.

In contrast with the high-input philosophy of the Green Revolution technological package, the efforts to improve the yields of upland crops in the semihumid and humid tropics are based on the low-input philosophy. Only the continuous cropping experiment of Yurimaguas (see Section 8.1) used high-input technology on tropical ultisols. After the results of this high-input experiment were known in 1981, concentration on further research into low-input systems has been strongly recommended (see Section 8.2).

Why concentration on low-input systems in tropical agriculture? At this point we need not discuss again the various socio-economic viewpoints for the simple reason that at least some of the scientists at the agricultural research centres in the tropics, who face the food crisis right there, would do everything possible to develop a high-input system, if they saw any chance of success in the field of upland cultivation.

As far as wetland cultivation is concerned, there is, in contrast, very intensive research on high-input systems. The activities of the International Research Institute in Los Baños, Philippines, are directed mainly at high-input systems for wet-rice cultivation on paddy soils. IITA, CIAT and YAES do the same whenever they come across wetlands or alluvial floodplains within their geographical research area. For upland crops, however, the preference remains with low-input systems.

In the case of the African main crops, such as cassava, yams, sweet potatoes and various food legumes, plant physiological characteristics are again not the reason why low-input systems are recommended since these plants show immediate positive responses to high input in locations where this technology is applicable. Up to now the problem that has not been solved is the transfer of fertilizer to the cultivated plants under most of the tropical soil and climate conditions. In Section 8.1 dealing with the Yurimaguas continuous cropping experiments, the obstacles have been explained in detail.

In brief, the success of high-input technology in extratropical and some semiarid outer tropical regions is based on the fact that in these environments climate and soil constitute a comparably effective and reliable nutrient-commuting system. It combines low leaching potential with high nutrient exchange capacity. In the tropical humid environments, on the other hand, there is a simultaneous occurrence of excessive leaching potential and extremely low nutrient exchange capacity. Being initially poor in nutrients the soil loses additional nutrient inputs due to its high permeability and low exchange capacity. Tropical rainfall conditions lead to severe leaching and the entire commuting system becomes extremely ineffective and unreliable.

Composting and mulching, which play a key role in low-input systems, are aimed at maintaining a certain organic matter level in order to heighten somewhat the low exchange capacity and reduce nutrient loss due to leaching.

The other possibility for more effective nutrient transfer is increase of the nutrient-catching ability of the root system of the cultivated plants. This is mainly applicable in long-living plants, such as trees, bushes or manioc, which develop a deeper-reaching root system, possess a higher density of fine roots and which can be inoculated with mycorrhizas as 'nutrient traps'.

An application of these principles and a practical approach to the design of low-input farming practices on relatively well-suited soils in tropical highlands is represented by Egger's ecofarming (Section 9.1).

234

9

Practical attempts at applying new farming techniques

9.1 The ecofarming approach

Since the mid-1970s, Kurt Egger, a botanist at the University of Heidelberg, has combined theoretical approach and practical application in agricultural development projects in East Africa, developing a remarkable alternative for improving agricultural production in the tropics which he calls site-adapted land management or 'ecofarming' (Egger, 1976).

The area of application and source of collecting experience is Rwanda, a mountainous region between 1500 and 1800 m of elevation, with the bimodal rain periods typical of the equatorial-type stations and soils that developed mainly over basaltic or plutonic rocks, which makes the area one of the ecologically exceptional regions (see Sections 5.1 and 5.4). Also exceptional is the high population density. Still, the principles of ecofarming developed there are of general validity and, *ceteris paribus*, transferable to other regions of the tropics.

The underlying principle of the ecofarming approach is to handle agriculturally used natural units as well as specific farm units as ecological systems, intended to render lasting productivity. Dating from times of less population pressure, there are many autochthonous systems in the tropics that display continuous productivity and are able to sustain populations at fairly constant levels. As long as population density is low, even shifting cultivation in its strictest sense is an ecologically stable autochthonous system, i.e. a closed system from which no biomass must be extracted.

Today's agricultural systems are aimed at the regular extraction of agricultural products through harvests, and in this sense, ecofarming has to be treated as an open system. In order to ensure continuous productivity, the output of the harvest products must be balanced by an input of internally non-renewable matter: the ecophysiological functional system of agricultural production is complemented by a system of abiotical input. This abiotical input consists mainly of chemical fertilizers and herbicides and mechanical tillage.

Accordingly, the search is for a biological outfit that guarantees a stable production level with a minimum input (input on a small scale is necessary for any system). In this sense, ecofarming is a low-input system that is subject to three economically limiting conditions:

1. Ensuring continuous productivity takes precedence over maximizing output.
2. Guaranteeing subsistence is more relevant than commercialization.
3. Utilization of internal resources is preferable to external input.

The biological outfit of the production system is based on a principle inherent in the natural biotic system of the tropics, namely, a highly active biological cycle with high biomass production (Egger and Rottach, 1986). Similar to the tropical forests, the agricultural production system needs a maximum amount of biomass, most of which is utilized within the system through litter-fall, decomposition and nutrient recycling, while the portion extracted from the system through harvest remains small. This is the exact opposite of the modern high-input systems where the non-usable biomass is kept at a minimum, for example the dwarf cereals and other HYVs fostered by the Green Revolution package (Egger and Martens, 1987), and also the opposite of the main policy of the continuous cropping approach used in the Yurimaguas project (see Section 8.1). Instead it is likened to a fully equipped agroforestry system (see Section 9.4).

Agricultural production with a large volume of biomass and biomass turnover requires multiple measures that, even though serving various purposes, are all aimed at keeping soil organic matter at the highest possible level.

The most important measures for optimizing the available biomass are:

1. Integration of different trees for economic use (fuel wood, timber, fruit, animal forage) and for ecological benefit (nutrient pump, nitrogen fixation, soil improvement through litter, shelter, possibly soil conservation).
2. Planting of hedges with various species for different purposes: shelter, soil conservation, forage, pruning, and biomass for mulch and compost.
3. Managed short-lived bush fallow. Experience has shown that even high amounts of compost, mulch or fertilizer do not have the increasing effect on crop yield that well-managed 1- or 2-year bush fallows have. These bush fallows consist of legumes (suitable for nitrogen fixation) of various species planted close together. Compared to herbaceous fallow plants, such as kudzu, the bush fallow has the advantage of making the nutrients from the deeper soil horizons available to the plants (nutrient pump). Further, after 1 or 2 years more biomass is available, and, finally, when the biomass is used as green manure, the lignified parts, taking longer to decompose, prolong the soil-improving effect. In addition, legumes are not as prone to excessive growth as herbaceous plants and the closely planted managed bush fallow serves the purpose of weed control. If necessary, leaves from the bush fallow may be used for compost or for feeding livestock, but the animals must not be allowed on the fallow land. The bush fallow must be a fixed component of the field crop rotation. Experiments should determine the most favourable ratio in the field–fallow rotation, but the period of bush fallow should always remain relatively short, since the largest biomass increase occurs

during the first stage of growth. At the end of the bush fallow period there should obviously be no burning, which would defeat the entire purpose. Instead, the material must be worked into the soil as green manure, a very labour-intensive process which might warrant implementation of mechanical tools.

4. Field cultures are composed of multistorey systems with different species which increase yields, offer security against infection and yield loss (integrated plant protection through multiple species), and ensure a continuous food supply. The crops are planted successively (multiple cropping) and spatially integrated (intercropping). Plant residues are used as mulch. Details of rotation and cultivation procedures are known from the common home garden practices.

5. An important element of the ecofarming concept is the manner in which livestock is raised. Extensive grazing must be replaced by stable feeding. Even though this means extra expenditure for fodder cultivation and feeding, the overriding consideration is again that of maximizing the utilization of organic matter, in this case animal excrement. Together with kitchen refuse and human excrement, it must be selectively applied as manure and thus be incorporated into the nutrient cycle of plant production.

6. To ensure the economic feasibility of the farm units, cash crops are to be included among the cultivated plants in order to be traded for family supplies.

7. In the interest of long-term system security, mineral fertilizer application is necessary to balance the export losses involved in the production of cash crops. This is especially true for low-base-status soils with scarce residual mineral content. Slow soluble mineral fertilizers, such as stone powder or rock phosphate, are preferable to easily soluble fertilizers. The specific minerals necessary vary with site and can be determined by monitoring the nutrient dynamics – if appropriate facilities are available. Nitrogen fertilizers are not recommended since the nitrogen regeneration is to be achieved by the internal biotic system itself.

For a critical evaluation of the ecofarming approach, the various components must be studied in the context of the technical know-how and especially the intense labour requirements. It is obvious that this alternative assumes good management by farmers who understand and have mastered the system. It also demands high labour input, since no machines are being used nor planned. The entire system is intended for small family-type farm units.

In fact, ecofarming has been developed for Rwanda, a country that musters the highest population densities supported by rainfed agriculture. At present, the average farm unit size in Rwanda is 1.4 ha (Kotschi, 1986) and is expected to decrease to 0.5–1.2 ha in the near future (Egger and Rottach, 1986). The entire system is designed to increase yields on the limited productive land, thus ensuring subsistence, averting hunger and enabling survival.

Since the actual population growth rate of Rwanda is 3 per cent, and 97 per cent of the country's economy is based on agriculture, the time will come when even the most labour-intensive and best functioning ecofarming system will collapse, since it is a low-input system. As to the densely populated exceptional regions of the tropics, the following statement by Egger (1982:125) is most pertinent: 'Sooner or later the present population development will override every, even the most subtle, development of production methods. The driving force behind every agricultural production increase, therefore, is to finally reach the goal of population stability.'

A short reflection may be added here. A few generations ago, the people in the ecologically exceptional regions of the tropics enjoyed much leisure time, despite relatively high population densities. Today, they must work very hard just to manage at subsistence levels and the quality of their lives has hardly improved. Development in the industrialized nations of the extratropics has been exactly the opposite: a few generations ago, everybody had to work hard. Today the quality of leisure time is a major issue.

9.2. The Transamazon Colonization Project

In 1971 the Brazilian postal service used two remarkable stamps. One showed a schematic picture of the new highways which were planned as access roads to the Amazonian rain forests. The other displayed a spade-armed colonist symbolizing the agro-economic evaluation of that particular part of Brazil. The stamps were part of the official campaign to support the Programme of National Integration (PIN) (Stephanes, 1972).

9.2.1 *The general framework of the colonization projects in Amazonia*

Having declared, in 1966, 'Amazonia Legal' with a surface area of 4.6 million km² as a special region for future development, the I. Plano Nacional de Desenvolvimento 1971–1974 (First National Plan for Development) destined 2 billion cruzeiros ($US 440mio.) for the Transamazon Highway ('Transamazonica'), the aerial photogrammetric survey (RADAM), and for colonization projects along the Transmazonica. The Brazilian government regarded the national integration of the almost undeveloped Amazonia as a prestige matter. In the transition from the wet-and-dry *campos cerrados* in the states of Goias, Matto Grosso and Pará to the permanently humid rain forest of the inner tropics, large haciendas for cattle raising were to be installed. The heart of the rain forest along the Transamazonica was reserved for agricultural colonization. Details of land division, agrarian settlements, infrastructural provisions, financial support and loans were planned with great thoroughness. Even details such as land-use systems and yield guidelines were set up (see Fig. 9.2). All measures were taken into consideration to guarantee the success of this national project. After more than 15 years the results of the efforts made under favourable administrative circumstances should provide valuable insights into the actual agro-economic potential in the rain-forest region of the inner tropics.

As part of the Amazonian Integrated Colonization Programmes (PIC) the Transamazon Highway Colonization Scheme was announced by the Garrastazú Medici

238

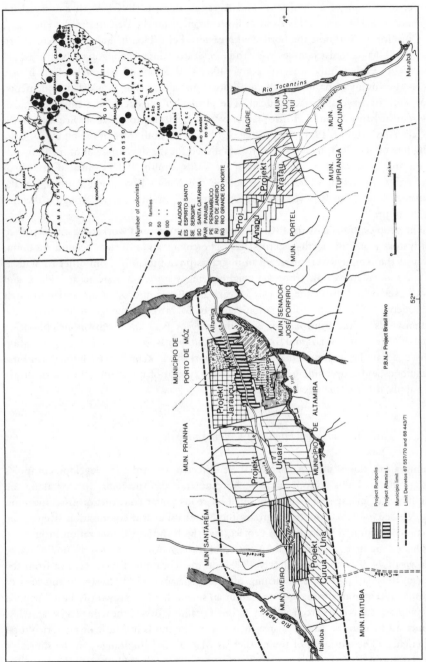

Fig. 9.1 The Amazon Colonization Project and origin of colonists (inset) (after Kohlhepp, 1976).

administration in 1970. The project was to be administered by the newly created National Institute of Colonization and Agrarian Reform (INCRA). It envisaged the opening of a 3300 km highway from Porto Franco, on the Belém–Brasilia Highway, to Cruzeiro do Sul, near the border with eastern Peru (Henshall and Momsen, 1974). Some 100 000 colonist families were to be relocated along the highway between 1972 and 1976. Each family was to receive 100 ha of land and a prefabricated house, along the highway or in one of the more than 66 *agrovilas* (nucleated settlements) that were to serve as pioneer nuclei. The plans also considered the enhancement of the semi-urban functions of three larger towns (*agrópolis*) to act as service centres for the *agrovilas* (Moran, 1981). In a fringe of 100 km on both sides of the highway, large long lots of 500–3000 ha (*glebas*) were established for the development of medium-size farms and cattle ranches. At the end of the 1970s, Smith (1982) estimated that along the highway stretch between Altamira, on the Xingú River, and Itaituba, on the Tapajóz River (Fig. 9.1) *glebas* occupied some 2.7 million ha, i.e. roughly the size of Belgium.

The people attracted by the PIC were mostly destitute rural workers or impoverished cultivators from the north-eastern states. The colonization nucleus around Altamira, west of the Xingú River, attracted immigrants predominantly from other rural areas of the same state (Pará), from the neighbouring states of Maranhão and Piauí, and to a lesser degree from the states of Ceará and Pernambuco, in the arid north-east (Mougeot, 1985). In the colonization centre of Marabá, on the Tocantins River, the colonists also came from *municipios* of the state of Pará and from the neighbouring state of Maranhão, Goiaz, and from states in southern Brazil. In these two colonization nuclei it can be assumed that the immigrants have kept the traditional land-use techniques and crop-tending practices that are typical of their tropical areas of origin and apply them in the newly opened lands.

9.2.2 *The Transamazon Highway Colonization Scheme*

Immigration flows into Amazonia were directed towards 15 development poles (Programa do Polos Agropecuãrios e Agrominerais da Amazonia) located along the interfluves of major rivers and connected by roads under construction (Kleinpenning and Volbeda, 1985). Among them are the Marabá–Tucuruí colonization stretch, the Altamira Integrated Colonization Project, and the Rondônia Colonization Project.

One of the first agrocolonization centres in eastern Amazonia was around Marabá, a town of 14 585 inhabitants (early 1970s) on the Tocantins River, 225 km from the Belém–Brasilia Highway's branching off the Transamazon Highway. In this area of latosols over granites from the Pre-Cambrian shield that had previously been intruded by swidden cultivators from villages in the Tocantins River, the *agrovilas* became less numerous than in the Altamira area. Most of the migrants had little or no agricultural experience. The existence of towns such as Marabá and Itupiranga (on the shores of the Tocantins River) assured a market for surplus staples. However, the ferrallitic soils proved to be of poor productivity, and by the end of the 1970s, disenchanted and bankrupt colonists, as well as growing numbers of landless migrants, flocked into Marabá which, by 1980, counted as many as 40 000 inhabitants (Smith, 1982).

A showcase of land opening and colonization in a previously undisturbed tropical setting is the Altamira Integrated Colonization Project. The choice of Altamira as model site was determined not only by the slow but steady colonization process in progress since the 1950s, but also by less severe local climatic conditions than in the rest of hot–humid eastern Amazonia. In addition, the apparent presence of terra roxa soils (alfisols), hailed 'as good as those of Paraná' (Moran, 1981, p. 286), made for a promising campaign. It turned out later, though, that the terra roxa soils occurred only in patches amid other soils ranging from infertile podzolic sands to oxisols and latosols.

Without considering the different soil conditions and topographic locations, colonists were allocated regular lots of 100 ha for the establishment of family farms. The dimensions of the holdings varied from 400 m of frontage by 2500 m of depth, or 500 m of frontage by 2000 m of depth. Only half of the 100 ha were allowed to be cleared for crop cultivation, while the rest was to be kept as 'forest reserve' (Kleinpenning, 1979). Planned *agrovilas* were intended to accommodate up to 64 families living in modest wooden houses and their functions were also elementary: primary schools, a general merchandise shop and a cafe. Only the larger *agrópolis* were planned to have major facilities such as a small hospital, government offices and vocational schools.

For the normal settler families the National Institute of Colonization designed a land-use model adapted to the hydroclimatic conditions of the Altamira project area (Fig. 9.2). Rice and maize as staple food were the first crops cultivated after the slash-and-burn clearing of a 4 ha plot. Immediately after the rice harvest beans and manioc were to be planted, followed by black pepper. This type of tillage is a permanent rainfed cultivation system (Ruthenberg, 1971, 1980; see Section 2.2) in which agriculture is practised year after year on the same well-tended plot. The land is allowed a short fallow period only during the dry spell of the year.

Even though the soil limitations in some areas of the Altamira PIC were known to the INCRA agents in charge of the project, the inflow of prospective colonists was so overwhelming that the officers were soon forced to abandon local counsel given by *caboclos* on soil quality and site convenience and had to allow the occupation of new lots by incoming farmers unfamiliar with the natural conditions of the area as long as they were confident with their own soil selection abilities (Moran, 1983b).

Farmers who, in the course of this indiscriminate procedure, ended up with infertile lots, found that their labour and capital investment were yielding increasingly meagre results and that their debts with the lender institutions were taking on frightening proportions. In fact, most of the agricultural production was destined for subsistence: rice, manioc, beans and maize. The rest was made up by modest cash crops: sugar-cane (for processing in a distillery located west of Altamira), cocoa and pepper. So unimpressive was the agrarian income generated by the farms that, in 1976, none of the Transmazon Highway settlements had reached the productive phase (Kleinpenning, 1979).

241

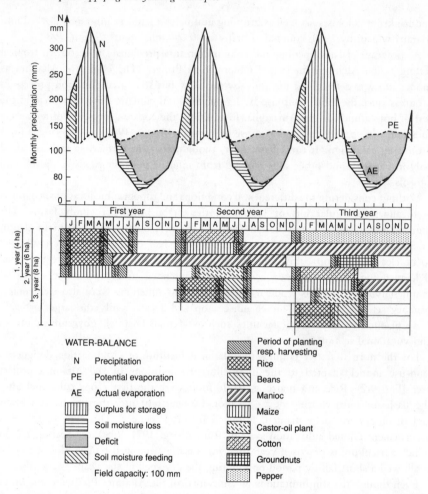

Fig. 9.2 Water balance and INCRA projected land use in the Amazon colonization area (after Kohlhepp, 1976).

Once settlements were established, the colonists began to face ecological and organizational problems. Not only did the differences in soil quality become apparent after only a few crops, but the lack of technical advice, and the non-existence of markets for the agricultural products made the situation critical. Settlers who lived in *agrovilas* but had farming land some kilometres away, started to built modest dwellings on that land and abandoned their assigned houses. Moreover, tropical diseases such as malaria, gastrointestinal infections and other ailments beleaguered the struggling settlers. And on top of this, plant diseases demonstrated that not only soil conditions and management limitations, but also environmental circumstances make the occupation of this area of Amazonia a very delicate enterprise.

242

A third effort to establish agrarian colonies along the Transamazon Highway evolved west of the town of Itaituba, a forlorn fishing town on the Tapajóz River, south of Santarém. Nearly 67 500 ha of *glebas* along the highway were made available by INCRA to colonists who mostly came from the states of Pará and Maranhão and from the vicinity of Santarém. As in the examples mentioned so far, the soils proved to be oxisols of low fertility that evinced reduced productivity after a few crop years. This, together with adverse weather conditions, pest and diseases led to the abandonment of lots and the establishment of the colonists in Itaituba (the town grew from 3782 inhabitants in 1970 to 15 000 in the early 1980s).

9.2.3 *Lessons from the colonization programme*

Varied assessments have been produced concerning the results of the colonization programmes along the Transamazon Highway. Considering the propositions and ideas advanced hitherto, the reader will probably suspect that our view of the projects destined to provide homesteads for former landless peasants on upland soils in equatorial forest regions is sceptical. Let us see what the experts have to say. Several substantial dissertations exist based on extensive field studies, and their observations, experiences and controversial conclusions form the bases of our discussion.

The anthropologist Emilio Moran writes: 'The most serious constraint to the development of the Amazon is not the absence of adequate soils, or is it the lack of labour, or even an insufficient amount of capital. Rather, the lack of managerial capacity at all levels of institutional functioning makes the process of Amazon development extremely hazardous from both an environmental and a social point of view' (Moran, 1981: 10).

It is particularly interesting to underline those parts of his argument in which natural conditions are implied. He argues that prominent governmental planning documents emphasize that the areas which have low soil fertility will be spared as forest reserves. In the surroundings of *agrovila* Vila Roxa (where he conducted his year-long study) nearly one-third of the colonists either chose or were assigned infertile plots. Yet, when the infertility became apparent, the credit banks would not condone the debts colonists had incurred in their efforts to force that land into crop-producing lots. The first administrators in Altamira sought to identify fertile land with the help of local *caboclos*, but when the original idea of making selective land assignments was abandoned due to the recognition of the soil's poverty and the large influx of colonists, the newcomers were taken to forest clearings and told to find a tract of land to their liking (Moran, 1981). Migrants from the north-east and from southern Brazil applied incorrect criteria in selecting their lots. Obviously, the consequences of not setting aside areas of low-fertility soil as forest reserves were not suffered by INCRA personnel, but the colonists who chose the wrong plots were saddled with debts. It was, indeed, a hazardous business finding the right plot, although around Altamira the chances of obtaining a good soil lot were 3 : 1.

The next imponderable pertains to the weather conditions and their consequences when related to misjudgements made by planning authorities. In the agricultural year 1971–72, farmers in the *agrovila* Vila Roxa produced rice at an average of 1053 kg ha^{-1}, and a few of them exceeded the projected 1500 kg ha^{-1}, as Moran reports. The results of the following years were beyond the farmers' control. The rice production failure of 1972–73, when yields averaged 634 kg ha^{-1}, was largely due to the utilization of a rice seed inappropriate to the area that had been promoted by INCRA officers. The fast-growing variety which reaches maturity in as little as 90 days was at its peak of fullness when the rains culminated in March, and the rice became waterlogged. Farmers who had not used this rice variety reached yields near the expected range. In 1973–74, low yields of 597 kg ha^{-1} were caused by the early arrival of the rainy season which impeded a good burn in the cleared fields. We concur with Moran in equating this to a drought disaster in an arid region and that it would be incorrect to blame the farmers for a climatic event.

Just as it is not correct to blame the farmers, it is equally unjust to blame the planning authorities for such 'acts of God'. If the rainy season 1972–73 had arrived and reached its culmination earlier, the early-maturity rice variety would have yielded reasonable amounts. If in the following year the rain had begun later in the season, the yields would also have been much better. The temporal variation of the rain onset and the interannual variability of precipitation amounts are climatic characteristics of the humid tropics that are beyond human control. Their effects can be catastrophic because, owing to the fragile soil conditions, the production system lacks any buffering capacity. If one had adequate soil conditions to rely upon, malfunctioning human interventions would not be so devastating as to endanger the very existence of the farmer.

Nigel Smith, a cultural geographer, states in the last chapter of his published dissertation *Rainforest Corridors*:

> Many individuals have benefited from the Transamazon scheme. Government personnel on the highway receive a handsome salary bonus for working at a hardship post; several have purchased ranches and farms along the road and in the vicinity of Marabá, Altamira and Itaituba. Most construction companies, hotels, restaurants, storekeepers, and landowners in the pre-existing towns have profited from the Transamazon. For the first time, several thousand peasants have titles to land that is rapidly increasing in value. Children can study in schools, albeit rudimentary ones, and families have access to free medical care, even if it is not always prompt. Some colonists, who arrived with few goods and little if any savings have managed to generate modest cash surpluses with which to improve their houses and purchase equipment.
>
> Nevertheless, the Transamazon has not achieved many of its objectives. The rapid turnover of lots indicates that the project has not created a favourable environment and growth of small-scale farms by formerly landless peasants' (Smith, 1982: 170).

Concerning land turnover, he reports that of the 170 lots he visited in 1979, only 59 per cent were still occupied by their original owners. Despite the fact that colonists are not officially allowed to sell their land, in practice they do. 'The price of a 100-ha parcel along the Transamazon depends on soil type, crops, and location. Lots of terra roxa bordering the main axis of the highway with a perennial water course sell for about U.S. \$35,000. Lots with sandy soil on side roads can be purchased for as little as \$4,000' (Smith, 1982, p. 171).

We believe that one of the best indicators of the scarcity and agricultural advantages of the terra roxa soils is their market price when compared with the acidic oxisols and ultisols of the same area.

According to Smith, several factors are responsible for prising settlers off their original lands. Disappointing agricultural yields, a confusing credit system, administrative shortcomings and disease have all contributed to the demise of new homesteaders. The prevalence of any of the aforementioned causes varies from case to case, but, in all cases, the least successful of the colonists are particularly prone to sell out.

Who purchases the land? Smith mentions two groups with different motivations and goals. Successful colonists acquire lots for their sons; entrepreneurs buy lots with the intention of investing capital and bringing both labourers and machinery to their new farms.

It is not difficult to discern which group purchases which land: the first group, being insiders, select land which is productive in the long run and where their sons will hopefully establish permanent homesteads. Entrepreneurs, on the other hand, are more likely to look for plots that will guarantee them a quick profit. The machinery and workers they bring are an indication that they slash the forests, take out the timber, burn the rest of the biomass, and plant the cleared plots with annual crops until the soil is depleted. Either way, agricultural productivity is thus likely to increase along the highway, at least in the short term (Smith, 1982: 171) but, what about the long-term effects? Deforestation and bare ground must be expected to be a forthcoming problem outside the terra roxa areas.

Throughout Smith's work, and in different contexts, the importance of the difference between terra roxa and other less fertile soils for the success of the settlers is evident. The soil quality is apparently the most decisive economic foundation in the colonization process. Other factors such as inadequate credits, fiscal incentives, planning shortcomings, impact of health problems, family size, and management ability recede to a secondary level. All of them are personally bound variables in time, and susceptible to alterations to a certain extent. Pertaining to the soil, however, there seems to be a permanent and irreversible dependency. Its gravity depends on the actual meaning and connotation of the terms 'fertile' and 'poor'. In order to discriminate between these two qualifications one would need to have access to data about the soil physical and chemical properties which, however, are nowhere to be found in Smith's work.

Philip E. Fearnside, an ecologist, conducted field research in the surroundings of *agrovila* Grande Esperanza, 50 km west of Altamira, for his dissertation, later published under the title *Human Carrying Capacity of the Brazilian Rainforest* (Fearnside, 1986). He claims that pieces of information from this area 'are more representative of the highway than are survey results in terra roxa areas. Most visitors to the Altamira

245

Table 9.1 Physical and chemical properties of soil samples in the surroundings of Grande Esperanza, Transamazonica (data from Fearnside, 1986)

Class	pH		Clay content		Carbon content		Phosphorus content	
	Range	Frequency (%)	Range (% of dry weight)	Frequency (%)	Range (% of dry weight)	Frequency (%)	Range (ppm)	Frequency (%)
1	< 4.0	33.0	0–14	21.6	< 0.50	1.9	0–1	83.8
2	4.0–4.4	30.2	15–29	25.3	0.50–0.86	31.0	2	8.3
3	4.5–4.9	15.3	30–44	23.6	0.87–0.99	5.1	3–4	5.7
4	5.0–5.4	12.5	45–59	21.2	1.00–1.49	40.8	5–6	2.1
5	5.5–5.9	5.3	60–74	7.5	1.50–1.99	17.2	7–9	0.1
6	6.0–6.4	3.6	75–89	0.8				
7	> 6.5	0.1						

246

area see only the agrovilas located on terra roxa at km 23 and km 90. Most researchers who have worked in the Altamira–Itaituba area have focused on these relatively small areas of terra roxa' (e.g. Homma, 1976; Homma *et al.*, 1978; Moran, 1975, 1976, 1981; Smith, 1976, 1978, 1982). The main agricultural research station in the area is also located on terra roxa' (Fearnside, 1986).

In the area researched by Fearnside the soils are very patchy, including some terra roxa (alfisol), the best soil type, and wider areas of poorer soils such as ultisols (yellow latosol). In his 'intensive study area' encompassing 23 600 ha with 236 lots, over 900 surface soil samples were taken at a minimum of 15 locations in each field. In addition, a series of 151 soil profiles were analysed. To describe the original soil condition under virgin forest, Fearnside uses soil pH, clay, carbon and phosphorous content. In Table 9.1 the corresponding data are summarized.

Considering the grain size composition, the soils vary from sandy to clay – loam soils. The upper horizons show as initial property a rather high content of organic matter (40.8 per cent have carbon content ranging from 1.00 to 1.49 per cent) due to their location under the undisturbed forest. Phosphorus is a particularly scarce element in most soils. Nearly two-thirds of the samples (63.2 per cent) are extremely acidic, with pH values under 4.4. Some 27.8 per cent are very acidic or acid, and only 3.7 per cent react neutral. Since only samples with almost neutral reactions can be classified as terra roxa (alfisols), the survey demonstrates that less than 4 per cent of the 23 600 ha of the study area can be considered as favourable soil types. Fearnside holds this percentage as representative for the whole area along the Transamazon Highway.

The areal estimation of soil distribution that follows, which also includes soil chemical characteristics and a qualification of the agricultural potential, has been taken from Fearnside's work (Table 9.2).

Concerning the percentage values of the transects, he adds:

> These soils are from the 799 km of the 1254 km section of the Transamazon Highway between Estreito and Itaituba for which soil identifications are reported by Falesi (1972). The percentage of more fertile soil (terra roxa) shown here is substantially more than exists in the Estreito–Itaituba stretch as a whole, since none of the 455 km (36 per cent of the total distance) not reported are terra roxa (Fearnside, 1986: 95).

The early estimates of the terra roxa total area have decreased steadily as the Amazon has become better prospected in recent years. In 1966, it was estimated to cover some 25 000 km^2, by 1974 this figure had dropped by more than half to 10 000 km^2, according to Falesi (1974). Most recent estimates suggest that terra roxas make for 'only about one five-hundreth of the area of Legal Amazon' (Fearnside, 1986: 227).

In a general survey of the Amazon basin soils, Cochrane and Sanchez (1982) make the rough estimate that the majority of the soils in the region are oxisols (45.5 per cent), followed by ultisols (29.5 per cent) and entisols (14.9 per cent). They also state: 'Unfortunately only about 6 per cent of the Amazon has well drained soils relatively high in native fertility.' To understand the implications of this statement, it should

247

Table 9.2 Major soil types of Transamazon Highway colonization areas

Soil type (Brazilian system)	USDA Classification*	% of transect†	Agricultural potential	pH (in H_2O)	Al^{3+}	P_2O_5	K^+	Ca^{2+}	Mg^{2+}	Na^+	N (%)	C (%)	CEC‡ (meq per 100 g)	Sample depth (cm)
						(meq per 100 g)								
Terra roxa	Alfisol	9.6	Good	5.8	0.00	0.15	0.29	5.31	0.81	0.05	0.21	1.56	9.60	0–20
Yellow latosol	Ultisol	18.2	Poor	3.9	1.56	0.22	0.03	0.04	0.06	0.03	0.13	1.24	8.85	0–30
Red-yellow podzolic§	Ultisol	38.6	Poor	4.1	2.89	0.21	0.06	0.10	0.09	0.05	0.11	1.09	6.77	0–20
Latcritic concretionary	Petroferric paleudult	6.9	Very poor	4.5	2.25	0.45	0.26	0.45	0.98	0.06	0.13	1.25	8.81	0–20
Quartzose sands¶	Quartzi-psamment	17.9	Very poor	4.8	0.40	0.23	0.03	0.10	0.02	0.04	0.04	0.22	1.84	0–20
Others‖	Others	8.8	Poor–very poor	4.2	6.1	0.06	0.20	1.1	1.6	0.05	0.17	1.19	15.9	0–28

Notes: These soils are from the 799 km of the 1254 km section of the Transamazon Highway between Estreito and Itaituba for which soil identifications are reported by Falesi (1972a). The three Transamazon Highway colonization areas at Marabá, Altamira and Itaituba are all within this area. The percentage of more fertile soil (terra roxa) shown here is substantially more than exists in the Estreito–Itaituba stretch as a whole, since none of the 455 km (36 per cent of the total distance) not reported are terra roxa. The percentage of terra roxa for the Transamazon Highway as a whole is even lower, since none of the remainder of the highway (west of Itaituba) has terra roxa (Brazil, EMBRAPA-IPEAN, 1974). The 76.8 km of terra roxa reported represents 2.6 per cent of the approximately 3000 km of the Transamazon Highway in Amazonia.

The soil chemical information given in the table is for the superficial layers of typical profiles. Data are from the following sources: terra roxa, yellow latosol, red-yellow podzolic, lateritic concretionary, and quartzose sands are from Falesi (1972a: I 36, 69, 168, 99, and 108, respectively); data for "others" are from Brazil, DNPEA (1973a: 57).

★ U.S. D Agr. (1960); Seventh Approximation classification equivalents from Beinroth (1975); Sánchez (1976); and Brazil, RADAM (1978 18: 271).

† Percentage of 799 km reported by Falesi (1972a). In some cases where more than one soil type were reported for a given highway segment, the stretch is apportioned equally between the types.

‡ Cation exchange capacity (sum of Ca^{2+}, Mg^{2+}, Na^+, K^+, H^+, and Al^{3+}).

§ Side-looking airborne radar survey (Brazil, RADAM 1974, vol. 5) classifies a number of areas as yellow latosol, which are closed as red-yellow podzolic by Falesi (1972a).

¶ Most of the quartzose sands (142.4 km or 17.82 per cent of the transect) are distrophic red and yellow sands. One small area (0.4 km, or 0.05 per cent of the transect) is white sand Recosol.

‖ Brunizem 2.4 km, or 0.30 per cent; cambisol (inceptisol) 30.9 km, or 3.87 per cent; grumosol (Vertisol) 7.0 km, or 0.88 per cent; slightly humid gley (tropaquept) 10.7 km, or 1.34 per cent; hydromorphic soils 18.0 km, or 2.25 per cent; alluvial soils, 1.0 km or 0.13 per cent.

Source: Fearnside (1986, p. 39).

be noted that in that 6 per cent (31 million ha) are included all possible fertile soils that can be found in Amazonia, such as well-drained alluvial soils, cambisols, vertisols and two varieties of terra roxa: *terra roxa estruturada* (Paleustalfs) and *terra roxa legitima* (Eutrothox). The same authors concede that the latter occur only in very restricted areas of the Amazon basin, for instance, in the vicinity of Altamira, and that only there 'permanent agriculture has a better chance of success' (Cochrane and Sanchez, 1982: 155 p. 155).

Condensing the findings of these experts, it seems that between 90 and 97 per cent of the Amazon colonization project area is covered by nutrient-poor, acidic soil groups whose agricultural feasibilities are discussed in detail in section 8.1. Up till now, there has been no technology capable of guaranteeing sustained land utilization under such soil conditions. So far it has not been possible to put into practice the original idea of creating continuous cropping systems in order to curtail slash-and-burn, and prevent progressive deforestation.

Gerd Kohlhepp, an economic geography specialist on tropical South America, observed from the beginning the development stages of the colonization project. In 1986 he concluded that the Transamazon Colonization Scheme was a failure. The original target of installing 100 000 settler families and the development of an integrated settlement system (*agrovila–agrópolis–rurópolis*) has virtually failed. Until 1975, only 7 per cent of the planned 100 000 families were officially settled and were provided with the promised government assistance. During the following years, only squatters settled without government support. As a result of the lack of co-ordination the planned *agrovilas–agrópolis–rurópolis* system collapsed, too. Apart from the various socio-economic reasons for this failure, the actual cause was the misleading estimation about the natural potential of the equatorial forest area (Kohlhepp, 1989: 64).

With regard to the whole region he stresses: 'To present agrarian colonization in the tropical rainforests as an alternative to agrarian reform is not only wrong in concept and from the point of view of agrarian policy but is also likely to have disastrous consequences for development strategies.' He underlines the judgement of Sioli, that any large-scale production and extraction of biomass are unfeasible and recommends:

> The Amazonian rainforests should no longer be an experimental area for ecologically and socially inappropriate 'development models' and a battle-ground for non-Amazonian problems and conflicts of interest which have been shifted to the periphery. Amazonia cannot serve as a spatial 'escape valve' for the rural and urban population pressure in other regions of the various countries nor as a 'granary' (Kohlhepp, 1989).

9.2.4 New concepts for Rondônia

During 1981, the Brazilian government approved a new 'Programa Integrado de Desenvolvimento do Noroeste do Brasil' (*POLONOROESTE*) (Integrated Development Programme for the North-west of Brazil). It was mainly concerned with the

Federal State of Rondônia. This region has a longer dry period than the central Amazon basin and the soils were originally believed to contain extensive terra roxas over ancient basic volcanites (*Atlas de Rondônia*, 1975). In the meantime it seems that, according to Moran (1983b: 15), only 10 per cent of Rondônia's soils fall into that category and that most of these are patches interspersed in large areas of oxisols and podzols.

Taking into consideration the experience gained in the Transamazon Colonization Project, a new development concept was formulated for the POLONOROESTE programme:

1. Forest reserves should be laid out in large *en bloc* areas;
2. The project should concentrate on smallholdings;
3. Colonization should occur in three different forms: as an integrated colonization project (PIC) similar to those along the Amazon Highway, as governmental supervised settlements, and as squatter dwellings;
4. The production of basic foodstuff, such as dry rice, manioc, maize and beans should be based on field–fallow rotation systems, while cash crops, such as coffee and cocoa, should rest on low-input systems;
5. A detailed survey of natural conditions should permit the adaptation of land use in micro-regions to relief, soil quality and water supply.

The problem here is to enforce these concepts with respect to the large immigration flow. In 1983, approximately 90 000 immigrants arrived; in 1984 the number rose to 130 000–140 000 people (Kohlhepp, 1985). Within the scope of the government programme of the INCRA, 40 000 families were settled by 1984, and by 1985 another 30 000 had been registered on the waiting list (Kohlhepp, 1985).

Soil limitations, in addition to low infrastructural facilities, lack of site-specific agrarian know-how (Muller, 1980), conflicts concerning landownership and land speculation, plague agricultural colonization in Rondônia as much as in the schemes of the Transamazon Highway, although the prospects are little more promising than in the inner tropical part of the rain forest.

As a corollary of the information presented, one must agree with Kohlhepp (1989: 61) that 'current agrarian colonization projects should be very strictly controlled and aimed at intensive, ecologically-compatible farming. The project areas should not be further extended because adequate potential for more intensive farming under better living conditions is available outside the Amazonian region'.

9.3 Transmigration in Indonesia

Ever since colonial times, Dutch authorities have attempted to resettle people from the overpopulated island of Java in the sparsely inhabitated neighbouring islands and to start active agricultural nuclei there. As early as 1905, farmer families were resettled around Lampung in south Sumatra and later in west Sumatra. Most of these efforts were aimed at easing population pressures in Java, but also at making use of areas that appeared to have agricultural potential or – as some cynical observers point out – to provide cheap labour for the plantations that were being established by the Dutch in Sumatra (Adiwilaya, 1970, quoted by Hardjono, 1977). With the declaration of independence and the establishment of a national government in Jakarta, the emphasis in the transmigration projects targeted Kalimantan, Sulawesi, Maluku and Irian Jaya in recent periods.

9.3.1 *Transmigration plans and number of people involved*

A Fifteen-Year Transmigration Plan, drawn up between 1947 and 1951 with the ambitious goal of relocating more than 48 million people, had to be abandoned and replaced by a new one, in 1953, when it was realized that very little had been achieved. The new plan sought the relocation of 100 000 people yearly, but only 40 000 were actually moved by the end of 1953 (Hardjono, 1977).

Ineffectiveness made it necessary to prepare a new Eight-Year Plan in 1961 which included provisions to transport, accommodate, resettle and feed the relocated people. Around the new settlements the land was to be cleared and irrigation facilities for growing rice – the main staple – were deemed necessary.

Then, before this plan was seen through, a new Three-Year Plan was developed in 1967 to deal with the mounting demographic pressures in Java and to enhance certain population-moving trends that had become apparent in a population census conducted in 1961. The plan emphasized the establishment of settlements in the wet coastal plain of south Sumatra and in central and south Kalimantan. When the construction of drainage and irrigation facilities in the tidal flats proved expensive and slow, a new shift in transmigration policies focused on settlement of non-tidal areas, and, to this effect, the transmigration ordinances of 1969–74 were issued (Hardjono, 1977).

During the period 1950–69 an estimated 358 321 people (excluding spontaneous migrants) were moved to Sumatra; 41 698 to Kalimantan; 14 868 to Sulawesi; and 8502 to Maluku and Irian Jaya (Hardjono, 1977). Their total amounts to less than half a million, which does not come even near the goal set by the transmigration authorities in the early 1950s. More recent data indicate that between 1973 and 1979 the total number of individuals moved by transmigration projects was 456 987 (Uhlig, 1984). According to Scholz (1980), 100 000 Javanese enter Sumatra each year. As a matter of fact, the Five-Year Plan of 1979–84 envisaged the relocation of 2.5 million people, a number which, considering a population total of nearly 85 million (1980) and the poor achievements of previous transmigration plans, appears illusory and insufficient to ease the population pressures in Java.

9.3.2   *Discussion of the different transmigration projects*

The official data and the governmental projects towards which resettled farmers were sent until 1973 are listed in Table 9.3.

The main and older projects on Sumatra are those around the area of Lampung. The relative vicinity of this region to central Java, with which it is connected by ferry, made this province the key target of transmigration. According to Scholz (1980), 65 per cent of all migrants moved there, thus swelling the number of inhabitants to 4 million in the early 1980s. Located almost on the Equator, southern Sumatra receives over 2000 mm of annual precipitation. Rainfall during the zenithal highs of March and September exceeds 300 mm, and during the less rainy periods it seldom drops below 125 mm. Strongly acidic ferrallitic soils have developed over igneous or metamorphic rocks and the dense rain-forest cover makes land opening difficult. Even more accentuated are the conditions on western Sumatra where annual rainfall commonly exceeds 3750 mm.

The land first chosen was reasonably fertile and well watered, and each colonist was given 1 ha. In river estuaries and along tributaries where tidal influence occurs farmers have established rice fields where the irrigation water flow is controlled by the tides.

Table 9.3   Transmigration projects, area, and numbers of settlers 1905–73 (Zimmermann, 1980)

Provinces of settlement	People 1905–41	People 1950–73	Total of people 1950–73	Area (ha)
Aceh		695	695	22 172
North Sumatra	11 426	10 582	22 008	6 477
West Sumatra	1 945	13 150	15 095	44 186
Riau		1 814	1 814	4 100
Jambi		9 771	9 771	4 000
Bengkulu	7 443	7 270	14 713	14 500
South Sumatra	25 153	146 858	172 011	57 700
Lampung	173 959	284 569	467 348	252 143
West Java		5 032	5 032	
West Kalimantan		13 824	13 824	69 200
Central Kalimantan		9 825	9 825	7 200
East Kalimantan	164	21 160	21 324	52 593
South Kalimantan	3 950	15 546	19 496	15 501
North Sulawesi		6 322	6 322	38 200
Central Sulawesi	146	17 144	17 290	43 071
South-east Sulawesi	984	7 291	8 275	2 176
South Sulawesi	13 464	13 283	26 747	18 538
Maluku		1 863	1 863	1 400
West Nusatenggara		654	654	
Irian Jaya		1 132	1 132	
Total	238 634	587 785	835 238	

With the help of these irrigation works wet rice has provided the farmers on tidal flats with some marketable surpluses. In places where irrigation was not readily available the relocated Javanese farmers had no choice but to practise dryland cropping, rotating maize, cassava, rice and soyabeans. Only on successful and accessible farms is their income supplemented by cash crops, such as coffee, pepper and cloves.

But as new colonists were moved to the Lampung area the settlements became overcrowded and in most areas the already small holdings were fragmented among family members. Since in the original land distribution no land had been reserved for natural expansion, farmers who could not maintain an adequate rotation cycle returned to subsistence level, growing increasingly more cassava as their dry rice and maize yields decreased (Hardjono, 1977). Consequently, much land changed hands as people left, or formerly cleared land was invaded by alang-alang grass (*Imperata cylindrica*), which, once established, is very hard to eradicate. In northern Lampung land is suitable for dry farming, and the transmigration authorities are hoping that the relocation projects there will not become mere 'cassava villages' as occurred in older parts of the Lampung relocation region.

In the settlement areas of South Sumatra the results of transmigration have also been mixed. In the Belitang project area, good soils and abundant water have rendered the wet-rice fields and rubber farms quite profitable. Around the Upang delta special settlements were started to cultivate rice on swamplands using tidal irrigation. Under favourable water conditions and in a soil that is rich in continental sediments, rice yields of up to 3.6 t ha^{-1} have been obtained without fertilization. If fertility can be maintained, the project should continue to prosper (Hardjono, 1977).

The province of west Sumatra is the third largest area in number of settlers and hectares opened. The project has not done well because of poorly selected sites. In 1973 a 100 000 ha transmigration site in Sitiung was selected as a research area within the Humid Tropics Programme of TropSoils (McCants, 1985). Sitiung is considered a representative transmigration site, and the Indonesian government has the intention of transferring the experience gained here to other sites. The *TropSoils Triennal Report 1981–1984* (McCants, 1985) provides valuable ecological details for a better understanding of the actual difficulties encountered up to now in the course of the resettlement of Javanese farmers. In the project area live 1500 indigenous families and 6000 transmigrated families. A modest home, 1.25 ha of recently cleared land, and a year's supply of food, fuel, other living essentials, seed and fertilizer awaited each family upon arrival. Land quality in the resettlement area is quite varied. The soils range from moderately fertile inceptisols on river terraces to highly leached and impoverished oxisols and ultisols on the dissected peneplain. Since the productive land on the river terraces has long been settled, the recent transmigrants are directed to the less desirable lands off the alluvial sediments. The mechanical clearing of the virgin rain forest poses a serious problem for the fragile oxisols and ultisols. After bulldozer clearing and burning, a typical farm plot has bare spots – sterile subsoil patches from which topsoil and organic matter have been stripped and/or eroded away – alternating with green strips that correspond to the ash lines of burnt trees. This variability presents a problem for the farmer. Bare spots produce nothing, sometimes not even cassava will grow there. Therefore, they are neglected, erosion progresses, and eventually they are abandoned altogether. If farmers do not learn to recognize the soil variability and lime or fertilizer are applied uniformly over a field, the bare spots receive too little and the green strips are overdosed. Immigrants from Java and Bali make the painful discovery that the knowledge and experience that worked so well on the rich volcanic soils of their homelands does not work in this oxisol and ultisol environment.

This summary presentation of actual experience in the context of the TropSoil projects in west Sumatra illustrates that resettlement under the premisses of the transmigration plans mentioned is a hazardous business. Those who encounter too many bare spots on their 1.25 ha plot will drop fast down to the lowest subsistence level, if they have not abandoned the land already. By now, the reclamation of the barren lands has become a major problem in the transmigration area of west Sumatra. Another problem is the development of sustainable, low-input farming systems. As in other parts of the tropics, careful attention to the existing organic matter and a thorough use of it, play a key role in such farming systems.

The projects at Bengkulu have fared no better. Located in rugged terrain, connected by poor roads, and plagued with land claims from the native inhabitants, many of the Javanese have abandoned their holdings either to return to Java or move to other ongoing projects on Sumatra. Those who chose to stay sell timber and charcoal and make simple furniture, which has led to a rapid deforestation of the surrounding valleys. Only in the Bukit Peninjauan project were the results encouraging due to the good rice soils and proximity to the market at Bengkulu some 23 km away (Hardjono, 1977).

Other projects in central Sumatra, such as Jambi and Riau, and in northern Sumatra (Aceh and North Sumatra) were not originally among the provinces earmarked for transmigration. However, the inflow of Javanese forced the implementation of the projects mentioned. Given the relatively high population density of northern Sumatra (104 persons per square kilometre, according to the census of 1980) and the existence of large plantations, there has been little incentive for planning settlements in these areas; still, many Javanese who came to work on the plantations during the first half of this century have made themselves independent in the rural areas of northern Sumatra.

Scholz (1980) points out repeatedly that on Sumatra the most successful projects have been those in which the smallholdings have based their agrarian economy on the exploitation of small tree crops (pepper, coffee and cloves) as cash crops mixed with food staples, as is typical of some agroforestry variations.

The transmigration projects on the island of Kalimantan were partially curtailed by the extensive unhealthy swamps on the southern and western coasts; thus, settlement attempts were made mostly in the eastern part. However, there are remarkable differences in the degree of success within the varied projects on the island. The most successful are those near the town of Samarinda, where Javanese settlers grow wet rice on excellent soils, developed on base-rich rocks, and find good markets in that town. Elsewhere in east Kalimantan the projects are located close to logging

255

companies which entice the colonists to leave their fields and work for them. Particularly convenient climate conditions concurred to make the colonization efforts more successful than on other islands. At Samarinda, annual rainfall varies around 1800 mm and around 100 mm during the late summer and early autumn months, making this region very similar to wet–dry zones of Africa and South America in their precipitation regime.

In South Kalimantan – the same as in South Sumatra – some colonization schemes are conducted on coastal plains, utilizing tidal irrigation with good results, although some problems with the acidity of the coastal peat soils have arisen. It is common that, as fertility begins to decline in these tidal flats, the colonists will abandon their allocated plots and look for work with timber companies. Soon, the tidal fields are overgrown with grass and weeds (Hardjono, 1977).

In the projects of central and west Kalimantan various difficulties were encountered. Swamps prove a hindrance to agricultural development and communication, but not so for logging enterprises which make use of the rivers. Remoteness, distance from markets, and abandonment have severely limited any progress. In west Kalimantan only a few projects on tidal lands have prospered, although again, distance from markets make the selling of agricultural surpluses and cash crops difficult.

Sulawesi (Celebes) is an island of uplands, cut by deep valleys. Since recent volcanism has not been frequent, except for small areas on the northern peninsula, most soils are infertile, having developed from non-basic parent rocks. The coastal plains are very reduced in size and tidal lands are scarce. Climatic conditions also make a difference on Sulawesi. Unlike other Indonesian islands, precipitation is irregularly distributed over the year and across the land. At Luwuk yearly precipitation is as low as 626 mm, while at Malili it reaches 3150 mm. The driest periods are autumn and early winter, a fact that allows the soils a certain resting period before the summer harvesting. Moreover, annual precipitation peaks between April and July, which makes for good watering conditions during the spring and summer growing season.

Despite the disadvantages, two of the most successful transmigration projects are located in the districts of Parigi and Luwu, in the southern part of the island. In the district of Luwu more than 20 000 people settled between 1969 and 1974. Although the settlements keep growing, agricultural production has not risen above subsistence level due to poor market accessibility. The decline of soil fertility after 2 or 3 years, is combated with some success by crop rotation and manuring. The latter stems from herds of cattle, usually water buffaloes, which are kept by the settlers in the Luwu region in addition to their agricultural activities.

Into the transmigration projects located in the mountainous and isolated parts of central Sulawesi less than 7000 people – most of them from Bali – were brought. The Balinese cultivate rice on terraced fields and sell it to rice-merchants who tour the Gulf of Tomini by boat. In the south-east and north of Sulawesi the projects mostly involve migrants from Java or Bali made homeless by natural catastrophes, such as the Balinese who were affected by the Mt Agung eruption. The villages are generally doing fine, for the Balinese are industrious and resourceful in their production of rice, corn, soyabeans and coconuts.

Other transmigration projects targeted the 'Eastern Provinces of Indonesia'. As such are understood the island of Seram and the greater part of the province of Maluku. Apart from the remoteness with respect to central Indonesia, the scarcity of roads and markets, the rapidly declining soil quality and the lack of irrigation infrastructures have curbed the progress of transmigration towards Maluku, as well as into Irian Jaya (the northern part of New Guinea).

Thus, with land still available in Sumatra, Kalimantan and Sulawesi, transmigration is unlikely to be pushed very hard into these outlying provinces, unless the current demographic growth continues in the heavily populated areas of central Indonesia.

9.3.3 *Lessons from the transmigration projects*

An assessment of the transmigration plans in Indonesia should address the following key issues; Have these policies achieved the goal of easing the population pressure in central Java? Under which environmental conditions did transmigration projects succeed, and under which conditions did they fail? Were the land-opening projects so satisfactory as to warrant further migratory plans?

The scholars and politicians involved in the study of the transmigration process in Indonesia all agree that the aim of easing population pressure in Java has not been achieved. For one, the number of outmigrants still lies far below the planned figures, and even considering the undeclared spontaneous colonists, the high growth rate of the Javanese population has overtaken the numbers of those who have left.

Thus, the emphasis in the purposes of the transmigration plans has changed. Instead of insisting on easing demographic pressures, the new priorities stress the development of new and productive agricultural regions in the empty spaces of the 'outer islands' as a way to enhance national security, and to provide higher living standards for the transmigrants and also for the local populations (Report 17 of Transmigration Area Development, TAD, Bappeda, 1982, quoted by Uhlig, 1984). The fact is that Java continues to be an overcrowded island which, nevertheless, is still able to meet the food demands of its large population.

As to the results in the new lands, they are mixed. If one considers that a good number of the Javanese who left did so because they were landless, quasi-landless, or heirs of undersized plots (Uhlig, 1984), experience with the newly allocated 1 or 2 ha showed that these holdings were still insufficient for upland cultivation. Inadequate waterworks, lack of communication, distance from markets, and crop decline in fields of low nutrient status have all conspired to curtail the progress of the transmigration projects. Thus, with the quality of the land deteriorating and the incomes insufficient to maintain large families, the trend of leaving cultivated lots and searching for new ones in lands recently opened or in the process of being opened, or of changing jobs altogether – either to work in logging companies or urban occupations – has upset the original goal of providing for each transmigrant farmer an improvement over his original living conditions.

To be sure, there are successful and productive settlements, particularly on tidal plains, in limited alluvial sedimentation areas, and in nutrient-rich soil locations, but these are exceptions (Hardjono, 1977). A remarkable example is that of the Balinese who settled in the mountainous region of central Sulawesi. The favourable soil conditions are apparently good enough to render successful their traditional excellent working habits and sense of co-operation.

The Lumpung province's transmigration projects show that the impact of an enormous inflow of migrants, added to existing natural limitations in productivity potential, is deleterious for the agricultural sector. Zimmermann (1980) reports that as many as 50 per cent of the farm population of Lampung who were given land in the early stages of the colonization have become landless again. Abandonment of the land, moving to new land-opening areas (usually as squatters or as spontaneous colonists), and engaging in deforestation have become discouraging practices in the present transmigration projects.

Of great significance is the fact that the new colonization seems to have fared better in the mountainous areas of Java itself. As reported by Uhlig (1984), successful agricultural settlements in east Java have been pushing up to elevations of 1600–1800 m, which is above the established farming areas dedicated to growing coffee, tea and chinchona bark. There 'even steep slopes are under permanent cultivation; they are practically terraced, but they lie on a porous, stable base of volcanic ash, which also provides a lasting supply of nutrients' (Uhlig, 1984, p. 78). With this statement the author lends support to an explanatory variable that he himself scorns, namely that on tropical soils of low nutrient status shifting cultivation is the only possibility, while permanent dryland cultivation can be conducted only on high-base-status soils. And this is exactly the case in large parts of Java, where soils originate from base-rich volcanic rocks, or have evolved on alluvial and coastal plains where the steady addition of new sediments assures a balanced supply of nutrients. In the other islands of Indonesia, the most successful agricultural colonization projects lie in river deltas – the Upang delta, the Indragiri River estuary in Riau province, and the delta, of the Barito River – or, on alluvial plains, such as the Bengkulu area of west Sumatra. Settlements that were established on volcanic deposits, such as in the Luwu region on Sulawesi and in the northern peninsula of Sulawesi, are also doing well.

Many of these examples prove sufficiently that wherever transmigration resulted in viable agricultural settlements, the conditions of the soil and the management of water were key elements ensuring their ultimate success.

It is enlightening that whenever the sociologist Hardjono refers to unsatisfactory projects, she invariably mentions, first, deteriorating soil conditions along with bureaucratic bottlenecks, lack of agricultural expertise and insufficient farmer preparation as the most common causes for the failures. Thereby, she recognizes that a natural condition – soil quality – is the primary controlling factor of the transmigration project's success or failure.

258

9.4   Agroforestry schemes as alternatives in tropical agriculture

Especially in the 1970s, 'agroforestry' was favoured as a sound concept for rational and successful land use in the humid tropics. In the foreword to *Agroforestry in the African Humid Tropics*, Gilles Lessard of the International Development Research Centre and Lee MacDonald of the United Nations University wrote: 'In the last five years there has been a virtual explosion of interest in agroforestry. The concept has spread from a few anthropologists, foresters and agricultural scientists to become a priority for a number of national and international agencies' (MacDonald, 1982). On the other hand, however, some authors have warned that agroforestry land–use systems should not be considered as a panacea 'to cure all evils of land management' (Budowski, 1982:13) and to meet the increasing food and fuel demands of humans in the tropics while, at the same time, minimizing the impact on the environment. In 1977 the *International Council for Research in Agroforestry* (ICRAF) was founded in an attempt to focus agroforestry research and development on actual problems and conditions of the tropical world (King and Chandler, 1978). Given the rather controversial views on the subject, an analysis of agroforestry as a viable land-use alternative in tropical agriculture is deemed necessary.

9.4.1   *The general aim of agroforestry*

Basically, agroforestry involves agricultural systems with a convenient ecological integration of trees with annual or perennial crops, grasses and animals when the latter are feasible with the purpose of obtaining 'multiple outputs of food products – plant and animal – and tree products that may range from food to fuel. . . . At the same time, the agroforestry system should be stabilizing in its impact on the environment and stable in its output of products' (Steppler, 1982:3).

From an ecological perspective, agroforestry systems tend to maximize certain environmental conditions that are advantageous to subsistence crops and large-scale agriculture in the humid and subhumid tropics. The newly developed term 'resource pools' applied to the zonal advantages refers particularly to light conditions, water budget and soil nutrients that are distributed horizontally and vertically and that tend to change with time. Sharing the resource pools in agroforestry systems entails competition as trees, shrubs, herbs and crops try to gain access to light, soil moisture and nutrients. Another aspect entails the exploitation of these resources in a partitioned fashion, such as root depths and perimeters, or location of deciduous trees in overstoreys. There are also complementary benefits when some species correct soil deficiencies for the benefit of others as in the case of nitrogen fixation by the roots of *Leucaena leucocephala*, *Acacia albidia*, *Gliciridia sepium* and *Alnus* (Buck, 1986).

Multistorey agroforestry systems provide shadow and nutrient-rich organic matter for the underlying cultivated crops. Coffee, as an understorey tree, is used in this way

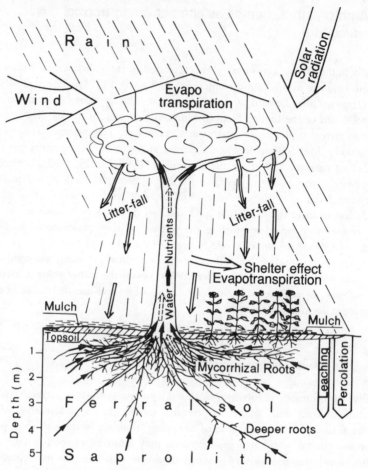

Fig. 9.3 Schematic overview of the functioning of an agroforestry system.

in some parts of Africa and South America (Huxley, 1985). Implicit in the concepts of resource sharing and multistorey agroforestry systems is the spacing between the different species involved. Given the different nutrient requirements of the integrated crops and the varied chemical soil properties it is understandable that there is not a universal formula for the interspacing of trees and agricultural crops. The interspaces are ruled by the complex set of interactions of soil and plant requirements in the 'tree–crop interface' (Huxley, 1985). The way in which tree–crop interface and resource pools operate is illustrated in Fig. 9.3. (A comprehensive overview on the importance of soils research in agroforestry appears in Mongi and Huxley, 1979.)

It is obvious that the spatial dispersion of trees and crops, and the proportion of different species that will interact in a system, can be controlled or manipulated by man. This human intervention requires awareness of the ecological mechanisms involved in the interface, a knowledge of the physiological needs of the plants, and wise economic considerations. Given the relatively low food production in most parts of the humid tropics, fast-dwindling land reserves due to progressive destruction of the remaining forests and shortage of domestic fuel wood, on one hand, and the rapid growing populations that impose greater demands for food and fuel in the future, on the other, attention has been given to the improvement of the bush fallow system by developing new land-use practices with the application of the agroforestry approach.

The ultimate goal of agroforestry is set high: it should potentially overcome the main constraints of land use in the humid tropics, applying an appropriate technology in areas with fragile ecosystems and subsistence farming (Steppler, 1982). The properties of the tropical ecosystems and the characteristics of the dominating field/fallow rotation systems as subsistence farming have been discussed in Chapters 3–5 and 2 respectively.

The possible combinations of arable crops and trees in agricultural units are manyfold. From a formal viewpoint, agroforestry techniques can be classified as agrosilvicultural when trees are mixed with perennial or temporal crops, silvopastoral when tree cultivation is combined with animal husbandry, and agro-silvo-pastoral when all three techniques are used together or in sequence (Nair, 1985). Less academic but more practical is a differentiation according to the techniques actually utilized. Besides the techniques suitable for smallholdings, agroforestry also includes the group of large-scale permanent tree-crop farming systems such as rubber, oil-palm, coconut, coffee, cocoa and tea monocultures, as well as the so-called *taungya* system. The problems faced by large land-use systems differ from those encountered by smallholdings.

9.4.2 *Permanent tree-crop farming*

The initial distinction must be made between large-scale commodity plantation operations and the smallholding sector with traditional multi-crop farming.

Agroforestry applied to large-scale tree crop plantations includes growing annual or perennial crops in the interrow areas between the long-term crop trees: 'intercropping' or 'row intercropping'.

On rubber plantations 'intercropping with food crops is possible in the early years of establishment but is not favoured in the large scale operations because of management problems' (Watson, 1982: 7). Experience in Indonesia, Ivory Coast and Brazil has highlighted the problem that 'the main crop becomes neglected in favour of the food crops. In the smallholding sector, however, intercropping has been a traditional practice with pineapples, bananas, cucurbits and other as cash crops, and with maize and upland rice for subsistence' (Watson, 1982: 7). Similar information is also available from West Africa, Malaysia, Thailand and Sumatra.

Intercropping oil-palm with food crops is quite possible and may be advantageous if the soil fertility does not impose limitations. Experiments under different soil conditions have indicated 'that intercropping of oil palm is perfectly feasible on good

261

soils, but that on poor soils there is a clear risk of nutrient exhaustion and eventual decline of productivity of the palms' (Watson, 1982: 8).

Coconuts, traditionally a subsistence crop of low soil impact, are particularly suitable for intercropping by small farmers. This type of intercropping has been practised in some parts of Indonesia, Malaysia, Sri Lanka and India. Coconuts are often grown on sandy coastal soils where little else grows. Under these circumstances intercropping presents similar limitations as those of the oil-palm, although coconut is more adaptable than the latter. With improving soil conditions, cropping intensity can be increased (Watson, 1982).

Cacao and coffee can be grown in the shade of mulch-providing timber trees. In plantation operations intercropping in the literal sense is not possible, since the cacao or coffee bushes stand too close together. In the smallholding sector, however, cacao and coffee are frequently integrated in the multistorey crop association.

For half-commercialized farms and especially for small-scale peasant holdings, intermixing crop trees in association with a variety of perennial or annual food crops is a characteristic pattern that differs from usual intercropping systems. Ruthenberg (1980) refers to this system as 'mixed cropping'. Around their homesteads farmers operate small garden-like plots, in which different crop trees form the upper storey (e.g. oil-palm, coconut or mango). The middle storey consists of citrus, breadfruit, bananas, coffee, cocoa or plantain, and in the ground storey maize, beans, pineapples, vegetables, peanuts or cocoyam are cultivated. The association may differ from region to region, but what is important is the multistorey arrangement. Very often ground cropping is practised on mounds or soil ridges because this procedure increases the soil volume. Tree products and crops from the intermediate storey may serve as cash crops. Pruning the trees ensures that adequate light reaches the ground while the organic residues together with those of plantain, banana and other plants are used for mulching. Additional kitchen waste, ashes from the fireplace, and human and animal excrement supply the necessary organic matter for making permanent farming possible on nutrient-poor ferrallitic soils with low exchange capacity. The main advantages of these compound gardens are the concentration of all soil-improving measures in a small area, the optimal utilization of nutrient and water supply by plants of varying root depths, reduced plant diseases and insect spread, and, finally, a reliable income and year-round subsistence crops. These tree-crop farming systems are, however, not true agroforestry systems, and very often the farmers are unable to meet their own demands for fuel wood and timber.

9.4.3 *Taungya farming*

To some extent the counterpart of permanent tree-crop farming systems are the land-use forms included in the concept of *taungya* farming. *Taungya* originally meant 'shifting field' in Burma, but today *taungya* system is applied to forest establishments connected with the production of food crops in the early years after tree planting. The origin of the term can be traced back to the Burmese Hill Farming Experiment. In the middle of the nineteenth century the forest administration used the system to establish teak plantations. After clearing the natural forest and before reforestation, the cleared land was first allotted to small farmers for a 1- or 2-year intermediate cultivation of food crops. After that, teak seedlings and rice seeds were distributed among the farmers. Under supervision they planted the teak seedlings and were allowed to do intercropping in the interrow areas between the trees for the following 2 or 3 years. In return they promised to look after the seedlings and help keep the plantation free of weeds. When shade from the developing tree canopy caused cessation of food cropping, the farmers were given another small piece of land to repeat the same procedure.

The advantage for the whole area is that the indiscriminate destruction of the forest and shifting cultivation are reduced. Instead, the potential shifting cultivators promote the reforestation of cleared land. Their immediate interest in this kind of agriculture is based on the fact that the soils in the plots made available to them are generally less leached and potentially more fertile than those cultivated up till then. This also explains why *taungya* has not succeeded where fertile agricultural land is available. For forest authorities, reafforestation costs can be reduced when food crops are grown in the first 1 or 2 years of tree establishment, as cited by Watson (1982) for Nigeria, Costa Rica and Surinam. The difficulties of the system lie mainly in proper management, the discipline of the farmers and continuous official supervision.

Today the *taungya* system has become a successful form of agroforestry not only in South-east Asia but, with land-specific variations, in most countries of the tropics (King, 1968; MacDonald, 1982; Ball and Umeh, 1982; Lundgren, 1978, 1979, is particularly explicit concerning the pedological and ecological aspects).

Obviously, the *taungya* system is operational only for as long as sufficient land is available for the itinerant cultivator; as soon as spatial constraints appear, *taungya* must adapt to the particular population density conditions. One such adaptation is the development of 'forest villages' in Thailand through which shifting cultivators are enticed to settle down in officially assigned villages. Land lots in the settlement are distributed to the families participating in the project, while the forest land around the villages is managed by official foresters and exploited by the villagers under supervision. For the first 3 years, the cultivators are allowed to raise agricultural crops within the tree spacings, thereafter reforestation with exploitable trees is conducted by the cultivators and the foresters (Boonkird *et al.*, 1984). The newly established farmers are then encouraged to grow long-term cash crops in the widened interrow spaces between the commercial tree species. This modality is an attempt to integrate traditional agriculture with an equally traditional commercial tree exploitation.

Taungya definitely represents an agroforestry form that contributes to rational and less destructive land use in the humid tropics. Nevertheless its long-term goal is reforestation, while gains in agricultural productivity are secondary. However, the tropical world of today faces the serious problem of progressive destruction of its remaining natural forests to open new agricultural land and pastures without achieving an improvement in the food supply for its fast-growing population. Therefore, a

263

stabilization of the food production process, an increase in food production outputs, and a guaranteed wood supply at the small farm level must be considered the main targets of agroforestry.

9.4.4 *Basic necessities of a real agroforestry system at smallholder level*

In order to reach the above-mentioned targets, it is mandatory to preserve the ecologically stabilizing effect of a natural tropical forest ecosystem for the agrarian land-use system as much as possible. 'If society demands that the high forest be removed, it must be replaced by economic crops that copy as closely as possible the characteristics of the natural forest, and that afford equal protection to the environment' (Bowers, 1982: 51).

From an ecological viewpoint, this requires the implementation of measures such as:

1. Maintaining as far as possible the amount of biomass return to the soil;
2. Optimal exploitation of the nutrient pool of organic matter;
3. Using the advantage of the high cation exchange capacity (CEC) residing in the organic matter;
4. Minimizing nutrient leaching by using the nutrient-trap effect of the plant root/mycorrhiza symbiosis;
5. Utilizing the ability of mycorrhiza to efficiently extract available nutrients from the soil minerals;
6. Using the deep-reaching tree roots as 'nutrient pumps' (Lundgren, 1978);
7. Prevention of erosion through multistorey plant associations to forestall extreme rainfall conditions;
8. Prevention of temperature extremes (elevated maxima accelerate the decomposition of organic matter).

These basic ecologic requirements demonstrate that in regions with generalized nutrient-poor ferrallitic soils an allocation (landscape pattern) of delimited areas destined for forest resources, food crop production and animal raising is not suitable.

The problem with transforming even partially a natural forest into a forest/food crop association is that the trees compete with the crop plants for light, nutrients (stored in stems and branches) and water in the soil. Additional disadvantages are the mechanical soil damage in connection with tree harvesting in certain localities, the increase of air moisture, and the decrease of air movement, which favours fungal diseases and proliferation of noxious insects.

A workable compromise could be found by means of tree-crop farming systems. But since these systems are not able to satisfy the food requirements of growing populations they should be designed in such a way as also to guarantee a substantial production of grain and tuber crops.

In the search for 'a way to place agroforestry as a major factor in the struggle to meet the increasing food and fuel demands of human beings while minimizing the impact on the environment' (Steppler, 1982: 5), an idealized model of action for agroforestry farming units can be constructed.

First, a gradual opening and thinning of the complex natural ecosystem with its many species must be conducted to guarantee sufficient space and light for the crop plants and to simplify the forest composition, so that – in the long run – the remaining stands become richer in ecologically and/or economically valuable species. Besides the economic viewpoint, peasant farmers have already recognized the superiority of certain species in regenerating soil nutrients and land productivity, as can be gathered from the various articles contained in the work edited by MacDonald (1982).

Clearing and thinning processes should create the best possible stands of managed seedlings and crops through a gradual opening of the forest canopy, removal of some trees and pruning of others. It is crucial for the success of these procedures that the resulting organic matter is not burned. As mulch, it protects the soil, reduces water loss, restricts nutrient waste due to the slow release of nutritional elements during the decomposition of the organic matter, and adds humic acids to the soil, thus increasing the CEC. Mulching is the most rational use of organic matter. Since the soil stays relatively loose under the mulch cover, tillage becomes unnecessary. Minimum tillage practices with abundant biomass residues on the soil have great ecological advantages and, in addition, reduce labour input.

The cleared areas between the remaining trees are used for simultaneous cultivation of varied crops (mixed intercropping), as is common in tropical subsistence farming, or for relay intercropping, where a second crop is planted after the first one has entered the reproductive growth phase but prior to harvest, as occurs with the maize–beans systems in tropical America. Planting plantains or bananas for food or cash may complete this type of viable agroforestry farming system.

The final result can be a smallholding with open stands of selected trees for timber and fuel wood in the top gallery, coconut or oil-palms with breadfruit in the second storey, beneath plantains, cocoa and coffee, and in some plots a bottom storey of maize, cocoyams, beans, vegetables, cassava, upland rice, sweet potato, etc. 'In the long run the aim would be not to get a maximum yield of one particular crop per unit area, but, rather, to maximize total production' (Bowers, 1982: 51) in a permanent cultivation system with resident peasant cultivators.

Supplementary use of modest amounts of inorganic fertilizer, to maintain the continuity of the system and to improve its revenues, may be possible when organic matter in the soil is kept at a level that guarantees sufficient exchange capacity. This implies, again, that the burning of organic matter should be prohibited by all means. The use of machinery is hardly possible in this system and manual labour from family members will remain its basis.

The gradual transformation of tropical forest plots into an agrarian landscape on the basis of agroforestry principles demands the utmost in know-how, labour and discipline. To begin with, the potential farmers are to be educated as to the immense and complicated workload which is required for many years to replace the practices involved in the slash-and-burn system. Gradual clearing of the dense

265

forest also confronts the farmers with all the unpleasant aspects of the rain forest, such as snakes, predators, plundering apes, or flocks of hungry birds. As soon as the planting starts, the struggle against the fast-growing weeds commences. Because of all these nuisances, there is always the danger of the peasants losing their patience and returning to the familiar clearing-by-fire method. Therefore, strict control and discipline are necessary.

For the farmers to accept the extremely hard work, an extensive programme of infrastructural development of the area must be undertaken by national or international institutions, so that the societal conditions become bearable.

The agroforestry system described can prevent the transformation of the 'green hell' of the tropical rain forest into a 'red desert'. However, the variety of species and the genetic pool will disappear and give way to an economic system which, in most of the newly cultivated land, will provide no more than a basis for subsistence farming. In this context, it is interesting to take note of a basic discussion of the possibilities of tropical agriculture by R.D. Bowers of the International Institute of Tropical Agriculture (IITA), Ibadan, Nigeria:

> Input : output ratios are generally unfavourable in the humid tropics. Low yields are a result of leached soils, heavy run-off, loss of nutrients, and weed competition. Drying and storage in a hot, humid environment present further problems. Generally, unfavourable input : output ratios are found where yields are low. Approximately the same inputs per hectare will produce about 2.5 tonnes of maize in the humid tropics as against 7 tonnes in the USA. In other words, only 0.15 ha of land in the US, as compared with 0.4 ha in the humid tropics, is required to produce 1 tonne of maize.
>
> Subsistence farming is the inevitable consequence of the unfavourable input : output ratios associated with the production of annual crops in the humid tropics. All other forms of annual crop production are economically impossible until such time as the agricultural scientists develop varieties that give input : output ratios comparable with those of similar crops grown in more favourable climates. . . . But subsistence farming has been shown to have already failed: it cannot produce sufficient surplus to feed the large towns; it cannot supply cheap basic foodstuffs; and it cannot retain the young people on the land. . . . The objective must be to give the people of the wet tropics crops that will outyield and undersell the main food crops grown in the temperate zones. . . . The tree crop alternative offers some hope. It is logical, in both economic and ecological terms. . . .
>
> One cannot know what the future holds but surely one must plan for several different possibilities. One of these is a world of increasing energy shortage, where energy related products become more and more expensive. In such a situation low energy-input tree crops have a potential that is unmatched (Bowers, 1982: 50).

The facts brought to light by this discussion of agroforestry systems on nutrient-poor ferrallitic soils of the humid tropics show that, as experts suggest, agroforestry systems in the humid tropics are likely to be most effective when they copy the natural forest as closely as possible. ('All the evidence indicates that the widespread clearing of tropical forest and the large-scale planting of annual crops leads to financial and ecological disaster' (Bowers, 1982: 51).) Copying the natural forest has three major implications: to turn away from the slash-and-burn practice for clearing and weed control, the impossibility of using machines for agricultural work, and an enormous investment of manual labour.

After the complete installation, the labour-intensive farming unit is submitted to the ecologically enforced restrictions of subsistence farming because the export of biomass via food crops or wood products has to be kept to a minimum in order to prevent a collapse of the production system. Let us not forget that the natural forest works on the basis of a closed nutrient cycle (Section 4.3). MacDonald (1982) alerts us to the danger resulting from the export of timber: with the wood calcium is removed, which plays a critical role in the nutrient cycle of humid tropical ecosystems (Sections 3.7 and 4.4).

All this underscores, once again, the ecological disadvantage of agricultural development in the humid tropics as compared with the situation in the middle latitudes, where, utilizing heavy machinery and intensive fertilizing, less people produce more food than is necessary for the inhabitants of the densely populated industrialized countries.

9.4.5 *Feasible agroforestry alternatives at smallholder level*

The implementation of a perfect agroforestry system for subsistence farmers can be approached only at selected locations and for restricted areas because financial, infrastructural and educational prerequisites are generally missing. The search for feasible alternatives then becomes mandatory. A simple and widely practised form is the 'alley cropping system' developed by the IITA in Ibadan, Nigeria (Wilson and Kang, 1980). Food crops, usually cereals and legumes, are grown in narrow alleys of 2 or 3 m width formed by fallow species (usually shrub or tree legumes). During the cropping period the fallow is suppressed by regular pruning (Getahun *et al.*, 1982), and the organic residues are worked in as mulch to improve the soil. Much depends on which fallow species are included in the alley cropping system. Juo and Lal (1977) found that *Leucaena leucocephala*, in naturally regenerating fallows, was an effective restorer of soil organic carbon and exchangeable cations in alfisols. On acid ultisols, maize–*Leucaena* alley cropping was less successful, and other tree species had to be tested.

Another form of utilizing leaves and twigs as mulch is the 'cut and carry' fallow management. The fallow species are grown on land unsuitable for arable cropping and the prunings are transported to the cultivated plots.

It must be noted that the use of shrubs or bushes solely for soil improvement, does not yet serve the purpose of real agroforestry. However, it avoids burning of the biomass and optimizes its use as mulch.

Efforts have been made to accelerate soil restoration through planted fallow after short-term cultivation (see Sections 8.2 and 8.4). Trees that are known for their effectiveness in restoring soil fertility are planted in the shortest possible time so that they also yield timber products for the requirements of the small farmers.

In the following sections a short review will be made of the different agroforestry schemes that are practised in the tropics.

9.4.6 *Agroforestry in tropical Africa*

The papers of the colloquium *Agro-forestry in the African Humid Tropics*, held in Nigeria, 1981 (MacDonald, 1982) offer a comprehensive view of the varied systems of agroforestry in the countries of tropical Africa.

Some authors, particularly Ball and Umeh (1982) and Getahun *et al.* (1982), report that mixed agriculture was practised in areas of western Africa long before agroforestry became a fashionable alternative to traditional rotational farming. The pertinence of a careful look at traditional agroforestry variations is stressed by the distinction made in south-eastern Nigeria between spontaneous fallow and planned fallow, when referring to fallows that are left to regenerate themselves versus others in which the agriculturalist promotes the growth of certain bushes or tree species that he perceives as beneficial for the edible crops he cultivates.

In Africa, regional differentiations are reflected in the combinations of trees and agricultural crops that prevail in particular landscapes. In the populated humid tropical belt of western Africa, particularly on the northern edge of the Gulf of Guinea where the density rises above 80 inhabitants per square kilometre, agroforestry is practised at the edge of the natural evergreen forests. The cultivators grow food staples such as manioc, yam, maize and plantains, in combination with cacao, bananas, coffee or oil-palm. In terms of both area covered and population involved, southern Nigeria is the heart of agroforestry in the Gulf of Guinea states. Ball and Umeh (1982) estimate that 9629 ha and 17 744 cultivators were involved in agroforestry in 1979. South-east Nigeria, spreading over an area of evergreen forests on ferrallitic soils and ferrisols, tends to combine productive trees, such as banana, cacao and oil-palms, with food staples and pastures, whereas western Nigeria (with slightly lower precipitation rates and less population density) specializes in exportable timber species (teak), mixed with cacao, bananas or oil-palm.

'Managed *taungya*' makes intensive use of certain tree species for protection against wind or excessive insolation, such as *Gmelina arborea*, one of the most utilized trees in African agroforestry, or *Terminalia superba* and *Albizzia spp.* Other species, such as the woody legume *Leucaena leucocephala* and *Gliricidia sepium*, help restore fertility to the soil. *Gmelina arborea* appears to be beneficial when planted at particularly convenient interspaces with yams and maize, but not in combination with manioc (Agbede and Ojo, 1982). *Gliricidia* increases the content of sodium, potassium, calcium and manganese in the soil. Concurrently, a measurable decrease in soil acidity has been observed when *Gliricidia* is associated with maize, yams, vegetables and manioc for subsistence purposes (Agboola *et al.*, 1982).

In general, agroforestry alternatives in the northern states of the Gulf of Guinea have been formulated as measures to stall excessive deforestation, and, even with incentives from national governments and assistance from international agencies, their success has been only partial. Experimentation with *Gmelina Albizzia* and *Terminalia* has proved successful at the level of small plots and experimental farms, but the individual cultivators have remained unenthusiastic, if not outright hostile, to the introduction of these new farming strategies.

In the countries of equatorial Africa, where dense rain forests still exist in large expanses, the lower population density still allows the practice of shifting cultivation; but the growing populations' demand for staples makes agroforestry measures advisable.

In the subhumid, semi-deciduous forest regions of Africa pastures become more abundant and agropastoral activities occupy a more dominant place than in the humid tropics. Among the commercial crops that are also grown by small agriculturalists are goundnuts, cotton and cacao. In Ivory Coast, where monocultures of cacao, coffee, coconuts and oil-palms are well developed, some of these crops are grown in association with forage plants, such as *Eleusine indica*, *Axonopus compressus* and *Panicum maximum*, a grain that grows especially well between palms and provides forage for cattle during the dry season. In central semiarid Senegal, agropastoral and silvopastoral strategies use the 'ronier' palm (*Borassus flabellifer*), baobab (*Adansonia digitata*) and *Acacia albidia* in association with millet, sorghum and peanuts.

In the tropical highlands of eastern Africa only Kenya practises *taungya*, in small farms called *shamba*, that are tended by traditional native cultivators. In the highlands, *Albizzia gummifera* is associated with pastures, whereas *Acacia albidia* does well with local grains in the arid northern region, and *Balanites aegyptica* combines favourably with pastures in the districts of Baringo and Samburu.

Considering Africa as a whole, intensive application of agroforestry measures appear to have had a better chance in Kenya than in other African countries due to the existing tradition of tree growing and forest conservation and not as a consequence of modern innovations (Owino, 1982). In general, agroforestry alternatives that have been attempted in Africa so far appear to have been most successful in districts where the cultural inclination and similar systems already existed among indigenous farmers, as exemplified by the experience of southern Nigeria. In other instances, and especially when agroforestry measures were devised by national governments to forestall further depletion of forests by itinerant agriculture, the response of the local population has been far from enthusiastic. In fact, many of the scientists and governmental representatives who report on national situations agree in citing the agriculturalists' rejection of imposed agroforestry measures as the major reason for the demise of such experiments in Africa. Thus, the future success of agroforestry in Africa remains uncertain, notwithstanding the efforts applied.

9.4.7 *Agroforestry in tropical South America*

In South America plan associations that resemble the contemporary and purposefully pursued agroforestry alternative have been in use since pre-Hispanic times by the Amerindians, and even the agricultural techniques employed by the Amazon basin Indians today qualify as agroforestry systems.

The agroforestry systems that function in the humid Amazonian lowlands are largely based on the mixing of tree species with assured cash value for their wood or their products (rubber in the case of *Hevea brasilensis*) or by the association of shade trees of potential timber value with tree crops such as cacao, pepper or coffee. Particularly convenient is the combination of cacao with *Erythrina*, a legume that provides much of the nitrogen demanded by the cacao trees, and the use of *Cordia alliodora* as a shade tree with good returns as a timber species. Pepper, which tolerates a maximum of 20 per cent shading, can be grown under *Erythrina* and *Gliricidia* trees that are easily pruned and provide additional income from their wood. Coffee has been traditionally associated with *Erythrina* and *Gliricidia*. Further use of native woody species like *Cordia goeldiana* or *Scizolobium amazonicum* are recommended as shade trees and protection against pests (Hecht, 1982b).

Agroforestry systems in the non-Brazilian segments of the South American tropical lowlands have also developed locally, although less elaborate and diverse. On the margins of the Amazon River, close to Iquitos, Peruvian Amazonia, different vegetal species have been grown in associations. Umari (*Paragueiba sericea*), uvilla (*Pourouma cecropiaefolia*) and Brazil nuts are grown for their fruits, and their wood is used for charcoal. In the shade of *Bactric gasipaes*, *Inga edulis* or cashew, food staples such as manioc, plantains and rice are cultivated. It is also common to find papaya, pineapple and passion fruit in the shade of Amazonian palms (Padoch *et al.* 1985). Multi-strata mixtures of perennial species, such as forage legumes (*Desmodium ovalifolium*) at ground level and *Canna edulis*, whose roots are eaten by hogs, form the basis for hog farming in Ecuador's Oriente (Bishop, 1982).

Away from the humid and warm environment of Amazonia, in South-east Bahia, better results in agroforestry strategies are achieved – in the combinations of cacao with rubber trees, clove with pepper, cacao with clove in the wake of decayed pepper plants (Alvim and Nair, 1986). In the drier environment of north-eastern Brazil, cultivation of perennial crops such as cashew, coconut, babassu palm (*Orbignya phalerata*), and the carnauba wax palm (*Copernicia prunifera*) in combination with natural pastures to which some herbaceous foreign species have been added, provide good grazing for sustainable silvopastoral systems (cattle, sheep and donkeys). In grazing areas the babassu palm provides shade for the cattle, while in agriculturally oriented places, it serves as shade for rice, maize, cassava and even bananas and plantains (May *et al.*, 1985). The cashew tree provides shelter for other productive crops such as sorghum, groundnuts and sesame (Johnson and Nair, 1985).

This short review of agroforestry in South America reveals that improvements in production and in ecological conditions are achieved only in those areas of the humid tropic margins where cash crops have been established for a long time. Inversely, in areas with recent forest clearings or colonizing thrusts into virgin lands, the initial successes are restricted to experimental plots or the agroforestry-like gardens of native cultivators, suggesting that agroforestry on that continent is successful only in small-scale farming and that its widespread diffusion is still questionable.

9.4.8 *Agroforestry in South-east Asia*

According to Adeyoju (1982), agroforestry in South-east Asia has been practised for over a century under different conditions and in various locations. Cultivation of sago palms and sago production practised in supplemented stands similar to the natural forest have been a traditional form of land use in Malaysia for thousands of years (Bruenig, 1984). The successful use of tree species and food staples or cash crops has been common in Sri Lanka since the nineteenth century and it is thought that the Sri Lankan Kandy gardens are the best examples of that farming system's potential for the humid tropics (Watson, 1982). Kandy gardens refer to small farms based on a close association of coconut, kitul and betel palms with cloves, cinnamon, nutmeg, citrus, mango, durian, jackfruit, rambutan and breadfruit, with a lower storey of bananas and pepper vines, and a peripheral ground storey of maize, cassava, beans, pineapples and others, often supplemented by an outside field of paddy rice (McConnell and Dharmapala, 1978).

In Indonesia, manioc, pepper and benzoin are grown under the canopy provided by coconut palms and plantains. In most parts of Sumatra, today more than half of the farming area is planted with tree and bush cultures, where rubber, coffee and spices such as cloves, cinnamon and pepper, prevail as cash crops. Tree and bush cultures in combination with fields of paddy rice dominate Sumatra's agrarian landscape today (Scholz, 1984). It must be stated, however, that most of the small farm enterprises have been laid out on volcanic soils where production conditions are much better than on the nutrient-poor ferrallitic soils. Outside volcanic environments, the rubber tree plays the most important role in the agroforestry enterprises of small farmers in Sumatra, Malaysia and Kalimantan, because it places very few demands on the soil (Scholz, 1984).

The *taungya* system of South-east Asia, which is to be considered as one of the most successful agroforestry systems, has already been described. Among the most common agroforestry products of South-east Asia are the association of commercial timber (particularly teak) or tree crops such as tea, cocoa, bananas, breadfruit, mangoes or kitul with groundnuts, pepper, maize, manioc or pineapples.

Reviewing the various applications of agroforestry principles in the tropics on different continents, it becomes evident that agroforestry schemes involving the production of edible crops and forest regeneration ventures are the most widespread. Even though both help improve food production in tropical countries, this contribution is only of secondary importance. Considering the scale of forest destruction through shifting cultivation, their positive effects are negligible.

At the small farmer level, agroforestry schemes spread over large areas only when the 'foreststand' consists of crop trees whose products can be exported and sold on the world market (e.g. Sumatra, Malaysia, West African coastal region). Generally, this form of land use is practised on rich tropical soils. Outside these soils agroforestry schemes appear only as small traditional compound gardens within the larger shifting cultivation areas. Genuine agroforestry schemes, where the natural forest is progressively converted – without burning – into an ecologically stable and permanent cultivation system for residing peasant cultivators at subsistence level exist only as research models.

9.4.9   *Facing the obstacles against the spread of agroforestry*

Many authors mention traditional attitudes and the dislike of planned farming alternatives as reasons for the poor response of subsistence cultivators and as the main obstacles to the spread of agroforestry schemes as appropriate alternatives for rotational fallow land-use systems. In our opinion, however, this does not address the crux of the problem, because one of the traditional land-use forms in the humid tropics, the compound garden, is largely in accordance with agroforestry principles. This proves that small farmers can manage a small-scale agroforestry system. However, a compound garden alone does not guarantee subsistence. Basic foodstuffs such as grain and tuber crops must be cultivated in outfields with the rotational bush fallow method when the natural forest has been removed by the slash-and-burn method at the beginning of cultivation, thus wasting a large part of the initially stored potential of nutrients. Here we are again confronted with the ecologically indispensable condition that there should be no burning of organic matter. Instead, the biomass should be added to the soil by mulching and natural decomposition processes. The labour expenditure necessary for this kind of biomass conversion is much higher than for the easier slash-and-burn practice.

The idea of spending a lot of manual labour on agroforestry endeavours and having to wait years before timber or crop trees can be harvested, is not at all appealing to subsistence farmers. Several African specialists agree that two of the major reasons why agroforestry systems have not caught on in their countries are the impatience of the small itinerant farmer and the naturally conservative character of native cultivators (Nadjombe, 1982). For as long as it is possible for itinerant small farmers or commercialized enterprises to seize areas of virgin tropical forest, the forest will be burned since this is the easiest method of opening agricultural land. The problem of lasting fertility is unknown and forest conservation is of no interest to these people. Thus, the paradox arises that the agroforestry strategies that have been strongly supported and encouraged by national and international agencies with the purpose of forestalling the disappearance of forests and raising the productivity of the agriculturally most disadvantaged natural regions of the world have been met with the least understanding and co-operation by precisely those cultivators who had most to gain. J. B. Ball and L. I. Umeh (1982) report that, between 1975 and 1980, the number of agriculturalists who engaged in *taungya*-like activities in Nigeria decreased from 24 607 to 17 744.

Throughout tropical Africa it can also be observed that agroforestry alternatives are mostly endorsed and practised by old farmers who have opted to stay in the countryside and have acquired a rather conformist attitude, while the younger people have lost hope that the socio-economic conditions in the countryside will improve and try their luck in the sprawling urban centres. Under these inauspicious circumstances the promotion of agroforestry does not hold much promise as a panacea for the situation of insufficient food production and forest destruction that characterizes the countries of the humid tropics.

Notwithstanding these real and objective difficulties that have arisen in the course of the implementation of agroforestry schemes in most countries, specialists at the service of world organizations such as FAO and the World Bank are strong supporters of agroforestry, which, after years of unsuccessful attempts and unprofitable investments, still represent for them the most promising means of effecting change in tropical agriculture (Catterson, 1982).

Experts such as Watson (1982) or Raintree and Warner (1986) support the implementation of 'integral' *taungyas*, meaning an agroforestry practice in which commercial forest species are grown at convenient interspaces so that cash crops and staples can be accommodated between tree rows. This modality allows crop diversification and provides shelter or supplementary nutrients for underlying plants, depending on the species that are combined, and adds the financial advantages of forestry income. Since subsistence agriculture has been the hallmark of itinerant cultivators and since cultural traits associated with this activity are difficult to abrogate in one generation, the allocation of permanent agricultural plots to grow food essential for the families involved must be included; in this way the cultivator and his family still feel in control of their lives. Population density increases are compatible with integral *taungya* since intensification of labour is needed for tending the timber species, the cash crop trees and the familial gardens. At lower agricultural land pressures there exists a much higher reliance on cash income and incentives of community development that may contribute to make integral *taungya* an attractive alternative for the independent-minded farmer.

In a final analysis of agroforestry as the way out of the bottleneck created by ecological limitations in the humid tropics it appears that much educating about the system's advantages will have to be done among those rural communities in which the development analysts see possibilities of success. Experience has shown that medium-size farmers who understand a little more about agricultural economics show more enthusiasm for agroforestry than subsistence cultivators. However, it is precisely the small and unsuccessful farmer who might benefit most from this system.

10

Summary and conclusions

The departure point of all the reflections presented in this book originates in the notorious disparity of the state of socio-economic development between the North and the South. The aim of the argumentation is to determine whether this disparity is rooted in ecological conditions.

One essential symptom of the disparity is insufficient food production in most tropical countries. Some of them show a high population density and a large individual per area ratio. In extended countries, however, this ratio is particularly small. Concerning the geographical distribution of these diverse demographic situations there are two remarkable facts. First, the countries which lie in the inner tropics and enjoy an optimal climate for plant growth exhibit, in general, smaller population densities than those located in the outer tropics where the climate is less favourable and more hazardous. Second, within the inner tropical belt there are clearly identifiable islands with extraordinary population densities. Since the population distribution during the course of history has mainly rested on agro-economic foundations, the problem arises of what has been the special form of food supply, the production potential and the development possibility of the corresponding land-use systems.

In most of the sparsely populated humid inner tropics the dominating form of rainfed agriculture is shifting cultivation in its broader connotation (Section 2.3). This type of land use is characterized by alternation of shorter cropping periods and longer periods of forest, bush or grassland fallows, which, in their extreme consequences, may entail long-term shifting of dwellings or farmsteads. The various principles of rotation are described in Section 2.2.

The main reason for abandoning a cultivated plot after only a short period of use is the rapid decline of yield after the first crop that follows slash-and-burn clearing of either the virgin or secondary forest, or the bush (Fig. 2.5; Section 2.6).

Obviously this form of land use is land-demanding and forest-squandering, and as soon as population density exceeds 20 people per square kilometre, the basic food requirements are barely met. International scientists categorize shifting cultivation not only as a backward stage of agricultural practice which increasingly fails to satisfy the requirements of higher production, but also the expression of a backward stage of culture in general (Section 2.7). Considering the shortcomings attributed to shifting

274

cultivation, two questions must be raised: why is rotational crop–fallow land usage with slash-and-burn methods of land clearing still so widespread in the tropics and why does it still prevail after more than 30 years of agrotechnical assistance from abroad?

The main postulates of the explanatory hypotheses on shifting cultivation permanence which ignore environmental causes are unsatisfactory (Section 2.7). *On the basis of ecological reasoning it can be proven that this particular mode of rotating food crop and forest or bush fallow is a specialized adaptation to the environmental conditions of the tropics which not even modern agrotechnology has been able to replace, at normal farm unit scale, by making use of fertilizers and intensive continuous cropping systems for the production of cereals or tubers as man's main staples.*

These constraints are primarily related to certain properties of the xanthic and orthic ferralsols (oxisol, ultisols, highly weathered ferrallitic soils, *sols ferralitiques fortement désaturés*, see Ch. 3) which dominate areally the humid tropics. *The ecologically decisive soil properties which act as limiting factors for plant growth, and which cannot be markedly altered or manipulated by man, are the poorness in (1) weatherable mineral content and (2) cation exchange capacity (CEC).* While the effect of nutrient poverty on plant growth is obvious, the low CEC values prove even more disastrous. In general, CEC acts in that it offers temporary lodging to nutrient cations, thus preventing their outwash. Soils with low values are ineffective in this sense. In the case of the application of water-soluble fertilizers only an insufficient amount of nutrient cations can be adsorbed and made available for plant growth.

The limiting conditions mentioned are caused by the intensive hydrolytic weathering under the hot and humid climate, the age of the soils, and the dominance of acidic parent rocks. *The third reason for the field/forest alternation which cannot be manipulated by man is the high mineralization rate of organic substances.* In view of the scarcity of nutrients and CEC in the mineral substance of the deeper soil it is the uppermost soil horizon (solum) with its organic matter which becomes the chief supplier of plant nutrients and exchange capacity. As soon as biomass input ceases after deforestation the high mineralization rate causes rapid depletion of the organic matter. The cultivated plot loses the most important source of nutrients and CEC after no more than two, at most three, growing periods. *The fourth reason is, that in the forest plus soil ecosystem most of the nutrient content – and almost all calcium – is stored in the phytomass and not in the soil itself* (Section 2.6 and Figs 2.6 and 2.7). Thus, utilization of the nutrient reserves contained in the phytomass of the natural vegetation and, once the reserves are exhausted, shifting to another location, become an ecologically imposed necessity.

Slash-and-burn is acceptable only with respect to labour saving and weed control; in terms of effective use of the nutrient reserve stored in the biomass it is a squandering practice. A considerable portion of nutrients contained in the ash does not enter the biomass through recycling. It is washed away by the frequent downpours (Section 1.2) and lost during the time interval between the maximum ash accumulation and the consumption by crops sown in the forest ash layer. In the long run slash-and-burn leads to a progressive impoverishment of essential elements in the ecosystem whenever the fallow period is shorter than the time necessary for full recuperation

275

of the natural vegetation. Unfortunately, this is happening already in most shifting cultivation areas.

The formidable challenge of overcoming the squandering system was met by the long-term continuous cropping experiment at the Yurimaguas Experimental Station (YAES) (Section 8.1). However, in the end it turns out that plots on tropical ultisols and oxisols can only be kept continuously productive by applying technological measures (constant monitoring of soil dynamics) and material means (fertilizer and lime) which, at present, cannot be made generally available under on-farm conditions in the humid tropics. Especially the amounts of limestone (3.5 t ha^{-1} every 3 years), required for lowering the soil acidity and aluminium toxicity and/or slightly increasing the CEC (Table 8.2), cannot be found in humid tropical lowlands for geological and climatic reasons (Section 8.1.4).

In those places where the new technology has been tried or where it will be attempted in the future, the application of water-soluble fertilizers on unprotected open plots remains a hazardous business because of the high frequency of risky downpours (Sections 1.2 and 8.1.2).

The apparent discrepancy between the nutrient poverty of low-base ferrallitic soils with their very limited agricultural production, on one hand, and the abundant vegetal life of lush rain forests with exceptionally high net biomass production (Section 1.3) on the other, can be explained by the closed nutrient cycle of the virgin forests. As long as the forest is undisturbed, no net loss of macronutrients through outwash occurs as proven by the fact that (1) the autochthonous black- and clear-water streams of the tropical lowland rain forests may best be likened to slightly contaminated distilled water, and, (2) the waters of small creeks in virgin forest areas contain even less macronutrients than rain-water. Nutrient loss from the undisturbed rain-forest biomes is basically prevented by the symbiotic relationship between trees and the mycorrhiza fungi which grow around the feeding rootlets and act as 'nutrient traps' (Section 4.4). Forest burning destroys this important link in the chain of the tightly closed nutrient cycle. Erosion and leaching of nutrients can then proceed without restraint. The ecosystem of the apparently lush tropical rain forest is thus, in essence, as vulnerable as a haemophiliac.

The lesson that has to be drawn from the ecological interrelationship within the humid tropics' biomes and all theoretical experiments is: *any available organic matter should be introduced in the framework of a most careful and waste-free recycling process.*

Low-input technology (Section 8.2), the ecofarming approach (Section 9.1) and the model for genuine agroforestry at smallholder levels (Sections 9.4 and 9.5) are alternatives aimed at making optimal use of the existing organic matter. The biological foundations of these production systems rest on an inherent biotic principle of natural systems in tropical forests. It implies a very active biological cycle that ensures high biomass production, most of which is utilized again within the system through nutrient recycling. The extraction of agricultural products via harvests is limited to those quantities for which the corresponding nutrient removal can be compensated by the abiotical input of chemical fertilizers. Application of herbicides and pesticides can help to control weeds and pests (Section 8.4).

All these systems demand great investment of manual labour. For this and other reasons, *input : output ratios associated with the production of annual crops are generally unfavourable. Limitation to subsistence farming is the necessary result (Section 9.4). In contrast to the situation in the extratropics where a small number of farmers using heavy machinery and intensive fertilization are capable of producing food in excess for a large number of consumers, the people of the humid tropics bear the heavy burden of an ecologically based disadvantage.*

Exceptional regions with high-density agrarian populations in the belt of tropical lowland rain forests are found:

(a) wherever the parent rocks are particularly base-endowed;
(b) where rock debris, rich in primary nutrient minerals, is supplied periodically or episodically to the soil;
(c) where the soils are relatively young; or
(d) where the efficiency of hydrolytic weathering and mineralization of organic matter are reduced.

The concurrence of two or three of these requisites creates optimal pedological conditions. Perhaps with goodwill and carried away by too optimistic expectations, personalities of international development agencies have cited these privileged areas as showcases for arguing that the whole humid tropics have a much higher population-supporting capacity than is realized at present. It is a matter of scientific integrity to point out the dangers of such a fallacy to prevent further attempts at resettling farmers in ecologically limited areas.

For geological reasons, the beneficial effects of old base-rich gabbroic bedrocks are restricted to particular loci amid the ancient crystalline shields that extend widely in tropical Africa and South America east of the Andes (for examples, see Section 7.3). The most important rock-based exceptional regions are found over basaltic parent rocks, which, aside from being rich in bases, are much younger than those from the parent rocks that dominate old crystalline shields and basements. Furthermore, most of them occur in elevated parts of the earth's crust where, for climatic reasons, hydrolytic weathering is less vigorous (for example Rwanda–Burundi, Sections 1.4 and 5.2).

Episodic or periodic additions of weatherable nutrient-rich primary minerals to the soil occur as volcanic ash in some limited areas (Java, for example) and, to a greater extent, in the floodplains and deltas of large 'white-water' rivers (Tables 4.3 and 5.8). The periodically rejuvenated fluvisols or inceptisols (Table 8.6) have since early times been ecologically preferred sites (Section 5.3). Once effective drainage and irrigation systems are established, these habitats emerge as agricultural corridors with extreme concentrations of people amid sparsely populated tropical uplands.

Optimal utilization of the ecologically advantageous alluvial accumulations has been achieved by wet-rice cultivation on paddy soils. Submergence and puddling creates, in the long run, an artificial hydromorphic soil (Section 5.5, Fig. 5.4). The paddy soils are effectively protected against nutrient loss through leaching by an almost impermeable floor. In the flooded and puddled soil matrix the decomposition processes of organic matter are brought about by anaerobic micro-organisms which use nitrates and oxidized

277

manganic and ferric compounds as oxygen sources. The nitrate reduction causes an undesirable nitrogen loss, which is partially compensated through microbial nitrogen fixation by blue-green algae and photosynthetic bacteria. *The ecologically decisive advantage results from the reduction of manganic and ferric compounds.* A 'self-liming' process diminishes the soil acidity to an almost neutral level, optimizes the CEC and the availability of most nutrients, and eliminates, at the same time, aluminium toxicity. *The main problem with paddy soils is the provision of nitrogen to the rice plants.* The low recovery rate of nitrogen fertilizers is the reason why production techniques of the Green Revolution type are of limited success in the traditional rice bowls of humid South-east Asia (Section 7.3).

The ecological advantages of tropical mountain areas are mostly related to the reduced hydrolytic mineral weathering and lessened mineralization rate of organic matter due to lower temperature and moisture levels (Section 5.4). This leaves more nutrients and clay minerals other than kaolinites in the soil and reduces the input of organic substances required for maintaining an acceptable level of humic acids. Another favourable property is the freshness of soils due to the continuous denudation of the upper horizons and the subsequent exposure of slightly weathered parent rocks. Changes in soil conditions induced by elevation and relief are often associated with the beneficial effects of base-rich underlying rocks (Sections 5.1 and 5.2). The altitude-dependent advantages begin at elevations between 1000 and 1200 m.

Archetypal regions of the tropics which combine the advantageous conditions mentioned above are the islands of Java, Madura and Bali (Sections 5.2 and 5.5). Numerous districts of these islands support, on agricultural bases, 800–1000 people per square kilometre, densities that are exceptional in the non-industrialized world. The fact that extreme contrasts and inequalities in population densities occur within the Indonesian state, opens up the possibility of proving the dependency of occupation intensity on natural resources (Sections 5.2 and 9.3).

In contrast to the tropics of South America, and especially of Africa, the ecologically exceptional regions of the South-east Asian tropics are numerous and rather extensive. This is related to the special geologic–tectonic outlay of South-east Asia (Section 5.6). Nevertheless, an examination of agricultural statistics and land-use maps reveals that even in South-east Asia the intensively cultivated areas amount to less than 20 per cent of the total land area. The favoured regions are interspersed within large areas where the usual conditions of the humid tropics prevail, and where the development of continuously productive land-use systems is as difficult as in Africa or South America, as demonstrated by the failures of the Indonesian transmigration projects (Section 9.3).

Concerning the humid inner tropics, the general conclusion that must be drawn from the ecological facts is as follows: it is not possible to turn the ferrallitic soils of rain-forest regions into continuously productive agricultural areas for staple food production. Field–fallow rotation has been and is the ecologically imposed adaptation to decisive soil qualities. The removal of the rain forest by slash-and-burn is but a waste of nutrient resources. Thorough recycling of organic matter is highly recommended.

Alternative proposals have to consider regional differentiations. For instance, in South

America attempts to establish agricultural colonies in the rain forest make little sense from an ecological viewpoint (Section 9.3). The expansion of agricultural areas in the Amazon basin must be restricted to alluvial plains along the white-water rivers, where large reserves of potential agricultural land and productivity still exist (Section 8.2). The *terra firme* rain forest of Amazonia should be spared from permanent occupation.

A different conclusion holds true for the South-east Asian tropics, where all ecologically advantageous areas have been exploited for generations with such intensity that some regions today are totally overpopulated. Directing the land-hungry masses to the forests is not a viable alternative, however. The emphasis must be put on a detailed survey on given natural resources and ecological conditions which alone can provide the foundations for site-specific planning of land use and colonization centring on low-input farming or ecofarming. General land surveys based only on airborne data cannot meet the goal (Section 9.3).

Of the three major regions of the world's tropics, *tropical Africa is the most problematic with respect to agricultural perspectives*. In the regions of natural rain forests and humid savannas, rotational field–fallow land-use systems on the basis of burning the biomass are already rapidly failing to satisfy the basic needs of a population which is as yet not particularly dense. As population growth continues, it will become necessary to abandon the practice of burning and to adopt one of the high manual labour input approaches in order to achieve a more efficient use of the available biomass. This can be reached by means of improved low-input systems (Section 8.2), ecofarming (Section 8.4) or genuine agroforestry systems (Section 9.4). On a local level – natural circumstances permitting – wetland rice cultivation could mean a further step in the direction of higher production but it would demand large investment and a radical change in the population's taste.

In *the wet-and-dry outer tropics* the problematic is quite different. Corresponding to declining precipitation and the prolonged dry periods, the reduced effectiveness of hydrolysis results in the formation of shallow soils, increased primary mineral content and rising proportions of clay minerals other than kaolinites, which, in turn, increase the CEC. A more effective CEC and the upward soil-water movement due to the intense evaporation during the dry season generally leave the soil with more favourable agrarian characteristics than in the ferrallitics of the inner tropics. Reduced weathering and changing degrees of soil moisture also lead to a greater differentiation of the chemical soil characteristics with respect to parent rocks and toposequences (Section 3.11).

The brown soils of the thorn savannas and the fersiallitic soils (mostly alfisols) possess a considerably higher potential fertility than the ferrallitics of the humid savannas and the leached ferrallitics of the rain forest (Section 6.1). Problems with the fersiallitic soils, especially with the alfisols, result from their physical structure, which makes them highly susceptible to soil erosion (Section 8.3.4).

Whenever climate-induced soil-chemical advantages are combined with positive attributes from the parent rocks, successful implementation of the Green Revolution technological package has been possible (Sections 7.1, 7.2 and 7.4). This holds

279

particularly true for the black and red soils over the basalts of the central Deccan Plateau in India.

However, the same climatic conditions that lead to the formation of favourable soil conditions prove agriculturally limiting, either by restricting the cultivation period to the rainy season or by favouring cropland erosion when the bare soils are exposed to heavy rain showers. Nevertheless, the semihumid and semiarid outer tropics reveal higher densities of agricultural use and population as compared with the humid inner tropics.

In view of the potentially fertile soils, the relatively intense agricultural use by concentrated population (Figs 2.13, 1.13 and 1.14) confronts two problems: (1) the danger of heavy soil erosion and (2) the lack of water management and irrigation installations.

The low storage capacity, rapid surface sealing and vulnerability to widespread soil erosion in sandy clay loam alfisols are a grave constraint to improving agricultural productivity. Regarding the erosion problem, many studies conducted by IITA and ICRISAT repeatedly recognized the overriding importance of covering the soil surface with crop residues or harvested organic matter and to apply no-tillage practices.

The shortness of the rainy season, the variability of rainfall (Figs 1.6 and 1.7) and the recurrent droughts (Section 6.2 and Fig. 6.4) call for careful water management and installation of irrigation systems. Presently, however, only a small percentage of the cultivated upland areas is irrigated cropland and dams for the purpose of water storage are remarkably few as compared with the semiarid Mediterranean subtropics.

The analysis of the situation, particularly in the semihumid Deccan Plateau of India, indicates that *the reasons for the lack of dams are based on unchangeable natural geographical conditions.* The climagenetic geomorphological zone of extreme peneplanation is characterized by wide through-valleys instead of the deeply incised valleys of the subtropical and extratropical regions. In addition, periodical extreme discharges (Figs 6.6 and 6.7) complicate even more the construction of the extended dams required by the wide trough-valleys (Section 6.4).

The need for change in attitude

With regard to the recognized North–South disparity, we are confronted with the fact that the largest segment of the tropics is subject to ecologically related handicaps that restricted the possibility for agricultural progress to an extent not found in the extratropics. *People outside the tropics must realize and acknowledge that they live in an ecologically favoured part of the world, whose potential has been much easier to develop. This imposes on them a moral obligation to help their disadvantaged fellow men in other less fortunate parts.*

In addition, everybody in the tropics as well as in the extratropics must learn that *the uncontrolled removal of biomass from the seemingly lush tropical rain forests – such as exports of timber and bulk wood products for example – will eventually lead to the exhaustion of this slowly-renewable natural resource.* We need to develop a better ecological insight and an

280

improved conscience towards conservation. Even more important, we must be ready to act accordingly.

Unfounded hopes and false expectations often do more damage than harsh but realistic measures, taken in full awareness of the extremely complex obstacles that tropical environments present to agricultural development approaches which have been adopted only on the basis of experience and success gained in temperate regions.

The farmers of the tropics must be helped to understand that any kind of biomass burning, whether forest, bush, grassland or crop residues, is a wasteful practice that sooner or later will destroy the very foundation of their habitat. Engineers and scientists must raise their consciousness and explore feasible methods of rational and efficient land use.

The time will come when even the most labour-extensive, fossil fuel consuming, and effectively managed low- or high-input farming systems will no longer be able to satisfy the needs of a fast-growing population in the presently cultivated regions of the tropics. Since deforestation cannot continue on the scale of today, *demographic stabilization must be attained (Section 7.4).*

Breaking the codes embedded in the scientific language of 'black box' agricultural technology, is a challenge we have felt necessary to accept for the benefit of those responsible individuals who have repeatedly asked themselves why, after all, the enormous investments in tropical development have failed. It was not so much the unveiling of certain persistent myths about tropical fertility – myths, by now, widely known and dismissed – that guided our critical review of innovative farming systems and of traditional land exploitation modes, but the attempt to explain how these myths could be perpetuated and the fundamental laws that govern the tropical environments ignored for so long.

Appendix 1

State	Total area (km²)	Arable land (km²)	Agricultural Land permanent crops (km²)	permanent pasture (km²)	Total (km²)	Population to be fed (in 1000)	Forest and woodland
Fed. Rep. Germany	248.580	72.400	2.130	45.660	120.190	61.048	73.280[1]
Great Britain	244.820	71.170	600	115.670	187.440	56.842	22.730
Guinea	245.860	15.000[1]	760[1]	30.000[1]	45.760	6.227	101.600[1]
Ivory Coast	322.460	28.900[1]	12.000[1]	30.000[1]	70.900	10.155	73.800[1]
Ghana	238.540	11.200[1]	17.000[1]	34.200[1]	62.400	14.052	84.200[1]
Nigeria	923.770	285.500[1]	25.350[1]	209.600[1]	520.450	98.578	149.000[1]
Madya Pradesh[2] (Mysore)[2]	443.446	200.980	—	15.920	216.900	52.179	53.240

Notes: 1) FAO estimate
2) Source: Stat. Bundesamt Länderbericht Indien 1986, Wiesbaden

(Source: FAO Production Yearbook 1986, Rome)

Appendix 2A   Soil types of South America

				Cation exchange meq (%)							Particle size analysis (%)				
Horizon	Depth (cm)	H (H₂O)	C (%)	CEC	TEB	% BS	Ca	Mg	K	Na	Stones	Coarse sand	Fine sand	Silt	Clay

1. Humic andosol (ectropic andosol; Chile (Cautin); *Nothofagus* forest; parent material; andesitic volcanic ash)

Horizon	Depth (cm)	H (H₂O)	C (%)	CEC	TEB	% BS	Ca	Mg	K	Na	Stones	Coarse sand	Fine sand	Silt	Clay
A₁	0–15	5.6	10.4	64	5.9	9	3.0	2.2	0.6	0.2	0	2	6	75	17
A₂	–45	5.6	3.2	46	0.4	1	0	0.1	0.2	0.1	0	2	7	72	19
AB	–60	5.6	3.2	51	0.2	1	0	0	0.2	0.1	0	3	8	71	18
BC (?)	–100	5.6	2.2	40	0.4	1	0.1	0.1	0.2	0.1	0	6	10	64	20

2. Chromic vertisol (Vertisol; Brazil (Bahia); *Xerophytic* shrub; parent material: calcareous clays)

Horizon	Depth (cm)	H (H₂O)	C (%)	CEC	TEB	% BS	Ca	Mg	K	Na	Stones	Coarse sand	Fine sand	Silt	Clay
A	0–20	8.1	0.4	39.8	39.8	100	36.4	2.8	0.2	0.4	4	13	25	11	51
BC₁	–70	7.9	0.3	34.3	34.3	100	32.8	0.8	0.1	0.6	4	18	19	13	50
BC₂	–132	8.1	0.3	38.9	38.9	100	36.0	1.6	0.1	1.2	3	12	24	12	52
C	–142	7.8	0.2	33.4	33.4	100	30.8	1.0	0.1	1.5	3	9	27	12	52

3. Humic gleysol (humic gley; Brazil (Minas Gerais); *Hydrophilic* grass; parent material: holocene deposits)

Horizon	Depth (cm)	H (H₂O)	C (%)	CEC	TEB	% BS	Ca	Mg	K	Na	Stones	Coarse sand	Fine sand	Silt	Clay
A₁	0–20	5.0	6.4	20.5	0.7	3	0.5		0.1	0.1	1	4	24	24	48
A₂	–55	4.8	3.0	17.9	0.5	3	0.4		0.05	0.1	1	5	16	23	56
ACg	–65	5.0	1.3	13.1	1.2	9	0.9	0.2	0.05	0	1	5	12	19	64
Cg₁	–85	4.9	0.5	9.5	2.2	23	1.6	0.1	0.05	0.1	1	4	4	16	76
Cg₂	–125+	5.3	0.1	10.3	6.5	64	5.4	1.6	0.05	0.1	0	1	1	25	71

4. Orthic acrisol (Red yellow podzolic; Brazil (Rio de Janeiro); *Humid tropical* savanna; parent material: granitic gneiss)

Horizon	Depth (cm)	H (H₂O)	C (%)	CEC	TEB	% BS	Ca	Mg	K	Na	Stones	Coarse sand	Fine sand	Silt	Clay
A	0–10	4.8	1.3	4.5	2.2	49	1.4	0.4	0.1	0.3	0	49	23	13	15
E	–30	4.6	0.7	3.7	1.3	35	0.8	0.2	0.05	0.3	0	48	16	12	24
EB	–45	4.6	0.5	4.2	1.0	24	0.5	0.2	0.05	0.2	0	43	11	12	32
B₁	–75	4.4	0.5	6.0	0.5	8	0.3	0.1	0.05	0.2	0	30	4	7	59
B₂	–155	4.6	0.2	4.0	1.3	28	0.4	0.7	0.05	0.2	0	22	10	30	38
BC	–195	4.6	0.1	3.9	0.9	23	0.2	0.6	0.1	0.1	0	32	15	25	28
C	–195+	4.7		8.0	1.2	15	0.1	0.8	0.2	0.2	0	39	19	22	20

5. Orthic ferralsol (Dark red latosol; Brazil (São Paulo); *former tropical* forest; parent material: shales and argillites)

Horizon	Depth (cm)	H (H₂O)	C (%)	CEC	TEB	% BS	Ca	Mg	K	Na	Stones	Coarse sand	Fine sand	Silt	Clay
A	0–30	4.7	1.6	11.5	4.1	35	2.1	1.1	0.8	0.1	0	3	13	8	76
BA	–60	4.8	1.0	8.2	2.5	30	1.4	0.7	0.4	0.1	0	2	10	11	77
B₁	–150	5.2	0.5	5.1	1.1	22	0.5	0.3	0.2	0.1	0	4	9	10	77
B₂	–210	5.3	0.2	5.3	1.3	25	0.6	0.4	0.2	0.1	0	2	8	11	79
BC	–260	4.8	0.2	5.6	0.9	16	0.4	0.2	0.2	0.1	0	2	10	11	74
C	–280+	4.6	0.1	6.7	1.1	16	0.4	0.5	0.1	0.1	0	11	9	25	55

6. Xanthic ferralsol (Yellow latosol; Brazil (Pará) *Humid tropical* forest; parent material: Pliocene lacustrine sediments)

Horizon	Depth (cm)	H (H₂O)	C (%)	CEC	TEB	% BS	Ca	Mg	K	Na	Stones	Coarse sand	Fine sand	Silt	Clay
A	0–2	4.0	3.6	14.9	2.2	15	0.9	1.0	0.3	0.1	0	4	11	10	75
							Total									
AB	–20	4.2	1.3	6.9	0.7	11	0.6		0.1	0	0	2	8	7	83
B₁	–60	4.7	0.7	4.6	0.6	14	0.5		0.1	0	0	1	6	5	88
B₁	–150	5.2	0.4	2.7	0.6	22	0.5		0.1	0	0	1	13	12	74
BC	–250	5.5	0.3	2.0	0.6	28	0.5		0.1	0	0	1	10	14	75

Appendix 2B   Soil types of Africa

Horizon	Depth (cm)	H (H₂O)	C (%)	CEC	TEB	% BS	Ca	Mg	K	Na	Stones	Coarse sand	Fine sand	Silt	Clay
					Cation exchange meq (%)							Particle size analysis (%)			

7. Eutric regosol (Raw mineral soil; Mauretania; *Saharian* climate; parent material: Aeolian sand)

Horizon	Depth (cm)	H (H₂O)	C (%)	CEC	TEB	% BS	Ca	Mg	K	Na	Stones	Coarse sand	Fine sand	Silt	Clay
A	0–10	6.7	0.03	3.28	1.87	57	1.00	0.72	0.11	0.04	0	57.2	40.6	0.6	0.9
C1	60–80	7.6	0.05	2.52	1.75	69	1.12	0.52	0.08	0.03	0	34.1	62.8	0.9	1.1
C2	150–170	7.5	0.04	2.80	1.57	56	1.00	0.48	0.05	0.04	0	38.2	57.6	0.8	1.1

8. Luvic arenosol (subarid red brown soil; Senegal; *Dry* savanna; parent material: Argillaceous sandstone)

Horizon	Depth (cm)	H (H₂O)	C (%)	CEC	TEB	% BS	Ca	Mg	K	Na	Stones	Coarse sand	Fine sand	Silt	Clay
A	0–25	6.9	0.20	7.4	2.24	30.2	1.64	0.50	0.04	0.06	0	32.0	61.0	2.6	5.7
AB	45–55	5.6	0.17	4.1	1.88	45.8	1.28	0.50	0.04	0.06	0	30.5	59.0	4.5	7.4
Bt1	100	5.3		2.96			2.10	0.74	0.04	0.08	0	29.5	55.5	2.1	14.4
Bt2	130			2.95			1.91	1.00	0.04	0.10					

9. Chromic vertisol; (Sudan *Semiarid*, dry savanna; parent material: Calcareous alluvial)

Horizon	Depth (cm)	H (H₂O)	C (%)	CEC	TEB	% BS	Ca	Mg	K	Na	Stones	Coarse sand	Fine sand	Silt	Clay
Ah1	0–1	6.8	1.29	45.2	40.4		22.6	16.5	1.3	tr.	0	16.1	23.3	10.7	49.8
Ah2	1–18	7.5	0.64	50.1	46.3		26.4	19.2	0.7	tr.	0	16.4	22.5	7.3	53.8
AC	18–60	7.9	0.48	50.0	42.6		23.7	18.1	0.7	0.1	0	14.6	21.7	9.6	54.1
CA	60–90	8.6		50.4	46.3		26.4	18.7	0.8	0.4	0	13.5	19.6	11.2	55.7
CA	90–120	8.7		45.8	45.1		25.3	18.6	0.8	0.4	0	15.2	20.6	8.8	55.4
CA	120–150	8.8		54.1	51.3		29.2	20.7	0.8	0.6	0	12.2	18.3	9.0	60.5
C	150–175	8.8		56.0	54.1		31.4	21.2	0.9	0.6	0	12.4	17.3	12.7	57.6
Cck	200–250	9.2		48.1	55.5		33.5	20.9	0.8	0.3	0	10.1	16.4	10.9	62.6

10. Ferralic cambisol (Haplothox; Sierra Leone; *Savanna*; Parent material: Granodiorite)

Horizon	Depth (cm)	H (H₂O)	C (%)	CEC	TEB	% BS	Ca	Mg	K	Na	Stones	Coarse sand	Fine sand	Silt	Clay
Ap	0–30	4.8		10.65	1.81	17	0.96	0.44	0.06	0.35	22.8	37.8	26.8	9.0	26.8
Bs	70–150	5.3		6.86	0.54	8	0.05	0.13	0.04	0.03	11.4	24.3	26.4	15.9	33.8
C	150–235	5.5		6.00	1.44	24	0.05	0.10	0.07	0.08	8.5	34.6	25.8	16.8	23.4

11. Eutric cambisol (Eutrophic brown soil; Ivory Coast; *Shrubby* savanna; parent material: Amphibolic schist)

Horizon	Depth (cm)	H (H₂O)	C (%)	CEC	TEB	% BS	Ca	Mg	K	Na	Stones	Coarse sand	Fine sand	Silt	Clay
Ah	0–10	6.6	4.96	29.3	30.9	100	18.61	11.45	0.80	0.04	0	9.6	11.9	43.8	34.4
Bw	40–50	5.8	0.47	14.3	12.9	90	7.58	5.28	0.04	0.04	7.3	22.7	17.9	32.0	23.4
BC	100–110	6.4		20.6	25.0	100	12.36	12.60	0.03	0.04	7.1	9.0	10.9	39.6	37.7

12. Ferric luvisol; (Leached ferruginous soil; Senegal; *Humid* savanna; parent material: Sandstone)

Horizon	Depth (cm)	H (H₂O)	C (%)	CEC	TEB	% BS	Ca	Mg	K	Na	Stones	Coarse sand	Fine sand	Silt	Clay
Ap1	0–6	6.5	1.05	4.80	3.50	73	2.25	1.05	0.15	0.05		26.1	47.1	4.0	9.4
Ap2	6–13	5.6	0.68	4.05	2.50	62	1.70	0.65	0.10	0.05		31.1	43.9	4.1	9.4
E	15–25	5.6	0.38	4.45	2.80	63	1.70	1.05	0.05	trace		23.3	38.6	4.9	21.5
Bt	40–60	6.2	0.33	5.15	3.65	71	2.10	1.50	0.05	trace		21.3	28.0	5.3	35.1
Btg	90–110	6.3	0.29	4.65	3.70	80	1.30	2.30	0.05	0.05		16.6	23.7	5.3	43.7
Bcs	130–150	6.3	0.20	4.85	3.60	74	1.70	1.85	0.05	trace		17.0	25.0	5.6	40.1

13. Plinthic ferralsol (Ivory Coast; *Secondary humid* forest; parent material: Schist and gneiss)

Horizon	Depth (cm)	H (H₂O)	C (%)	CEC	TEB	% BS	Ca	Mg	K	Na	Stones	Total		Silt	Clay
Ah	0–3	4.0	6.0	13.7	4.0	30	1.3	1.6	0.92	0.24	24	62.7		3.2	32.0
BA	3–15	4.4	1.8	9.8	1.0	10	0.2	0.5	0.22	0.11	70	61.0		4.3	32.1
Bo1	15–33	4.6	1.0	5.8	0.7	13	0.1	0.3	0.16	0.06	65	53.7		4.1	41.2
Bu2	33–60	4.6	0.9	5.8	0.5	9	0.1	0.2	0.14	0.06	38	42.4		6.0	50.7
2Bs	60–98	4.7	0.7	5.4	0.4	7	0.1	0.1	0.11	0.05	24	24.5		21.9	53.2
2Bsm1	98–142	4.8	0.4	6.1	0.6	9	0.1	0.2	0.18	0.07	15	37.7		15.9	44.0
2Bsm2	142–258	4.7	0.3	4.9	0.4	8	0.1	0.1	0.13	0.05		28.0		17.8	49.5

Appendix 2C   Soil types of South-east Asia

Horizon	Depth (cm)	H (H_2O)	C (%)	Cation exchange meq (%)							Particle size analysis (%)				
				CEC	TEB	% BS	Ca	Mg	K	Na	Stones	Coarse sand	Fine sand	Silt	Clay
14. Humic andosol (Philippines; *Trop. Monsoon* (2000 mm annual); parent material: Fine volcanic ash)															
Ah	0–30	5.3	12.0	61.43	9.54	15.5	7.76	1.44	0.17	0.17	—		40.80	48.7	10.5
AB	30–58	5.5	5.2	42.36	2.41	5.7	1.44	0.59	0.20	0.29	12.72		40.88	40.0	6.4
Bw1	58–88	5.7	1.6	33.79	2.59	7.7	1.63	0.59	0.25	0.12	8.88		51.52	33.2	6.4
Bw2	88–120	6.0	1.5	33.06	3.73	11.3	2.50	0.80	0.26	0.17	9.01		64.39	22.0	4.6
BC	120–150	5.9	1.2	33.92	3.14	9.3	2.18	0.59	0.25	0.12	—		44.90	45.7	9.4
15. Calcic cambisol (Thailand; *Tropical monsoon* (1450 mm); parent material: Mixed basic rocks)															
Ap	0–29	7.8	1.5	84.0	78.8	94	73.6	4.1	0.7	0.4		12.0		36.0	52.0
BA	29–49	7.2	0.7	84.1	70.9	84	66.6	3.6	0.4	0.3		11.0		36.0	53.0
Bw	49–82	7.3	0.6	79.6	72.7	91	68.5	3.3	0.5	0.4		13.0		29.0	58.0
C	82–150+	8.0	0.3	38.0	52.4	138	50.5	1.4	0.2	0.3		29.0		38.5	32.5
16. Ferric luvisol; Sabah; *Tropical monsoon* (1800 mm annual); parent material: Quaternary basalt)															
Ah	0–2	7.3	7.18	39.74		88	27.81	5.15	1.81	0.28		7.7	18.7	51.3	22.3
AB	2–17	6.3	1.16	14.35		69	7.96	1.20	0.60	0.19		6.1	11.5	32.5	49.9
BA	17–55	6.2	0.37	10.81		67	6.17	0.69	0.19	0.16		4.7	8.2	24.3	62.8
Bt1	55–150	6.0	0.30	8.01		61	3.05	1.35	0.40	0.08		3.1	5.5	20.6	70.8
Bt2	150–200	5.8	0.34	8.58		52	2.93	0.68	0.77	0.06		19.0	13.3	14.6	52.4
17. Ferric acrisol (Thailand; *Tropical monsoon* (2100 mm); parent material: Old alluvial deposit)															
Ap	0–8/12	5.4	1.07	5.7	1.6	28	1.0	0.3	0.1	0.2		59.1		28.0	12.9
E	8/12–40	6.4	0.63	6.4	2.3	36	1.7	0.3	0.1	0.2		10.3	45.7	28.0	16.0
EB	40–53	5.7	0.36	4.8	1.6	33	1.0	0.4	0.04	0.2		55.0		27.3	17.6
BE	53–72	5.2	0.29	5.8	1.0	17	0.5	0.2	0.1	0.2		10.6	41.8	25.7	21.9
Bt	72–119	5.2	0.26	5.2	0.5	10	0.2	0.1	0.04	0.2		52.7		26.9	20.4
Btg	119–147	5.0	0.22	6.0	0.6	8	0.2	0.1	0.1	0.2		10.1	40.3	27.4	22.2
BCg	147–153+														
18. Orthic ferralsol (Thailand; *Tropical monsoon* (1500 mm); parent material: Old alluvial deposit)															
Ap1	0–5	5.9	1.51	9.6	6.5	59	4.8	1.2	0.3	0.2		64.0		31.5	4.5
Ap2	5–10	4.9	1.05	6.3	4.1	41	2.8	1.0	0.1	0.2		61.0		31.0	8.0
AB	10–15	4.7	0.70	6.0	2.3	28	1.3	0.7	0.1	0.2		64.0		28.5	7.5
BA	15–40	4.2	0.47	4.9	1.1	14	0.5	0.3	0.1	0.2		61.0		22.5	16.5
Bws1	40–70	4.2	0.25	4.3	0.7	11	0.4	0.1	0.1	0.2		55.5		22.5	22.0
Bws2	70–90	4.7	0.19	3.9	0.7	12	0.4	0.1	0.1	0.2		52.0		26.0	22.0
Bws3	90–140+	4.7	0.15	4.0	0.8	12	0.4	0.2	0.1	0.2		47.5		26.5	26.0

Appendix 3A   Dams in West Africa and the Sudan

Name of project (river)	Max. height (m)	Length (m)	Storage (10⁶ m³)	Irrigated area (ha)	Date
Sudan					
Sannar Dam (Blue Nile)	—	3 200	980	42 000	1925
Gebel Aulia (Nile)	—	—	—	—	—
Khashm El Gibra (Atabra)	50	3 846	1 300	250 000	1964
Er Roseires Dam (Blue Nile)	59	16 000	2 930	340 000	1966
Nigeria					
Kainji Dam (Niger)	65.53	8 500	15 000		
Jebba Dam (Niger)	—	—	1 043	—	—
Shiroro Gorge (Kaduna)	—	—	—	—	—
Ghana					
Akosombo Main Dam (Volta River)	112	640	148 000	—	1965
Akosombo Saddle D.	—	355	—	—	1965
Barikese Dam (Ofin River)	12	467	35	—	1968
Ivory Coast					
Ayame I (Bia)	26	310	1 250	—	1959
Kan (Kan)	14	718	2.8	—	1964
Ayame II (Bia)	29	310	8	—	1964
Kossov (Bandama)	51	1 431	25 000	—	—
Liberia					
Mt Coffee (Paul River)	—	—	—	—	1966
Sierra Leone					
Guma Dam	68	275	22	—	1963
Guinea					
Kaledam (Samou)	—	—	14	—	1963
Kinkon Dam (Kokulu)	—	—	—	—	—
Suapiti Dam	120	1 070	11 000	—	—

Source: INCOLD (1973).

Appendix 3B   Dams in the Deccan states of India

Name of project (river)	Max height (m)	Concrete and masonry		Earth		Storage (10^6 m^3)		Irrigated area (ha)	Date	
		Length (m)	Quantity (10^6 m^3)	Length (m)	Quantity (10^6 m^3)	Gross	Live		Start	Completion
Andhra Pradesh										
Kadam (Godavari)	40.6		0.432	2 102	1.33	137	215	34 400	1949	IV Plan
Nagarjunasagar (Krishna)	124.7	1 450	5.605	3 414	2.35	6 797	11 558	830 000	1956	End of IV Plan
Pochampad (Godavari)	42.7	958	0.668	13 651	8.20	2 299	3 170	230 000	1963	V Plan
Gotta Dam (Vamsadhara)	34.8	?	0.150	2 042*	1.44	?	456	?		
Orissa										
Hirakud (Mahanadi)	61	4 801	1.14	20 661	17.08	5 822	8 100	242 820	1948	1957
Salandi (Salandi)	51.8		0.201	818*	2.91	557	566	67 383	1961	IV Plan

287

Appendix 3B Continued.

Name of project (river)	Dams									Date	
	Max height (m)	Concrete and masonry		Earth		Storage (10^6 m^3)		Irrigated area (ha)		Start	Completion
		Length (m)	Quantity (10^6 m^3)	Length (m)	Quantity (10^6 m^3)	Gross	Live				
Mysore											
Hidkal (Ghataprabha)	50		0.99	8 841*	7.28	614	660	101 175		1949	1956
Tungabhadra (Tungabhadra)	49.4	2 441*	0.988		0.17	3 324	3 767	37 637		1945	1956
Bhadra (Bhadra)	71.6	440*	0.787		0.39	1 789	2 023	99 015		1947	End IV Plan
Kabini (Kabini)	59.5		0.386	2 701*	1.67	435	544	51 874		1959	V Plan
Almatti Dam (upper Krishna)	34.8	1 631*	0.746		0.19	2 090	2 350	242 820			
Siddapur Dam (upper Krishna)	23.6	6 951*	0.317		3.64	700	791			1964	V Plan

Appendix 3B Continued.

Name of project (river)	Max height (m)	Concrete and masonry Length (m)	Concrete and masonry Quantity (10^6 m³)	Earth Length (m)	Earth Quantity (10^6 m³)	Storage (10^6 m³) Gross	Storage (10^6 m³) Live	Irrigated area (ha)	Date Start	Date Completion
Maharastra										
Vir (Nira)	34.7	3 607*	0.741		1.21	266	328	26 710	1957	1965/66
Bagh (Bagh)	22.9		0.062	1 052*	1.02	187	203	33 671	1958	IV Plan
Girna (Girna)	55	963*	0.292		1.98	526	611	57 208	1958	IV Plan
Itiadoh (Garvi)	30		0.032	602*	0.87	414	470	46 136	1958	End of IV Plan
Khadakwasla (Ambi)	58		0.07	832*	2.93	275	312	22 298	1957	End of IV Plan
Mula (Mula)	46.6		0.178	2 820*	7.38	609	737	65 561	1959	IV Plan
Yeldari (Purna)	51.4		0.200	4 786*	2.20	814	962	61 514	1957	IV Plan
Siddashwar (Purna)	38.3		0.140	6 306*	0.82	74	247			
Pawna (Pawna)	42.9		0.180	1 700*	1.80	271	305	189 695	1964	End of V Plan
Ujjani (Bhima)	51.8	2 332*	0.760		0.15	1 439	3 113	141 645	1964	V Plan
Jayakwadi (Godavari)	36.6		0.420	9 904*	11.78	2 069	2 605	?		
Krishna (Krishna)	35.0		0.07	3 342*	1.03	129	175			Not fixed
Warna (Warna)	58.4		0.12	1 474*	7.73	996	2 467	99 058	1964	End of V Plan
Madhya Pradesh										
Barna (Barna)	46.6		0.143	396*	0.23	407	486	66 435	1960	End of IV Plan
Tawa (Tawa)	51.0		0,636	1 823*	3.11	2 087	2 311	303 560	1962	V Plan
USA										
Hoover (Colorado)	221	379*	3.36*				36 700			
Grant Coulee	168	1 272*	8.02*				11 743			
Switzerland										
Dixence	285	695	5.95				400			

*For the total length of the dam.

Source: India, Ministry of Irrigation and Power (1967).

Appendix 4 Soil types of India

Horizon	Depth (cm)	pH (H₂O)	C (%)	CEC	TEB	%BS	Cation exchange meq (%)					
							Ca	Mg	K	Na	Al	H
Eutric fluvisol (Bangladesh; non-calcareous alluvium (periodically irrigated); parent material: mixed alluvium)												
Ap	0–12		0.86	13.4								0.7
Cg1	12–20		0.46	10.1								1.2
Cg2	20–55		0.92	13.09	11.89	91	10.2	1.4	0.16	0.13		1.2
Cg3	55–73			11.63	10.43	90	8.8	1.5	0.06	0.07		
Eutric cambisol (Bangladesh; non-calcareous brown floodplain soil; parent material: mixed alluvium)												
Ap	0–12	5.9	1.0	7.3	4.1	57	2.9	0.9	0.20	0.14		3.2
B	12–32	5.9	0.74	7.7	4.1	53	3.4	0.5	0.14	0.08		3.6
BC	32–60	5.9	0.27	3.8	1.4	37	0.8	0.4	0.09	0.08		2.4
2C	60–137	6.2										
Dystric nitosol, (Bangladesh; deep red-brown terrace soil; parent material: Pleistocene/Tertiary clay)												
Ah1	0–4	6.1	2.12	10.32	5.83	56	3.26	2.02	0.50	0.05		4.49
Ah2	4–7	5.2	1.66									
BA	7–15	5.2	1.32	10.84	4.11	38	2.10	1.40	0.50	0.11		6.73
Bt1	15–31	5.2	1.02									
Bt1	31–47	5.5										
Bt2	47–70	5.6										
Bt2	70–95	5.7		9.75	1.54	16	1.02	0.34	0.15	0.03		8.21
C1	135–155	5.5		11.37	1.96	17	1.02	0.85	0.06	0.03		9.41

Appendix 4 Continued.

Horizon	Depth (cm)	pH (H₂O)	C (%)	CEC	TEB	%BS	Cation exchange meq (%)					
							Ca	Mg	K	Na	Al	H
Chromic luvisol (India; deep red loam; parent material: gneiss)												
Ah	0–13	6.7	0.30	6.5			0.100					
BA	13–56	6.5	<0.1	9.1			0.100					
Bt1	56–92	6.5	<0.1	8.5			0.105					
Bt2	92–150	6.6		9.6			0.120					
BC	150–183	6.6		11.3			0.135					
Pellic vertisol (India; deep black soil; parent material: basalt)												
Ap	0–2.5	7.9	0.76	70.11	67.84		57.0	5.70				
Ah	2.5–23	7.85	0.58	71.82	67.84		59.0	3.32				
ACh1	23–69	7.80	0.58	69.62	65.64		56.0	7.60				
ACh2	69–91	7.80	0.58	68.41	61.32		51.5	6.65				
Cck1	91–132	7.9	0.15	60.29	45.79		36.0	8.07				

Bibliography

Adeyoju, S. K. (1982) Agroforestry, legislation, policies and forest usages, in MacDonald, L. H., (ed.) *Agro-forestry in the African Humid Tropics*, The United Nations University, Tokyo, pp. 18–23.

Agbede, O. O. and **Ojo G. O. A.** (1982) Food crop yield under Gmelina plantations in Southern Nigeria, in MacDonald L. H. (ed.) *Agro-forestry in the African Humid Tropics*, The United Nations University, Tokyo, pp. 79–82.

Agboola, A. A., Wilson, G., Getahun, A. and **Yamoah, C. F.** (1982) Gliricidia sepium: a possible means to sustained cropping, in MacDonald, L. H. (ed.) *Agro-forestry in the African Humid Tropics*, The United Nations University, Tokyo, pp. 141–3.

Allan, W. (1965) *The African Husbandman*, Oliver & Boyd, Edinburgh.

Allen, B. L. (1977) Mineralogy and soil taxonomy, in Dixon, J. B. and Wee, S. B. (eds) *Minerals in Soil Environments*, Soil Science Society of America, Madison, Wisconsin, pp. 771–96.

Alvim, R. A. and **Nair, P. K.** (1986) Combination of cacao with other plantation crops: an agroforestry system in Southeast Bahia, Brazil, *Agroforestry Systems*, **4**(1) 3–15.

Anderson, J. M. and **Swift, M. J.** (1983) Decomposition in tropical forests, in Sutton, S. L., Whitmore, T. C. and Chadwick, A. C. (eds) *Tropical Rain Forest: Ecology and Management*, Blackwell Scientific Publications, Oxford, pp. 287–309.

Andreae, B. (1981) *Farming, Development and Space. A World Agricultural Geography*, Walter de Gruyter, Berlin and New York.

Anon/Klinge (1972a) Regenwasseranalysen aus Zentralamazonien, ausgeführt in Manaus, Amazonas, Brasilien, von Dr Harald Ungemach, *Amazoniana*, **III**, 186–98.

Anon/Klinge (1972b) Die Ionenfracht des Rio Negro, Staat Amazonas, Brasilien, nach Untersuchungen von Dr Harald Ungemach. *Amazoniana*, **III**, 175–85.

Atlas van Tropisch Nederland (1938) uit geven door het Koninklijk Nederlandsch Aardrijkskundig Genootschap, Amsterdam.

Aubert, G. (1965) Classification des sols. Tableaux des classes, sous classes, groupes, sous groupes des sols, utilisés par la Section de Pédologie de l'ORSTOM. *Cahiers ORSTOM*, Série Pédologie 3, pp. 269–88.

Ball, J. B. and **Umeh, L. I.** (1982) Development trends in Taungya systems in the moist lowland forest of Nigeria between 1975 and 1980, in MacDonald L. H. (ed.)

Agro-forestry in the African Humid Tropics, The United Nations University, Tokyo, pp. 72–8.

Bandy, D. E. and **Benites J. R.** (1983) *Lowland Rice Production on Alluvial Soils, Agronomic–Economic Research*, 1980–1981 Technical Report, North Carolina State University, Raleigh, NC, pp. 101–10.

Bandy, D. E., Mesia, R., Nicholaides, J. J., Hernandez and **Coutu, A. J.** (1980) Yurimaguas Small Farmer Extrapolation Program, in Nicholaides, J. J. III (ed.) *Agronomic–Economic Research on Soils of the Tropics*, 1978–1979 Technical Report, North Carolina State University, Raleigh, NC., pp. 198–218.

Bandy, D. E. and **Nicholaides, J. J. III** (1983) Composting and mulching, *Agronomic–Economic Research*, 1980–1981 Technical Report, North Carolina State University, Raleigh, NC, pp. 70–6.

Bandy, D. E. and **Sanchez, P. A.** (1985) Managed kudzu fallow, in McCants, Ch. B. (ed.) *TROPSOILS*, Triennial Technical Report 1981–1984, North Carolina State University, Raleigh, NC, pp. 159–61.

Basso, E. B. (1973) *The Kalapalo Indians of Central Brazil*, Holt, Rinehart and Winston, New York.

Bauer, P. T. and **Yamey, B. S.** (1957) *The Economics of Underdeveloped Countries*, Cambridge University Press, Cambridge.

Bayliss-Smith, T. B. and **Wanmale, S.** (eds) (1984) *Understanding Green Revolutions: Agrarian Change and Development Planning in South Asia*, Cambridge University Press, Cambridge.

Beckerman, S. (1983) Does the swidden ape the jungle? *Human Ecology*, **11**(1), 1–12.

Benchetrit, M., Cabot, J. and **Durand-Dastrès, F.** (1971) *Géographie zonale des régions chaudes*, Fernand Nathan, Paris.

Benites, J. R., Arevalo, L., Bandy, D. E., Rios, O., Tejada, E., Hormia, K. and **Rachiumi, A.** (1985) Intensive management of alluvial soils, in McCants, Ch. B. (ed.) *TROPSOILS*, Triennial Technical Report 1981–1984, North Carolina State University, Raleigh, NC, pp. 199–205.

Bennett, M. K. (1962) An agroclimatic mapping of Africa, *Food Research Institute Studies*, **III** (3), 195–216, Stanford University Press.

Biel, E. (1929) Die Veränderlichkeit der Jahressumme des Niederschlags auf der Erde, *Geographische Jahresberichte aus Österreich*, **14/15**, 151–60.

Birkeland, P. W. (1974) *Pedology, Weathering and Geomorphological Research*, Oxford University Press, Oxford.

Birkeland, P. W. (1984) *Soils and Geomorphology*, Oxford University Press, Oxford.

Bishop, J. P. (1982) Agroforestry systems for the humid tropics east of the Andes, in Hecht, S. B. (ed.) *Amazonia. Agriculture and Landuse Research*, CIAT, Cali, pp. 403–13.

Blum, W. E. H. and **Magalhaes L. M. S.** (1987) Restriçoes edáficas de solos na bacia sedimentar Amazonica à utilizaçao agraria, in Kohlhepp, G. and Schrader, A. (eds) *Hombre y Naturaleza en la Amazonía*, Tübinger Geographische Studien, No. 95, pp. 83–92, Tübingen.

Blüthgen, J. and **Weischet W.** (1980) *Allgemeine Klimageographie*, 3rd edn, de Gruyter, Berlin and New York.

Bocquier, G. (1973) *Genèse et évolution de deux toposéquences de sols tropicaux du Tschad*, Mémoires ORSTOM (Paris), No. 62.

Bohn, H. L., McNeal, B. L. and **O'Connor, G. A.** (1979) *Soil Chemistry*, John Wiley & Sons, New York.

Bonneau, M. and **Souchier, B.** (1982) *Constituents and Properties of Soils*, Academic Press, London.

Boonkird, S. A., Fernandez, E. C. and **Nair, P. K.** (1984) Forest villages: an agroforestry approach to rehabilitating forest land degraded by shifting cultivation in Thailand. *Agroforestry Systems*, **2**(1), 87–102.

Boserup, E. (1965) *The Conditions of Agricultural Growth: The Economics of Agrarian Change under Population Pressure*, Aldine Publications, Chicago.

Boserup, E. (1981) *Population and Technological Change: A Study of Long-Term Trends* The University of Chicago Press, Chicago.

Bowen, G. D. (1980) Mycorrhizal roles in tropical plants and ecosystems, in Mikola, P. (ed.) *Tropical Mycorrhiza Research*, Clarendon Press, Oxford, pp. 165–90.

Bowers, R. D. (1982) Agricultural tree crops as a no-tillage system, in MacDonald, L. H. (ed.) *Agro-forestry in the African Humid Tropics*, The United Nations University, Tokyo, pp. 49–51.

Braak, C. (1929) *Het Klimaat van Nederlands-Indië.* Verhand Koninklyk Magnestischen Meteorologisch Observatorium Batavia vol II.

Braun, H. (1974) Shifting cultivation in Africa. (Evaluation of questionnaires), in *FAO: Shifting Cultivation and Soil Conservation in Africa,* FAO Soils Bulletin 24, Rome, pp. 21–36.

Bray, F. (1986) *The Rice Economies. Technology and Development in Asian Societies*, Blackwell, Oxford.

Bridges, E. M. (1978) *World Soils*, Cambridge University Press, Cambridge.

Brinkmann, W. L. F. (1985) Studies on hydrobiogeochemistry of a tropical lowland forest system. *GeoJournal*, **11**(1), 89–101.

Brinkmann, W. L. F. and **Dos Santos, A.** (1970) Natural waters in Amazonia. III, *Acta Amazonica*, **2**, 443–8.

Brookfield, H. C. (1962) Local study and comparative methods: an example from Central New Guinea. *Annals of the Assoc. of Amer. Geog.*, **52**, 242–54.

Brown, G., Newman, A. C. D., Rainer, J. H. and **Weir, A. H.** (1978) The structure and chemistry of soil clay minerals, in Greenland, D. J. and Hayes, M. H. B. (eds) *The Chemistry of Soil Constituents*, Wiley, Chichester, pp. 29–178.

Bruenig, E. F. (1984) Nutzbarmachung des tropischen Regenwaldes: Eldorado oder Pandora? *Geographische Rundschau*, **36**(7), 352–8.

Buck, M. (1986) Concepts of resource sharing in agroforestry systems, *Agroforestry Systems*, **4**(3), 191–203.

Büdel, J. (1982) *Climatic Geomorphology*, Princeton University Press, Princeton.

Budowski, G. (1982) Applicability of agro-forestry systems, in MacDonald, L. H. (ed.) *Agro-forestry in the African Humid Tropics*, The United Nations University, Tokyo, pp. 13–16.

Bunting, B. T. (1969) *The Geography of Soil.* Hutchison, London.

Buol, S. W. and **Couto, W.** (1978) Fertility management interpretation and soil surveys of the tropics, in Drosdoff, M., Daniels, R. D., and Nicholaides J. J. (eds.), *Diversity of Soils in the Tropics*, American Society of Agronomy, Madison, Wisconsin, pp. 65–75.

Buringh, P. (1977) Food production potential of the world, *World Development*, **5**(5–7), 477–85.

Butzer, K. W. (1976) *Geomorphology from the Earth*, Harper & Row, New York.

Carlson, D. G. (1982) Famine in history: with a comparison of two modern Ethiopian disasters, in Cahill, K. M. (ed.) *Famine*, Orbis Books, Maryknoll, NY, pp. 5–16.

Carol, H. (1973) The calculation of theoretical feeding capacity for tropical Africa, *Geographische Zeitschrift*, **61**(1), 80–97.

Catterson, T. M. (1982) Agro-forestry production systems: putting them into action, in MacDonald, L. H. (ed.) *Agro-forestry in the African Humid Tropics*, The United Nations University, Tokyo, pp. 128–33.

Chatterjee, P. S. (1973) India, in *World Atlas of Agriculture*, vol. 2, Instituto Geografico de Agostini, Novara, pp. 155–219.

Church, R. J. H. (1974) *West Africa. A Study of Environment and of Man's Use of it*, Longman, London and New York.

CIAT (Centro Internacional de Agricultura Tropical), *Annual Reports; CIAT Highlights (1970–1986)*, Cali, Colombia.

CIMMYT (Centro Internacional de Mejoramiento de Maiz y Trigo) (1985) *Wheats for More Tropical Environments*, Proceedings of the International Symposium, 24–28 Sept. 1984, Mexico, DF. CIMMYT, Mexico.

Clarke, J. S. (1971) *Population Geography in Developing Countries*. Pergamon Press, Oxford.

Clarke, W. C. (1976) Maintenance of agriculture and human habitats within the tropical forest ecosystem, *Human Ecology*, **4**(3), 247–60.

Cochrane, T. T. (1984) Amazonia: a computerized overview of its climate, landscape and soil resources, *Interciencia*, **91**(5), 298–306.

Cochrane, T. T. and **Sanchez, P. A.** (1982) Land resources, soils and their management in the Amazon region: a state of knowledge report, in Hecht, S. B. (ed.) *Amazonia: Agriculture and Landuse Research*, CIAT, Cali, pp. 137–209.

Combe, J. (1982) Agroforestry techniques in tropical countries: potentials and limitations, *Agroforestry Systems*, **1**(1), 13–27.

Committee for the World Atlas of Agriculture (ed.) (1973) *World Atlas of Agriculture*, Instituto Geografico de Agostini, Novara.

Conklin, H. C. (1957) *Hanonoo Agriculture in the Philippines*, FAO Forestry Development Paper No. 12, Rome.

Cotton, C. A. (1961) Theory of savanna planation, *Geography*, **46**, 89–96.

Dalrymple, D. G. (1978) *Development and Spread of High-Yielding Varieties of Wheat and Rice in the Less Developed Nations*, US Dept of Agriculture Office of International Cooperation and Development, Foreign Agriculture Economic Report 95, 6th edn, Washington, DC.

Dalrymple, D. G. (1979) The adoption of high-yielding grain varieties in developing nations, *Agricultural History*, **53**, 704–26.

Da Silva, A. R. (1985) Breeding and disease problems confronting the successful cultivation of wheat in the cerrados of Brazil, in *Wheats for more Tropical Environments*, Proceedings of the International Symposium, 24–28 Sept. 1984, Mexico, DF, CIMMYT, Mexico, pp. 122–4.

De Boer, H. J. (1950) On the relation between rainfall and altitude in Java, Indonesia, *Chronica Naturae*, **106**, 424–7.

De Datta, S. K. (1978) Fertilizer management for efficient use in wetland rice soils, in *Soils and Rice* (IRRI), Los Baños, pp. 671–701.

295

De las Salas, J. P. (ed.) (1979) *Agroforestry Systems in Latin America*, Proceedings of Workshop held in Turrialba, Costa Rica, March 1979, CATIE, Turrialba, Costa Rica.

De Schlippe, P. (1956) *Shifting Cultivation in Africa. The Azande System of Agriculture*, Routledge & Paul, London.

De Vleeschauwer, D., Lal, R. and **de Boodt, M.** (1979) Influence of soil conditioners on water movement through some tropical soils, in Lal R. and Greenland, D. J. (eds) *Soil Physical Properties and Crop Production in the Tropics*, John Wiley & Sons, London, pp. 149–58.

Denevan, W. M. (1971) Campa subsistence in the Gran Pajonal, eastern Peru, *The Geographical Review*, **61**(4), 496–518.

Denevan, W. M. (1976) *The Native Population of the Americas in 1492*, The University of Wisconsin Press, Madison.

Denevan W., Treacy, J. M., Alcorn, J. B., Padoch, C., Denslow, L. and **Flores, C.** (1985) Indigenous agroforestry in the Peruvian Amazon: Bora Indian management of swidden fallows, in Hemming, J. (ed.) *Change in the Amazon Basin*, vol. I, Manchester University Press, Manchester, pp. 137–55.

D'Hoore, J. L. (1964) *Soil Map of Africa, 1 : 5 Million*, Comité de Coopération Technique en Afrique au sud du Sahara, Lagos.

Dickinson, R. E. (ed.) (1987) *The Geophysiology of Amazonia. Vegetation and Climate Interactions*, John Wiley & Sons, New York.

Domroes, M. (1968) Über die Beziehung zwischen äquatorialem Konvektionsregen und der Meereshöhe auf Ceylon, *Archiv Meteorologie, Geophysik, Bioklimatologie*, B **16**, 164–73, Vienna.

Duchaufour, P. (1982) *Pedology: Pedogenesis and Classification*, Allen & Unwin, London.

Dudal, R. (1976) Inventory of the major soils of the world with special reference to mineral stress hazards, in Wright, M. J. (ed.) *Plant Adaptation to Mineral Stress in Problem Soils*, Proceedings of a Workshop at National Agricultural Library, Beltville, Maryland, Cornell University Agricultural Experimental Station, pp. 3–13.

Dudal, R. (1980) Soil related constraints to agricultural development in the tropics, in International Rice Research Institute and New York State College of Agriculture and Life Sciences, Cornell University (eds) *Priorities for Alleviating Soil Related Constraints to Food Production in the Tropics*, Los Baños, the Philippines, pp. 23–37.

Eckholm, E. P. (1976) *Losing Ground*, W. W. Norton, New York.

Eder, J. F. (1977) Agricultural intensification and the returns to labour in the Philippine swidden system, *Pacific Viewpoint*, **18**, 1–21.

Egger, K. (1976) Ausbeutung oder Kooperation – Landbau in ökologischer Verantwortung, *Scheidewege*, **6**, 194–208.

Egger, K. (1982) Ökologische Alternativen im tropischen Landbau–Notwendigkeit, Konzeption, Realisierung, in: Havlik, D. and Mäckel, R. (eds) *Fortschritte landschaftsökologischer und klimatologischer Forschung*, Freiburger Geographische Hefte 18, Freiburg Geograph. Institut, pp. 119–32.

Egger, K. and **Martens, B.** (1987) Theory and methods of ecofarming and their realization in Rwanda, East Africa, in Glaeser, B. (ed.) *The Green Revolution Revisited. Critique and Alternatives*, Allen & Unwin, London and Boston, pp. 150–73.

Egger, K. and **Rottach, P.** (1986) Methoden des Ecofarming in Rwanda, in Rottach, P. (ed.) *Ökologischer Landbau in den Tropen*, 2nd edn, C. F. Müller, Karlsruhe, pp. 229–49.

El-Swaify, S. A. (1977) Susceptibilities of certain tropical soils to erosion by water, in Greenland, D. J. and Lal, R. (eds) *Soil Conservation and Management in the Tropics*, John Wiley & Sons, London, pp. 71–7.

El-Swaify, S. A., Moldenhauer, W. C. and **Lo, W.** (eds) (1985) *Soil Erosion and Conservation*, Soil Conservation Society of America, Ankeny, Iowa.

El-Swaify, S. A., Singh, S. and **Pathak, P.** (1987) Physical and conservation constraints and management components for SAT alfisols, in International Crop Research Institute for the Semiarid Tropics (ICRISAT), *Alfisols in the Semiarid Tropics*, Proceedings of Consultative Workshop, Dec. 1983, Patancheru, Andhra Pradesh, pp. 33–48.

Fairbridge, R. W. (ed.) (1968) *The Encyclopedia of Geomorphology*, Reinhold, New York.

Falagi, O. and **Lal, R.** (1979) Effect of aggregate size on mulching, erodibility, crusting, and crop emergence, in Lal, R. and Greenland, D. J. (eds) *Soil Physical Properties and Crop Production in the Tropics*, John Wiley & Sons, London, pp. 87–94.

Falesi, I. C. (1972) *Solos da Rodovia Transamazônica*, Boletim Tecnico No. 55, Instituto de Pesquisas Agropequárias do Norte, Belém.

Falesi, I. C. (1974) *O solo da Amazônia e sua relaçao com a definiçao de sistemas de produçao agricola*, Reunião do Grupo Interdisciplinar de Travalho sobre Diretrizes de Pesquisa Agricola para Amazonia, EMBRAPA/IICA, Documento 2, Brasilia.

FAO (1974) *Shifting Cultivation and Soil Conservation in Africa*, Papers presented at the FAO/SIDA/ARCN regional seminar, Ibadan 1973, Rome.

FAO (1986) *FAO Production Yearbook 1986*. Food and Agriculture Organization of the United Nations, Rome.

FAO/Unesco (1971) *Soil Map of the World 1 : 5 Million*, vol. IV, *South America*, Unesco, Paris.

FAO/Unesco (1977) *Soil Map of the World 1 : 5 Million*, vol. VI, *Africa*, Unesco, Paris.

FAO/Unesco (1978) *Soil Map of the World 1 : 5 Million, vol. VII, South East Asia*, Unesco, Paris.

FAO/Unesco (1982) *Potential Population Supporting Capacities of Lands in the Developing World*, Technical Report of Project 'Land Resources for Populations of the Future', Rome.

Fearnside, P. M. (1985) Deforestation and decision-making in the development of Brazilian Amazonia. *Interciencia*, **10**(5), 243–7.

Fearnside, P. M. (1986) *Human Carrying Capacity of the Brazilian Rainforest*, Columbia University Press, New York.

Fearnside, P. M. (1987) Rethinking continuous cultivation in Amazonia. The 'Yurimaguas Technology' may not provide the bountiful harvest predicted by its originators, *BioScience*, **37**, 209–14.

Fittkau, E. J. (1971) Ökologische Gliederung des Amazonasgebietes auf geochemischer Grundlage, *Münsterer Forschungen für Geologie und Paläontologie*, **20/21**, 35–50.

Fittkau, E. J., Illies, J., Klinge, H., Schwabe, G. H. and **Sioli, H.** (eds) (1968) *Biogeography and Ecology in South America*, W. Junk, The Hague.

Fittkau, E. J. and **Klinge, H.** (1973) On biomass and trophic structure of the Central Amazonian rain forest ecosystem, *Biotropica*, **5**, 2–14.

Fitzpatrick, E. A. (1971) *Pedology*, Oliver & Boyd, Edinburgh.

Flohn, H. (1968) Ein Klimaprofil durch die Sierra Nevada de Merida (Venezuela), *Wetter und Leben*, **20**, 181–191.

Frank, E. (1987) Delimitaciones al aumento poblacional y desarollo cultural en las culturas indígenas de la Amazonia antes de 1492, in Kohlhepp, G. and Schrader, A. (eds) *Hombre y Naturaleza en la Amazonía*, Tübinger Geographische Studien No. 95, pp. 109–23.

Freeman, J. D. (1955) *Iban Agriculture. A Report on the Shifting Cultivation of Hill Rice by the Iban of Sarawak*, Colonial Study No. 18, Her Majesty's Stationery Office, London.

Furch, K. (1976) Haupt- und Spurenmetallgehalte zentralamazonischer Gewässertypen. Neotropische Ökosysteme, *Biogeographica*, **VII**, 27–43.

Furch, K. (1984) Water chemistry of the Amazon basin: the distribution of chemical elements among freshwaters, in Sioli, H. (ed.) *The Amazon. Limnology and Landscape Ecology of a Mighty Tropical River and its Basin*, W. Junk, Dordrecht and Boston, pp. 167–99.

Ganssen, R. and **Hädrich, F.** (1965) *Atlas zur Bodenkunde*, Bibliographisches Institut, Mannheim.

Getahun, A., Wilson, G. F. and **Kang, B. T.** (1982) The role of trees in farming systems in the humid tropics, in MacDonald, L. H. (ed.) *Agro-forestry in the African Humid Tropics*, The United Nations University, Tokyo, pp. 28–35.

Giessner, K. (1985) Hydrologische Aspekte zur Sahel-Problematik, *Die Erde*, **116**, 137–57.

Glaeser, B. (ed.) (1984) *Ecodevelopment – Concepts, Projects, Strategies*, Pergamon, Oxford.

Glaeser, B. (ed.) (1987) *The Green Revolution Revisited. Critique and Alternatives*, Allen & Unwin, London and Boston.

Golley, F. B. and **Medina F. B.** (eds) (1983) *Tropical Rain Forest Ecosystems. Structure and Function*, Elsevier, Amsterdam.

Goodland, R. J. A. and **Irwin H. S.** (1975) *Amazon Jungle: Green Hell to Red Desert?* Elsevier, Amsterdam.

Goodland, R. J. A., Watson, C. and **Ledec G.** (1983) *Environmental Management in Tropical Agriculture*, Westview Press, Boulder.

Gourou, P. (1966) *The Tropical World. Its Social and Economic Conditions in its Future Status*, 4th edn (5th edn ed. 1980), Longman, London.

Grainger, A. (1983a) Improving the monitoring of deforestation in the humid tropics, in Sutton, S. L., Whitmore, F. C. and Chadwick, A. C. (eds) *Tropical Rain Forest: Ecology and Management*, Special Publication No. 2 of the British Ecological Society, Blackwell Scientific Publications, Oxford.

Grainger, A. (1983b) The state of the world's tropical forests, *The Ecologist*, **10**, 6–54.

Greenland, D. J. (1974) Evolution and development of different types of shifting cultivation, in *Shifting Cultivation and Soil Conservation in Africa*, Papers presented at the FAO/SIDA/ARCN Regional Seminar, Ibadan, Nigeria, 2–21 July 1973. FAO Soils Bulletin 24, pp. 5–13.

Greenland D. J. (1975) Bringing the Green Revolution to the shifting cultivator, *Science*, **190**, 841–4.

Greenland, D. J. (1981) *Characterization of Soils in Relation to their Classification and Management for Crop Production: Examples from Some Areas of the Humid Tropics*, Oxford University Press, New York.

Greenland, D. J. and **Lal, R.** (eds) (1977a) *Soil Conservation and Management in the Tropics*, John Wiley & Sons, London.

Greenland, D. J. and **Lal R.** (1977b) Soil erosion in the humid tropics: the need for action and the need for research, in Greenland, D. J. and Lal, R. (eds) *Soil Conservation and Management in the Tropics*, John Wiley & Sons, London, pp. 216–65.

Gregory, S. (1969) Rainfall reliability, in Thomas, M. F. and Whittington, G. W. (eds) *Environment and Land Use in Africa*, Methuen, London, pp. 57–82.

Grenzebach, K. (1984) Entwicklung kleinbäuerlicher Betriebsformen in Tropisch Afrika, *Geographische Rundschau*, **36**(7), 368–76.

Grigg, D. (1974) *The Agricultural Systems of the World. An Evolutionary Approach*, Cambridge University Press, Cambridge.

Grigg, D. (1979) Ester Boserup's theory of agrarian change, *Progress in Human Geography*, **3**(1), 64–84.

Grigg, D. (1984) The agricultural revolution in Western Europe, in Boyliss-Smith, T. B. and Wanmali, S. (eds) *Understanding Green Revolutions*, Cambridge University Press, pp. 1–17.

Grigg, D. (1985) *The World Food Problem, 1950–1980*, Oxford University Press; Oxford.

Hardjono, J. M. (1977) *Transmigration in Indonesia*, Oxford University Press, Kuala Lumpur and London.

Harley, J. L. and **Smith, S. E.** (1983) *Mycorrhizal Symbiosis*, Academic Press, London.

Harris, D. (1971) The ecology of swidden cultivation in the upper Orinoco rain forest, *The Geographical Review*, **61**(4), 475–95.

Harris, D. R. (ed.) (1980) *Human Ecology in Savanna Environment*, Academic Press, London.

Hastenrath, S. (1967) Rainfall distribution and regime in Central America, *Archiv Meteorologie, Geophysik, Bioklimatologie*, B **15**, 201–41.

Hastenrath, S. (1968) Zur Vertikalverteilung des Niederschlags in den Tropen, *Meteorologische Rundschau*, **4**, 113–16.

Hauck, F. W. (1974) *Introduction to Shifting Cultivation and Soil Conservation in Africa*, FAO Soils Bulletin No. 24, pp. 1–4, Rome.

Hay, R. L. and **Jones, B. F.** (1972) Weathering of basaltic tephra on the island of Hawaii, *Geol. Soc. Amer. Bull.* **83**, 317–32.

Hayman, D. S. (1980) Mycorrhiza and crop production, *Nature*, **287**, 487–8.

Hecht, S. B. (ed.) (1982a) *Amazonia: Agriculture and Land Use Research*, Centro Internacional de Agricultura Tropical (CIAT), Cali.

Hecht, S. B. (1982b) Agroforestry in the Amazon basin: practice, theory and limits to a promising land use, in Hecht, S. B. (ed.) *Amazonia: Agriculture and Land Use Research*, CIAT, Cali, pp. 331–71.

Hemming, J. (ed.) (1985) *Change in the Amazon Basin*, vol. 1, *Man's Impact on Forests and Rivers*; vol. 2, *The Frontier after a Decade of Colonization*, Manchester University Press, Manchester.

Henshall, J. and **Momsen, R. P.** (1974) *A Geography of Brazilian Development*, G. Bell & Sons, London.

Hernandez, D. and **Coutu, A. J.** (1983) Economic analysis of agronomic data, in Nicholaides III, J. J. *et al.* (eds) *Agronomic–Economic Research on Soils of the Tropics, 1980–81 Technical Report*, North Carolina, State University, Raleigh, NC, pp. 111–18.

Herrera, R., Merida, T., Stark, N. and **Jordan, C. F.** (1978) Direct phosphorus transfer from leaf litter to roots, *Naturwissenschaften*, **65**, 208–9.

Higgins, G. M., Kassam, A. H. and **Naiken, L.** (eds) (1982) *Potential Population Supporting Capacities of Land in the Developing World*, Report of Project INT/75/P13, FAO, Rome.

Hodder, B. W. (1980) *Economic Development in the Tropics*, 3rd edn, Methuen, London.

Homma, A. K. (1976) *Programaçao das Atividades Agropecuarias sob condiçoes de risco nos lotes de nucleo de colonizaçao de Altamira*, Universidade Federal de Viçosa, Minas Gerais.

Homma, A. K., Viegas, R., Graham, J., Lemos, J. and **Mendeslopes, J. dos.** (1978) *Identificaçao de sistemas de produçao nos lotes de nucleo de colonizaçao de Altamira, Para*, Embrapa-Cpatu, Belem.

Huxley, P. A. (1985) The tree crop interface – or simplifying the biological environmental study of mixed cropping agroforestry systems, *Agroforestry Systems*, **3**(4), 251–66.

IITA (International Institute of Tropical Agriculture) (1973) *IITA 1973 Report. Farming Systems Program*, IITA, Ibadan.

IITA (1986) *Resource and Crop Management Program at IITA*, IITA, Nigeria.

INCOLD, (International Commission on Large Dams) (1973) *World Register of Dams*, Unesco, Paris. 2nd updating, Paris 1977.

India, Ministry of Irrigation and Power (1957) *Hirakud Projects*, Glasgow Printing Co, New Delhi.

Instituto Brasileiro de Geografia e Estadistica (1975) *Atlas de Rondonia*, Instituto Brasileiro de Geografia e Estadistica, Rio de Janeiro.

IRRI (International Rice Research Institute) (1978) *Soils and Rice*, IRRI, Los Baños, Laguna, the Philippines.

IRRI (1980) *Soil-related Constraints to Food Production in the Tropics*, Los Baños, Laguna, the Philippines.

International Wheat Council (1980–81) *Review of the World Wheat Situation*, London.

Irion, G. (1976) Mineralogisch–geochemische Untersuchungen in der pellitischen Fraktion amazonischer Oberböden und Sedimente. Neotropische Ökosysteme, *Biogeographica*, **7**, 7–25.

Ishikawa, S. (1967) *Economic Development in Asian Perspective*, Kinokuniya Bookstore, Tokyo.

Janke, B. (1972) *Naturpotential und Landnutzung im Nigertal bei Namey/Republic Niger*, Jahrbuch der Geographischen Gesellschaft Hannover, 189pp.

Janos, D. P. (1983) Tropical mycorrhizas, nutrient cycles and plant growth, in Sutton, S. L., Whitmore, T. C. and Chadwick, A. C. (eds) *Tropical Rain Forest:*

Ecology and Management, Special Publication No. 2 of the British Ecological Society, Blackwell Scientific Publications, Oxford, pp. 327–45.

Jenny, H., Gessel, S. P. and **Bingham, F. T.** (1949) Comparative study of decomposition rates of organic matter in temperate and tropical regions, *Soil Science*, **68**, 419–32.

Johnson, D. V. and **Nair, P. K.** (1985) Perennial crop-based agroforestry systems in Northeast Brazil, *Agroforestry Systems*, **2**(4), 281–92.

Jordan, C. F. (1982) The nutrient balance of an Amazonian rain forest, *Ecology*, **63**, 647–54.

Juo, A. S. R. (1981) Chemical characteristics, in Greenland, D. J. (ed.) *Characterization of Soils in Relation to their Classification and Management for Crop Production: Examples from Some Areas of the Humid Tropics*, Oxford University Press, New York, pp. 51–79.

Juo, A. and **Lal, R.** (1977) The effect of fallow and continuous cultivation on the chemical and physical properties of an alfisol, *Plant and Soil*, **47**, 567–84.

Kassam, A. H. (1979) *Multiple Cropping and Rainfed Crop Productivity in Africa*, Consultant's Working Paper No. 5, FAO/UNFPA Project INT/75/P13, FAO, Rome.

Katzer, F. (1897) *Das Wasser des unteren Amazonas*, Sitzungsberichte der Böhmischen Gesellschaft.

Kawaguchi, K. and **Kyuma, K.** (1977) *Paddy Soils in Tropical Asia. Their Material, Nature, and Fertility*, University Press of Hawaii, Honolulu.

Khomvilai, S. and **Blue, W. G.** (1976) Effects of lime and potassium sources on the retention of potassium by some Florida mineral soils, *Soil and Crop Science Society of Florida, Proceedings*, **36**, 84–9.

King, K. F. S. (1968) *Agro-silviculture (the Taungya-System) Bulletin*, Department of Forestry, University of Jbadan, Nigeria.

King, K. F. S. and **Chandler, M. T.** (1978) *The Wasted Lands – the Programme of Work of the International Council for Research in Agroforestry*, ICRAF, Nairobi.

Kio, P. R. O. (1982) Forest conservation strategies for tropical Africa, in MacDonald, L. H. (ed.) *Agro-forestry in the African Humid Tropics*, The United Nations University, Tokyo, pp. 36–40.

Klatt, A. R. (1985) Introduction to the Symposium, in CIMMYT *Wheat for More Tropical Environments*, 24–28 Sept. 1984, Mexico, pp. 21–3.

Kleinpenning, J. M. G. (1979) *An Evaluation of the Brazilian Policy for the Integration of the Amazon Basin (1964–1976)*, Publicatie 9, Geografisch en Planologisch Instituut, Nijmvegen, Netherlands.

Kleinpenning, J. M. G. and **Volbeda, S.** (1985) Recent changes in population size and distribution in the Amazon region of Brazil, in Hemming, J. (ed.) *Change in the Amazon Basin*, vol. 2: *The Frontier after a Decade of Colonization*, Manchester University Press, Manchester, pp. 6–36.

Klinge, H. (1972) Biomasa y materia orgánica del suelo en el ecosistema de la pluviselva Centro-Amazónica, in IV. Congreso Latino-Americano de la Ciencia del Suelo, Table 6, Maracay, Venezuela.

Klinge, H. (1973a) Root mass estimation in lowland tropical rain forests of Central Amazonia, Brazil. I. Fine root masses of a pale yellow latosol and a giant humus podzol, *Tropical Ecology*, **14**(11), 29–38.

Klinge, H. (1973b) Root mass estimation . . . II. 'Coarse root mass' of trees and palms in different height classes, *Annales Academia Brasilera de Ciencias*, **45**(3/4), 595–609.

Klinge, H. (1976a) Nährstoffe, Wasser und Durchwurzelung von Podsolen und Latosolen unter tropischem Regenwald bei Manaus/Amazonien. Neotropische Ökosysteme. *Biogeographica*, **7**, 45–58.

Klinge, H. (1976b) Bilanzierung von Hauptnährstoffen im Ökosystem tropischer Regenwald (Manaus). Neotropische Ökosysteme. *Biogeographica*, **7**, 59–76.

Klinge, H. (1977) Preliminary data on nutrient release from decomposing leaf litter in a neotropical rain forest, *Amazoniana*, **6**, 193–202.

Klinge, H. and **Ohle, W.** (1964) Chemical properties of rivers in the Amazonian area in relation to soil conditions, *Verh. Internationaler Verein für Limnologie*, **XV**, 1067–76.

Klinge, H. and **Rodrigues, W. A.** (1972) Biomass estimation in a central Amazonian rainforest, *Acta Scientifica*, **24**, 225–37.

Kohlhepp, G. (1976) Planung und heutige Situation staatlicher kleinbäuerlicher Kolonisationsprojekte an der Transamazônica, *Geographische Zeitschift*, **64**, 171–211.

Kohlhepp, G. (1979) Brasiliens problematische Antithese zur Agrarreform: Agrarkolonisation in Amazonien, in Elsenhans, H. (ed.) *Agrarreform in der Dritten Welt*, Campus, Frankfurt-on-Main, pp. 471–504.

Kohlhepp, G. (1985) Agrarkolonisationsprojekte in tropischen Regenwäldern. Amazonien als Beispiel und Warnung, *Entwicklung und ländlicher Raum*, **19**, 13–18.

Kohlhepp, G. (1989) A challenge to science and regional development policy – reflections on the future development of Amazonia, *Applied Geography and Development*, **33**, 52–67, Institute for Scientific Cooperation, Tübingen.

Kohlhepp, G. and **Schrader, A.** (eds) (1987) *Homem e Natureza na Amazonia. Hombre y Naturaleza en la Amazonia*, Simposio Internacional e Interdisciplinar, Blaubeuren 1986, Tübinger Geographische Studien No. 95, Geographisches Institut, Tübingen.

Kotschi, J. (1986) Ökologischer Landbau als ein Instrument landwirtschaftlicher Entwicklung, in Rottach, P. (ed.) *Ökologischer Landbau in den Tropen: Ecofarming in Theorie und Praxis*, Muller, Karlsruhe, pp. 95–106.

Kowal, J. M. and **Kassam, A. H.** (1977) Energy load and instantaneous intensity of rainstorms at Samaru, Northern Nigeria, in Greenland, D. J. and Lal, R. (eds) *Soil Conservation and Management in the Tropics*, John Wiley & Sons, London, pp. 57–70.

Kunstadter, P., Chapman, E. C. and **Sabhasri, S.** (eds) (1978) *Farmers in the Forest. Economic Development and Marginal Agriculture in Northern Thailand*, The University Press of Hawaii, Honolulu.

Laatsch, W. (1957) *Dynamik der Mitteleuropäischen Mineralböden*, Theodor Steinkopf Verlag, Leipzig.

Lagemann, J. (1977) *Traditional African Farming Systems in Eastern Nigeria. An Analysis of Reaction to Increasing Population Pressure*, Weltforum Verlag, Munich.

Lagemann, J., Flinn, J. C. and **Ruthenberg, H.** (1976) Land use, soil fertility and agricultural productivity as influenced by population density in Eastern Nigeria, *Zeitschrift für ausländische Landwirtschaft*, **15**, 206–19.

Lal, R. (1976a) *Soil Erosion Problems on an Alfisol in Western Nigeria and their Control*, IITA Monograph No. 1, Ibadan, 126pp.

Lal, R. (1976b) No-tillage effects on soil properties under different crops in Western Nigeria, *Soil Sci. Journal, Society of America*, **40**, 762–8.

Lal, R. (1977a) Analysis of factors affecting rainfall erosivity and soil erodibility, in Greenland, D. J. and Lal, R. (eds.) *Soil Conservation and Management in the Tropics*, John Wiley & Sons, London, pp. 49–56.

Lal, R. (1977b) The soil and water conservation problem in Africa: ecological differences and management problems, in Greenland, D. J. and Lal, R. (eds.) *Soil Conservation and Management in the Tropics*, John Wiley & Sons, London, pp. 143–9.

Lal, R. (1979a) Physical characteristics of soils of the tropics: determination and management, in Lal, R. and Greenland, D. J. (eds) *Soil Physical Properties and Crop Production in the Tropics*, John Wiley & Sons, London, pp. 7–44.

Lal, R. (1979b) Modification of soil fertility characteristics by management of soil physical properties, in Lal, R. and Greenland, D. J. (eds) *Soil Physical Properties and Crop Production in the Tropics*, John Wiley & Sons, London, pp. 397–405.

Lal, R. (1979c) The role of physical properties in maintaining productivity of soils in the tropics, in Lal, R. and Greenland, D. J. (eds) *Soil Physical Properties and Crop Production in the Tropics*, John Wiley & Sons, London, pp. 3–6.

Lal, R. (1980) Physical and mechanical characteristics of Alfisols and Ultisols with particular reference to soils in the tropics, in Theng, B. K. G. (ed.) *Soils with Variable Charge*, Department of Scientific and Industrial Research, Lower Mettle, New Zealand, pp. 253–74.

Lal, R. and **Greenland, D. J.** (eds) (1979) *Soil Physical Properties and Crop Production in the Tropics*, John Wiley & Sons, London.

Landauer, K. and **Brazil, M.** (eds) (1990) *Tropical Home Gardens*. Selected papers from an international workshop held at the Institute of Ecology, Padjadjaran University, Dec. 1985, Bandung, Indonesia. United Nations University, Tokyo.

Lanly, J. P. (1982) *Tropical Forest Resources*. FAO Forestry Paper 30, Rome.

Lanly, J. P., Singh, Karn Deo and **Janz, Klaŭs** (1991): FAO's 1990 reassessment of tropical forest cover, *Nature and Resources*, **27**, (2) 21–6.

Ledger, D. C. (1969) The dry season flow characteristics of West African rivers, in Thomas, M. F. and Whittington, G. W. (eds) *Environment and Land Use in Africa*, Methuen, London, pp. 83–102.

Leopold, L. B. (1951) Hawaiian climate. Its relation to human and plant geography, *Meteor. Monogr. Am. Meteor. Soc.*, **1**, 1–6.

Lieth, H. (1978) Biological productivity of tropical lands, *Unasylva*, **28**, 24–31.

Linnemann, H., de Hoogh, J., Keyzer, M. A. and **van Heemst, H. D. J.** (1979) *Model of International Relations in Agriculture (MOIRA)*, Report of the Project Group, Food for a Doubling World Population, Elsevier, Amsterdam.

Lockwood, J. G. (1976) *World Climatology. An Environmental Approach*, E. Arnold, London.

Louis, H. (1964) Über Rumpfflächen- und Talbildung in den wechselfeuchten Tropen, besonders nach Studien in Tanganyika, *Zeitschrift für Geomorphologie*. New series, special issue, 43–70.

Lundgren, B. (1978) *Agroforestry in West Africa: An Appraisal of some IDRC-supported Research Projects in Ghana and Nigeria*, Swedforst Consulting AB, Solna, Sweden.

Lundgren, B. (1979) Research strategy for soils in agroforestry, in Mongi, H. O. and Huxley, P. A. (eds) *Soil Research in Agroforestry*. Proceedings of an Expert Consultation held at ICRAF, Nairobi, March 1979, pp. 523–35.

Lundgren, B. (1985) Global deforestation. Its causes and suggested remedies, *Agroforestry Systems*, **3**, 91–5.

Mabberly, D. J. (1983) *Tropical Rain Forest Ecology*, Blackie, Glasgow.

McCants, C. B. (ed.) (1985) *TropSoils*, Triennial Technical Report 1981–1984, Soil Science Department, North Carolina State University, Raleigh, NC.

McCollum, R. E. (1985) Continuous cropping: conventional tillage (Yurimaguas Experimental Station, Peru), *Tropsoils* **Y-104**, 2–8.

McCollum, R. E. and **Pleasant, J. M.** (1985) Weed control for low input systems, in McCants, Ch. B. (ed.) *TROPSOILS*, Triennial Technical Report 1981–1984, Soil Science Department, North Carolina State University, Raleigh, NC, pp. 163–4.

McConnell, D. J. and **Dharmapala, K. A. E.** (1978) *The Forest Garden Farms of Kandy FAO Agricultural Diversification Project*, Farm Management Report No. 7, Peradeniya, Sri Lanka.

MacDonald, L. H. (ed.) (1982) *Agro-forestry in the African Humid Tropics*, Proceedings of a workshop held in Ibadan, Nigeria, 27 Apr. – 1 May 1981, The United Nations University, Tokyo.

McDowell, L. R., Conrad, J. H., Ellis, G. L. and **Loosli, J. K.** (1983) *Minerals for Grazing Ruminants in Tropical Regions*, University of Florida, Dept. of Animal Science, Center for Tropical Agriculture, Gainesville, Fl.

McGrath, D. G. (1987) The role of biomass in shifting cultivation, *Human Ecology*, **15**(2), 221–42.

McVey, R. T. (ed.) (1963) *Indonesia*, Yale University Press, New Haven.

Mahar, D. (1988) *Government Policies and Deforestation in Brazil's Amazon Region*, World Bank Publications Department, Washington.

Mann, C. E. (1985) Selecting and introducing wheats for the environments of the tropics, in CIMMYT, *Wheats for More Tropical Environments*, Proceedings of the International Symposium, 24–28 Sept. 1984, Mexico, DF, pp. 24–33.

Manner, H. I. (1981) Ecological succession in new and old swiddens in montane Papue New Guinea, *Human Ecology*, **9**(3), 359–77.

Manshard, W. (1974) *Tropical Agriculture. A Geographical Introduction and Appraisal*, Longman, London.

Martinez, J. J. (1985) Selection and introduction of wheat types for subtropical conditions in Mexico, in CIMMYT, *Wheats for More Tropical Environments*, Proceedings of the International Symposium, 24–28 Sept. 1984, Mexico, DF, pp. 54–6.

May, P., Anderson, A., Frazao, J. and **Balick, M.** (1985) Babassu palm in the agroforestry systems in Brazil's mid-north region, *Agroforestry Systems*, **3**(3), 275–96.

Meggers, B. J. (1971) *Amazonia: Man and Culture in a Counterfeit Paradise*, Aldine Publishers, Chicago.

Melin, E. and **Nilsson, H.** (1954) Transport of labelled phosphorus to pine seedlings through the mycelium of *Cortinarius glaucopus*, *Svensk Botanic Tidskrift*, **48**, 555–8.

Melin, E. and **Nilsson, H.** (1955) Ca45 used as an indicator of transport of cations to pine seedlings by means of mycorrhizal mycelium, *Svensk Botanic Tidskrift*, **49**, 119–22.

Melin, E. and **Nilsson, H.** (1957) Transport of C14-labelled photosynthate to the fungal associate of pine mycorrhiza, *Svensk Botanic Tidskrift*, **51**, 166–86.

Mensching, H. (1986) *Die Sahelzone. Naturpotential und Probleme seiner Nutzung*, Aulis Verlag Deubner, Cologne.

Mensching, H. and **Ibrahim, F.** (1977) The problem of desertification in and around arid lands, *Applied Science and Development*, **10**, 7–43, Tübingen.

Mikola, P. (ed.) (1980) *Tropical Mycorrhiza Research*, Clarendon Press, Oxford.

Millot, G. (1970) *Geology of Clays. Weathering, Sedimentology, Geochemistry*, Chapman & Hall, London.

Millot, G. (1979) Clay, *Scientific American*, **240**, 109–18.

Ministry of Irrigation and Power (1957) *The Hirakud Dam*, New Delhi.

Mohr, E. C. J. (1930) *Tropical Soil Formation Processes and the Development of Tropical Soils. With Special Reference to Java and Sumatra* (Translated by Rob E. Pendleton), College of Agriculture. University of the Philippines, Laguna.

Mohr, E. C. J. (1938/44) *The Soils of the Equatorial Regions with Special Reference to the Netherlands East Indies*, Amsterdam 1938, Translated by R. L. Pendleton, Edward Busther-Inc, Lithoprint Ann Arbor, Michigan.

Mohr, E. C. J., van Baren, F. A and **van Schuylenborgh, J.** (1972) *Tropical Soils. A Comprehensive Study of their Genesis*, 3rd edn, Mouton/Ichtiar B. van Hoeve, The Hague.

Mongi, H. O. and **Huxley, P. A.** (eds) (1979) *Soils Research in Agro-forestry*, Proceedings of an Expert Consultation, International Council for Research in Agroforestry (ICRAF), Nairobi.

Monteith, J. L. (1979) Soil temperature and crop growth in the tropics, in Lal, R. and Greenland, D. (eds) *Soil Physical Properties and Crop Production in the Tropics*, John Wiley & Sons, London, pp. 249–62.

Moormann, F. R. (1981) Representative toposequences of soils in southern Nigeria and their pedology, in Greenland, D. J. (ed.) *Characterization of Soils in Relation to their Classification and Management For Crop Production: Examples From Some Areas of the Humid Tropics*, Oxford University Press, New York, pp. 10–29.

Moran, E. F. (1975) *Pioneer farmers of the Transamazon Highway: Adaptation and agricultural production in the lowland tropics*, PhD dissertation, University of Florida, Gainesville.

Moran, E. F. (1976) *Agricultural development in the Transamazon Highway*, Latin American Studies Working Papers, Indiana University, Bloomington.

Moran, E. F. (1977) Estrategias de sobrevivencia: O uso de recursos ao longo da Rodovia Transamazonica, *Acta Amazonica*, **7**(3), 363–79.

Moran, E. F. (1981) *Developing the Amazon*, Indiana University Press, Bloomington.

Moran, E. F. (1983a) Government-directed settlement in the 1970s: an assessment of Transamazon Highway colonization, in Moran, E. (ed.) *The Dilemma of Amazonian Development*, Westview Press, Boulder, pp. 297–317.

Moran, E. F. (1983b) Growth without development: Past and present development efforts in Amazonia, in Moran, E. (ed.) *The Dilemma of Amazonian Development*, Westview Press, Boulder.

Moran, E. (ed.) (1983c) *The Dilemma of Amazonian Development*, Westview Press, Boulder.

Morgan, W. B. (1969a) Peasant agriculture in tropical Africa, in Thomas, M. F. and Whittington, G. W. (eds). *Environment and Land Use in Africa*, Methuen, London, pp. 241–72.

Morgan, W. B. (1969b) The zoning of land use around rural settlements in tropical Africa, in Thomas, M. F. and Whittington, G. W. (eds) *Environment and Land Use in Africa*, Methuen, London, pp. 301–19.

Morgan, W. B. and **Pugh, J. C.** (1969) *West Africa*, Methuen, London.

Mougeot, L. J. A. (1985) Alternative migration targets and Amazonia's closing frontier, in Hemming, J. (ed.) *Change in the Amazon Basin*, vol. 2: *The Frontier after a Decade of Colonization*, Manchester University Press, Manchester, pp. 51–90.

Muller, C. (1980) Frontier-based agricultural expansion: the case of Rondonia, in Barbira-Scazzocchio, B. (ed.) *Land, People and Planning in Contemporary Amazonia*, pp. 141–51.

Nadjombe, O. (1982) Agrosilvicultural practices in Togo, in MacDonald, L. H. (ed.) *Agro-forestry in the African Humid Tropics*, The United Nations University, Tokyo, pp. 78–80.

Nair, P. K. (1985) Classification of agroforestry systems, *Agroforestry Systems*, **3**(2), 83–90.

National Research Council (1982) *Ecological Aspects of Development in the Humid Tropics*, National Academy Press, Washington, DC.

Netting, R. M. (1968) *Hillfarmers of Nigeria, Cultural Ecology of the Kofyar of the Jos Plateau*, University of Washington Press, Seattle.

Nicholaides, J. J. III, Bandy, D. E., Sanchez, P. A., Benites, R. J., Villachica, J. H., Couto A. J. and **Valverde, C. S.** (1985a) Agricultural alternatives for the Amazon Basin, *Bio Science*, **35**(5), 279–85.

Nicholaides, J. J. III, Couto, W. and **Wade, M. K.** (eds) (1983a) *Agronomic–Economic Research on Soils of the Tropics*, Technical Report 1980–1981, Soil Science Department, North Carolina State University, Raleigh, NC.

Nicholaides J. J. III, Piha, M. J. and **Vilavicencio, M.** (1985b) Minimum tillage × residue management × potassium rates with Al-tolerant cultivars, in McCants, Ch. B. (ed.) *TROPSOILS*, Triennal Technical Report 1981–1984, Soil Science Department, N. Carolina State University, Raleigh, NC, pp. 153–5.

Nicholaides J. J. III, Sanchez, P. A., Bandy, D. E., Villachica, J. H., Coutu, A. J. and **Valverde, C. S.**, (1983b) Crop production systems in the Amazon Basin, in Moran, E. F. (ed.) *The Dilemma of Amazonian Development*, Westview Press, Boulder, pp. 101–54.

Nieuwolt, S. (1977) *Tropical Climatology: An Introduction to the Climates of the Low Latitudes*, John Wiley & Sons, London.

Norman, D. W., Simmons, E. B. and **Hays, H. M.** (1982) *Farming Systems in the Nigerian Savanna. Research and Strategies for Development*, Westview Press, Boulder.

Norman, M. J. T. (1979) *Annual Cropping Systems in the Tropics. An Introduction*, University Presses of Florida, Gainesville, Fl.

Nye, P. H. and **Greenland, D. J.** (1960) *The Soil under Shifting Cultivation*, Commonwealth Bureau of Soils, Technical Communication No. 51, Farnham Royal, Bucks.

Odum, H. T. (1970) *A Tropical Rain Forest*, US Atomic Energy Commission, Washington, DC.

Ofori, C. S. (1974) Shifting cultivation – reasons underlying its practice, in *Shifting Cultivation and Soil Conservation in Africa*, FAO Soils Bulletin No. 24, Rome, pp. 14–20.

Okigbo, B. N. (1979) Farming systems and soil erosion in West Africa, in Greenland D. J. and Lal, R. (eds) *Soil Conservation and Management in the Tropics*, John Wiley & Sons, London, pp. 151–63.

Okigbo, B. N. and **Lal, R.** (1979) Soil fertility maintenance and conservation for improved agroforestry systems in the lowland humid tropics, in Mongi, H. O. and Huxley, P. A. (eds) *Soil Research in Agroforestry*, Proceedings of an Expert Consultation, ICRAF, Nairobi, pp. 41–78.

Olson, I. S. (1963) Energy storage and balance of producers and decomposers in ecological systems, *Ecology*, **44**, 322–31.

Owino, F. (1982) Agroforestry developments in Kenya: prospects and problems, in MacDonald, L. H. (ed.) *Agro-forestry in the African Humid Tropics*, The United Nations University, Tokyo, pp. 137–9.

Padoch, C., Chota Inuma, J., de Jong, W. and **Unruh, J.** (1985) Amazonian agroforestry: a market oriented system in Peru, *Agroforestry Systems*, **3**(1) 47–58.

Palmer, I. (1976) *The new rice in Indonesia*, UNRISD, Genoa.

Paramananthan, S. (1978) Rice soils of Malaysia, in IRRI, *Soils and Rice*, Los Baños, Laguna, the Philippines, pp. 87–98.

Parsons, J. J. (1980) Europeanization of the savanna lands of northern South America, in Harris, D. (ed.) *Human Ecology in Savanna Environments*, Academic Press, London, pp. 267–92.

Pathak, P., Miranda, S. M. and **El-Swaify, S. A.** (1985) Improved rainfed farming for the semiarid tropics: implications for soil and water conservation, in El-Swaify, S. A. Moldenhauer, W. C. and Lo, A. (eds) *Soil Erosion and Conservation*, Soil Conservation Society of America, Ankeny, Iowa, pp. 338–54.

Patrick, W. H. Jr. and **Reddy, C. N.** (1978) Chemical changes in rice soils, *Soils and Rice*, pp. 361–76.

Pelzer, K. J. (1948) *Pioneer Settlements in the Asiatic Tropics*, American Geographical Society Special Publication No. 29, New York.

Pelzer, K. J. (1964) Physical and human resource patterns, in McVey, R. T. (ed.) *Indonesia*, Yale University Press, New Haven, pp. 1–23.

Pelzer, K. J. (1978) Swidden cultivation in Southeast Asia: Historical, ecological and economic perspectives, in Kunstadter, P. *et al.* (eds) *Farmers in the Forest. Economic Development and Marginal Agriculture in Northern Thailand*, The University Press of Hawaii, Honolulu, pp. 271–86.

Penck, A. (1924) Das Hauptproblem der physischen Anthropogeographie. *Sitzungs-bericht der Preuss. Ak. d. Wiss., Phys. Math. Kl.* **24**, 249–57. Reprinted in Wirth, E. (ed.) *Wirtschaftsgeographie*, Darmstadt, 1969, pp. 157–80.

Pendleton, R. L. (1930) *Tropical Soil Forming Processes and the Development of Tropical Soils, with Special Reference to Java and Sumatra*. College of Agriculture, University of the Philippines, Laguna.

Pimentel, D., Hurd, L. E., Bellotti, A. C., Forster, M. J., Oka, J. N., Sholes, O. D. and **Whitman, R. J.** (1973) Food production and the energy crises, *Science*, **182**, 443–5.

Popenoe, H. (1957) The influence of shifting cultivation cycle on soil properties in Central America, *Proceedings of the 9th Pacific Science Congress, San Francisco*, vol. 7, pp. 72–7.

Popenoe, H. (1960) Some soil cation relationships in an area of shifting cultivation in the humid tropics, *7th Internat. Congress of Soil Science*, vol. 2, pp. 303–11.

Posey, D. A. (1985) Native and indigenous guidelines for new Amazonian development strategies: understanding biological diversity through ethnoecology, in Hemming, J. (ed.) *Change in the Amazon Basin*. vol. 1, Manchester University Press, Manchester, pp. 156–81.

Pullan, R. A. (1969) The soil resources of West Africa, in Thomas, M. F. and Whittington, G. W., (eds) *Environment and Land Use in Africa*, Methuen, London, pp. 147–91.

Raintree, J. B. and **Warner, K.** (1986) Agroforestry pathways for the intensification of shifting cultivation, *Agroforestry Systems*, **4**, 39–54.

Ramdas, L. A. and **Dravid, R. K.** (1936) Soil temperatures in relation to other factors controlling the disposal of solar radiation at the earth's surface, *Proceedings of the National Institute of Science India*, vol. II, pp. 131–43.

Rappaport, R. (1967) *Pigs for the Ancestors. Ritual in the Ecology of a New Guinea People*, Yale University Press, New Haven.

Redhead, J. F. (1980) Mycorrhiza in natural tropical forests, in Mikola, P. (ed.) *Tropical Mycorrhiza Research*, Clarendon Press, Oxford, pp. 127–42.

Redhead, J. F. (1982) Ectomycorrhizae in the tropics, in Dommergues, Y. R. and Diem, H. G. (eds) *Microbiology and Tropical Soils and Plant Productivity*, Clarendon Press, Oxford, pp. 253–70.

Reichholf, J. H. (1990) Der Tropische Regenwald. Die Ökobiologie des artenreichsten Naturraums der Erde, Deutscher Taschenbuch Verlag, München.

Revelle, R. *et al.* (1967) *The World Food Problem*, A Report of the President's Advisory Committee, vol. 2, The White House, Washington.

Revelle, R. (1976) The resources available for agriculture, *Scientific American*, **235**, 165–78.

Richards, P. W. (1952) *The Tropical Rain Forest*, Cambridge University Press, London.

Richey, J., Salati, E. and **dos Santos, U.** (1985) The biogeochemistry of the Amazon River: an update, *Mitteilungen des Geol. Pal. Inst. der Univ. Hamburg*, **5**, 245–57.

Rodin, L. E. and **Basilevič N. J.** (1968) World distribution of plant biomass. Functioning of terrestrial ecosystems at the primary productivity level, *Proceedings of the Copenhagen Symp. (Unesco)*, pp. 45–50.

Rottach, P. (ed.) (1986) *Ökologischer Landbau in den Tropen: Ecofarming in Theorie und Praxis*, Muller, Karlsruhe.

Rouvé, G. (1965) Ein Überblick über die Wasserwirtschaft Indiens, *Die Wasserwirtschaft*, **55**, 396–404.

Ruddle, K. and **Manshard, W.** (1981) *Renewable Natural Resources and the Environment. Pressing Problems in the Developing World*, Natural Resources and Environment Series, United Nations University, vol. 2, Tycooly Int. Publishing, Dublin, 396pp.

Ruthenberg, H. (1971, 1980) *Farming Systems in the Tropics* (3rd edn 1980), Clarendon Press, Oxford.

Ruttan, V. W. (1977) The Green Revolution: seven generalizations, *Int. Development Rev.*, **4**(1), 15–23.

Sanchez, P. A. (1976) *Properties and Management of Soils in the Tropics*, John Wiley & Sons, New York and London.

Sanchez, P. A. (1983) Nutrient dynamics following rainforest clearing and cultivation, *Proceedings of the International Workshop on Soils*, Townsville, Queensland, Australia, pp. 12–16.

Sanchez, P. A. (1985) Fertilizers make continuous cropping possible in the Amazon, *Better Crops International*, **1**(1), 12–15.

Sanchez, P. A., Ara, M. A., Schaus, R., Reategui, K., Ayarza, M. and **Bextley, R.** (1985a) Grass–legume mixtures under grazing, in McCants, Ch.B. (ed.) *TROPSOILS*, Triennial Technical Report 1981–1984, Soil Science Department, North Carolina State University, Raleigh, NC, pp. 169–72.

Sanchez, P. A., Bandy, D. E., Ara, A. M., Schaus, R., Reategui, K. Toledo, J., Sanchez, C. and **Ivazeta, H.** (1985b) Pasture germplasm adaption to acid soils of the humid tropics with minimum input, in Mc Cants, Ch. B. (ed.) *TROPSOILS*, Triennial Technical Report 1981–1984, Soil Science Department, North Carolina State University, Raleigh, NC, pp. 165–8.

Sanchez, P. A., Bandy, D. E., Villachica, J. H. and **Nicholaides, J. J. III** (1981) Continuous cultivation of annual crops, *Agronomic–Economic Research on Soils of the Tropics*, 1980–1981 Technical Report, Soil Science Department, North Carolina State University, Raleigh, NC, pp. 11–15.

Sanchez, P. A. Bandy, D. E., Villachica, J. H. and **Nicholaides J. J. III** (1982) Amazon basin soils: management for continuous crop production, *Science*, **216**, 821–7.

Sanchez, P. A. and **Buol S. W.** (1975) Soils of the tropics and the world food crisis, *Science*, **188** 598–603.

Sanchez, P. A. and **Cochrane, T. T.** (1980) Soil constraints in relation to major farming systems in tropical America, in IRRI, *Priorities for Alleviating Soil-related Constraints to Food Production in the Tropics*, Los Baños, Laguna, the Philippines.

Sanchez, P. A. and **Nicholaides, J. J. III** (1985) Humid tropics program Peru, Brazil. An overview, in McCants, Ch.B. (ed.) *TROPSOILS*, Triennial Technical Report 1981–1984, Soil Science Department, North Carolina State University, Raleigh, NC, pp. 89–104.

Sanders, E. R. (1972) *Quality of the Surface Waters of the U.S.*, US Geological Survey, Water Supply Paper, Washington, DC.

Sapper, K. (1939) *Die Ernährungswirtschaft der Erde und ihre Zukunftsaussichten für die Menschheit*, Ferdinand Enke Verlag, Stuttgart.

Scheffer, F. and **Schachtschabel, P.** (1976) *Lehrbuch der Bodenkunde*, Ferdinand Enke Verlag, Stuttgart.

Schmidt, E. and **Mattingly, P.** (1966) Das Bevölkerungsbild Afrikas um das Jahr 1960, *Geographische Rundschau*, **18**, 447–57.

Schmidt, G. W. (1972a) Chemical properties of some waters in the tropical rainforest region of Central Amazonia along the new road Manaus-Itacoatiara, *Amazonia*, **3** (2).

Schmidt, G. W. (1972b) Amounts of suspended solids and dissolved substances in the middle reaches of the Amazon over the course of one year, *Amazonia*, **3** (2).

Schmidt, R. D. (1952) Die Niederschlagsverteilung im andinen Kolumbien, in Troll, C. (ed.) *Studien zur Klima- und Vegetationskunde der Tropen*, Bonner Geographische Abhandlungen 9, pp. 99–119.

Schmidt-Lorenz, R. (1971) *Böden der Tropen und Subtropen. Handbuch der Landwirtschaft und Ernährung in den Entwicklungsländern*, vol.2, Eugen Ulmer, Stuttgart, pp. 44ff.

Schmithüsen, J. (1976) *Atlas zur Biogeographie*. Bibliogr. Institut, Mannheim, Vienna and Zurich.

Scholz, U. (1980) Land reserves in Southern Sumatra/Indonesia and their potentialities for agricultural utilization, *GeoJournal*, **4**(1), 19–30.

Scholz, U. (1984) Ist die Agrarproduktion der Tropen ökologisch benachteiligt? *Geographische Rundschau*, **36**(7), 360–6.

Scholz, U. (1986) Deforestation in the Asian Tropics – causes and consequences. Asien, *Deutsche Zeitschrift für Politik, Wirtschaft und Kultur*, DGA, No. 21, pp. 1–29.

Seavoy, R. E. (1973) The shading cycle in shifting cultivation, *Annals Assoc. Amer. Geographers*, **63**(4), 522–8.

Seubert, C. F., Sanchez, P. A. and **Valverde, C.** (1977) Effects of land clearing methods on soil properties and crop performance in an Ultisol of the Amazon jungle of Peru, *Tropical Agriculture Trimester*, **54**, 307–22.

Sherman, G. D. (1952) The genesis and morphology of alumina-rich laterite clays, in *Problems of clay and laterite genesis*, American Institute of Mining and Metallurgical Engineers, New York, pp. 154–61.

Singer, R. and **Araujo Aguiar, I.** (1986) Litter decomposing and ectomycorrhizal basidiomycetesin in igapo forest, *Plant Systematics and Evolution*, **153**, 107–17.

Sioli, H. (1954) Beiträge zur regionalen Limnologie des Amazonasgebietes. II. Der Rio Arapiuns. Limnologische Untersuchung eines Gewässers des Tertiärgebietes, *Archiv für Hydrobiologie*, **49**, 448–518.

Sioli, H. (1965) Zur Morphologie des Flußbettes des Unteren Amazonas, *Naturwissenschaften*, **52**(5), 104–10.

Sioli, H. (1967) Studies in Amazonian waters. Atlas do Simpósio sobre Biota Amazonica, *Limnologia*, **3**, 9–50.

Sioli, H. (1975) Tropical rivers as expressions of their terrestrial environments, in Golley, F. B. and Medina F. B. (eds) *Tropical Ecological Systems. Trends in Terrestrial and Aquatic Research*, Springer, New York, pp. 275–88.

Sioli, H. (1983) *Amazonien. Grundlagen der Ökologie des größten tropischen Waldes*, Naturwissenschaftliche Rundschau Paperback, Stuttgart.

Sioli, H. (ed.) (1984) *The Amazon. Limnology and Landscape Ecology of a Mighty Tropical River and its Basin*, W. Junk, Dordrecht-Boston.

Smith, N. J. H. (1976) Brazil's Transamazon highway settlement scheme: Agrovilas, agropoli and ruropoli, *Proceedings of the Association of American Geographers*, **8**, 553–66.

Smith, N. J. H. (1978) Agricultural productivity along Brazil's Transamazon Highway, *Agro-Ecosystems*, **4**, 415–32.

Smith, N. J. H. (1982) *Rainforest Corridors: The Transamazon Colonization Scheme*, University of California Press, Berkeley.

Soil Science Department, North Carolina State University (ed.) (1976–83) *Agronomic – Economic Research on Soils of the Tropics*, Technical Reports, Raleigh, NC.

Sombroek, W. G. (1966) *Amazon Soils*, Centre of Agricultural Publications, Wageningen.

Sombroek, W. G. (1984) Soils of the Amazon region, in Sioli, H. (ed.) *The Amazon: Limnology and Landscape Ecology*, Monographiae Biol. 56, pp. 521–35.

Sommer, A. (1976) Attempt at an assessment of the world's tropical moist forests, *Unasylva*, **28**(112/113), 5–24.

Sopher, D. E. (1980) Indian civilization and the tropical savanna environment, in Harris, D. (ed.) *Human Ecology in Savanna Environments*, Academic Press, London, pp. 185–208.

Spencer, J. E. (1966) *Shifting Cultivation in South-Eastern Asia*, University of California, Berkeley, Publications in Geogr. 19.

Stark, N. and **Jordan, C. F.** (1978) Nutrient retention by the root mat of Amazonian rain forest, *Ecology*, **59**, 434–7.

Stephanes, R. (1972) *O programa de integraçao nacional e a colonizaçao da Amazonia*, 2nd edn, INCRA, Brasilia.

Steppler, A. (1982) An identity and strategy for agro-forestry, in MacDonald, L. H. (ed.) *Agro-forestry in the African Humid Tropics*, The United Nations University, Tokyo, pp. 1–5.

Stocks, A. (1983) Candoshi and Cocamilla swidden in Eastern Peru, *Human Ecology*, **11**(1), 69–84.

Sutton, S. L., Whitmore, T. C. and **Chadwick, A. C.** (eds.) (1983) *Tropical Rain Forests: Ecology and Management*, Special Publication No. 2 of the British Ecological Society, Blackwell Scientific Publications, Oxford.

Swift, M. J., Heal, O. W. and **Anderson, J. M.** (1979) *Decomposition in Terrestrial Ecosystems*, University of California Press, Berkeley.

Swift, M. J. and **Sanchez, P. A.** (1984) Biological management of tropical soil fertility for sustained productivity, *Nature and Resources*, **20**(4), 1–9.

Thomas, M. F. (1974) *Tropical Geomorphology*, John Wiley & Sons, London.

Thomas, M. F. and **Whittington, G. W.** (eds) (1969) *Environment and Land Use in Africa*, Methuen, London.

Transmigration Area Development Project (TAD), East Kalimantan, Indonesia (1982) *Transmigration Feasibility Study*, vol. I, Samarinda, Kalimantan.

Trewartha, G. T. (1972) *The Less Developed Realm: A Geography of its Population*, John Wiley & Sons, New York.

Trojer, H. (1959) Fundamentos para una zonificación meteorológica y climatológica del trópico y especialmente de Colombia, *Bul. Inform. Centro Nac. de Invest. de Café*, **10**, 289–373.

Troll, C. and **Paffen, K. H.** (1964) Karte der Jahreszeitenklimate der Erde, *Erdkunde*, **18**(5): 5–28.

Turner, B. L., Hanham, R. Q. and **Portararo, A. V.** (1977) Population pressure and agricultural intensity, *Annals of the Assoc. Amer. Geographers*, **67**(3), 384–96.

Uhlig, H. (ed.) (1984) *Spontaneous and Planned Settlement in Southeast Asia*, Giessener Geographische Schriften, vol. 58, Institute of Asian Affairs, Hamburg.

Unesco (1971) *Discharge of Selected Rivers of the World*, vol. II, *Monthly and Annual Discharges Recorded at Various Selected Stations*, Unesco, Paris.

311

Van Baren, F. A. (1961) The pedological aspects of the reclamation of tropical, and particular volcanic, soils in humid regions, in *Tropical Soils and Vegetation*, Unesco, Paris, pp. 5–67.

Vickers, W. (1983a) Tropical forest mimicry in swiddens: a reassessment of Geertz's model with Amazonia data, *Human Ecology*, **11**(1), 35–46.

Vickers, W. (1983b) Development and Amazonian Indians: the Aguarico case and some general principles, in Moran, E. (ed.) *Dilemma of Amazonian Development*, Westview Press, Boulder, pp. 25–50.

Waddell, E. W. (1972) *The Moundbuilders: Agricultural Practices, Environment and Society in the Central Highlands of New Guinea*, University of Washington Press, Seattle.

Walter, H. (1977) *Vegetationszonen und Klima*, Eugen Ulmer, Stuttgart.

Ward, G. (1973) Papua and New Guinea, in *World Atlas of Agriculture*, vol. 2, *Asia and Oceania*, pp. 654–61.

Watson, G. A. (1982) Tree crop farming in the humid tropics: some current developments, in MacDonald, L. H. (ed.) *Agro-forestry in the African Humid Tropics*, The United Nations University, Tokyo, pp. 6–12.

Watters, R. F. (1960) The nature of shifting cultivation, *Pacific View* **1**, 59–99.

Watters, R. F. (1971) *Shifting Cultivation in Latin America*, Forestry Development Paper No. 17, FAO, Rome.

Webster, K. H. and **Wilson, P. N.** (1980) *Agriculture in the Tropics*, Longman, London.

Wedepohl, K. H. (1969) *Handbook of Geochemistry*, Springer-Verlag, Berlin and New York.

Weischet, W. (1966) La diversa distribución vertical de la precipitación pluvial en las zonas tropicales y extratropicales; sus razones y efectos geográficos, *Union Geogr. Int. Conf. Regional Mexico 1966*, vol. III, pp. 457–78.

Weischet, W. (1969) Klimatologische Regeln zur Vertikalverteilung der Niederschläge in Tropengebirgen, *Die Erde*, **100**, 287–306.

Weischet, W. (1977) *Die ökologische Benachteiligung der Tropen*, B. G. Teubner, Stuttgart (2nd edn 1980).

Weischet, W. (1978) *Die Grüne Revolution. Erfolge, Möglichkeiten und Grenzen in ökologischer Sicht*, Paderborn, Fragenkreise 23519, Ferdinand Schöningh, Paderborn.

Weischet, W. (1985a) Ecological considerations concerning the unsatisfactory development of the rural economy in the tropics, *Applied Geography and Development*, **26**(1), 7–32. Also *Scient. Rev. on Arid Zone Research*, vol. 6 (1988), Scient. Publishers, Joghpur, India. pp. 197–226.

Weischet, W. (1985b) Landwirtschaftliches Produktionspotential der immerfeuchten Tropen im Lichte der ökologischen Grundlagen, *Perspektiven der Zukunft* (Tübingen), **70/71**, 30–8.

Weischet, W. (1988) Neue Ergebnisse zum Problem Dauerfeldbau im Bereich der feuchten Tropen, *Tagungsbericht und wissenschaftliche Abhandlungen* 46, Deutscher Geographentag München, Franz Steiner Verlag, Stuttgart, pp. 66–86.

Weischet, W. (1990) Das Klima Amazoniens und seine geoökologischen Konsequenzen, in Hoppe, A. (ed.) *Amazonien. Versuch einer interdisziplinären Annäherung.* Berichte Naturforsch, Gesellschaft Freiburg Br. **80**, Freiburg pp. 59–91.

Went, F. W. and **Stark, N.** (1968a) Mycorrhiza. *Bio Science*, **98**, 1035–9.

Went, F. W. and **Stark, N.** (1968b) The biological and mechanical role of soil fungi, *Proc. of the Nat. Acad. of Sciences USA*, **60**, 497–504.

Wilson, G. F. and **Kang, B. T.** (1980) Developing stable and productive biological cropping systems for the humid tropics, *Proc. Conf. on Biol. Agric.*, Wye College, London, August 1980, London.

Young, A. (1972) *Slopes*, Oliver and Boyd, Edinburgh.

Young, A. (1976) *Tropical Soils and Soil Survey*, Cambridge University Press, Cambridge.

Zimmermann, G. R. (1980) 'Landwirtschaftliche Involution' in staatlich gelenkten indonesischen Transmigrationsprojekten, in Roll, W., Scholz, U. and Uhlig, H. (eds) *Symposium: Wandel bäuerlicher Lebensformen in Südostasien*, Giessener Geographische Schriften, vol. 48, Giessen, pp. 121–9.

SUBJECT INDEX

Acids,
  fulvic, 83, 105
  humic, 83, 105
  organic, 83, 84
Acre, 141
Acrisols, 80, 119
AEC, 95
Agreste, 25
Agriculture,
  altitudenal limit, 9
  itinerant, 49, 55
Agroforestry
  general aim, 259
  main target, 263
  model of action, 264, 265
  variation, 255
Agrópolis, 250
Agrosilviculture, 261
Agro-silvo-pastoral, 261
Agrovilas, 24, 240, 250
Alfisols, 80, 100, 121, 167, 223, 227, 241, 247
Alley cropping, 267
Allophanes, 90, 96, 122
Alluvial accumulations, 146
Alluvial plains, 258
Alluvial soils, 191, 247
  chemical characteristics, 154, 219
Altiplano, 9
Aluminium toxicity, 86, 87, 110, 111, 205
Amazon
  flow circumstances, 147
Andesite, 94

Andosols, 122
Anion exchange capacity, *see* AEC
Arenosols, 168

Bardobes, 122
Basalt, 93, 94, 97, 141
  chemical composition, 145
Basement, 138, 139
Base saturation, 86
Beef programme, 232
Biomass, 60, 61, 103
  net production, 103, 104
Black cotton soils, 122, 168, 192
Black earths, 122, 168
Breeding programme, 228-230
Brown tropical soils, 121
  eutrophic, 120
  red-brown, 121
Buffering capacity, 86, 205
Burmese Hill Farming Experiment, 263
Burning, 215, 217, 265, 272
Bush-fallow
  managed short-lived, 236
Buttressed trunks, 133

Caboclo, 55
Cambisols, 80, 120, 121, 168, 193, 249
Campo cerrado, 21
Capillary rise, 81
Carrying capacity, 162, 245
Cash crops, 165
  income, 164
Cassava, 232
  villages, 254